Sustainable Development: Science, Management and Environment

Sustainable Development: Science, Management and Environment

Edited by **Kane Harlow**

New York

Published by Callisto Reference,
106 Park Avenue, Suite 200,
New York, NY 10016, USA
www.callistoreference.com

Sustainable Development: Science, Management and Environment
Edited by Kane Harlow

International Standard Book Number: 978-1-63239-580-1 (Hardback)

Printed in the United States of America.

Contents

Preface

This book aims to highlight the current researches and provides a platform to further the scope of innovations in this area. This book is a product of the combined efforts of many researchers and scientists, after going through thorough studies and analysis from different parts of the world. The objective of this book is to provide the readers with the latest information of the field.

The technological improvement of our civilization has established a consumer society growing more rapidly than the planet's resources allow, with the needs for resource and energy rising exponentially in the past century. For the assurance of a secured future for humans, an enhanced comprehension of the environment as well as technological solutions, behaviors and mindsets is important, in accordance with modes of development that the ecosphere of earth can support. Sustainable development provides an approach that would be useful to blend with the managerial strategies and assessment tools for decision and policy makers at the regional planning level. This book focuses on sustainability and its application in life sciences, business and management, and environment.

I would like to express my sincere thanks to the authors for their dedicated efforts in the completion of this book. I acknowledge the efforts of the publisher for providing constant support. Lastly, I would like to thank my family for their support in all academic endeavors.

Editor

Part 1

Social Sustainability and Life Science

1

People, Places and History – Towards the Sustainability of Social Life in Traditional Environments[1]

Oscar Fernández
University of León
Spain

1. Introduction

The study of the historical centres of cities has not attracted much attention, perhaps because they have been considered a "consolidated urban phenomenon", or maybe for a number of other reasons. Nevertheless, they are places of great interest, liable to far-reaching transformations and innovations, with much social interaction, which varies over time, for they are generally inhabited by one group of individuals, the residents, and used and enjoyed by others, depending on their functionality, which is usually one of tourism, leisure and culture. And the most interresting point, as I shall try to point out, is that it is a phenomenon with characteristics and peculiarities common to many historical centres of medium-sized cities of southern Europe.

This article examines the characteristics, problems and solutions offered by historical centres in different cultural contexts. Before going into a general analysis, I offer a study of the case of the historical centre of the city of León, in Spain, which will allow us to reflect later and make proposals for the sustained development of such places.

León is a medium-sized city with a complex structure with a dominating urban characteristic reaching beyond the actual municipal boundaries. It forms, together with its suburbs, a small conurbation of nearly 180,000 inhabitants. It has two thousand years of history and is very attractive because of its situation in north-west Spain, an area where different peoples have settled and different cultures have developed over time. It is a centre of such important routes as the Pilgrim Route to Santiago, the Silver Route and the *Mesta* route of sheep transhumance. All this contributes to its cultural value and heritage of different periods and styles. Notable examples of this heritage are the 1st-century Roman city walls, St Isidore's Basilica and the Royal Burial Vault of the Kings and Queens of León (12th century), the Gothic Cathedral (begun in 1255) and the 18th-century Renaissance San Marcos building, now a *parador* (state-run hotel). The old city, the walled area, is the Roman and Medieval enclave, with a history going back over two thousand years. The quarter is characterized by a great morphological, functional and social variety. It retains a major role

---

[1] This study is part of a project financed by the Spanish Ministry of Science and Technology (ref. FFI 2009-12186).

in the life of the city owing both to its geographical location in the centre of the city and to its historical and cultural significance. The gradual process of degradation undergone in recent years has made it less competitive in comparison with other areas and new neighbourhoods of the city, which are more dynamic, despite the considerable efforts of the city council over the last ten years to maintain public spaces, buildings, lighting and other improvements, with the help of European financing. In this article, we shall analyse the problem of historical centres, general to many Spanish and Italian cities, and others of southern Europe.

Later, through the case study of the historical centre of the city of León, I shall concentrate on one of the most urgent problems for its people: habitability, housing and other indicators of sustainability. We consider that it is also possible to speak of the construction of sustainable cities through the situation of historical centres, for despite having their own life, they also form part of the rest of the city. By applying the criterion of sustainability, we shall discover how to achieve the city of tomorrow, the city of progress, while at the same time preserving and safeguarding our cultural heritage brought to us by the city's history.

The central methodological position of the discipline of anthropology is based on participatory observation, with in-depth interviews yielding qualitative information. We also use quantitative data obtained from a habitability survey, using indicators of sustainability, carried out in the city of León in 2004 under the auspices of the Re Urban Mobil Project and within the framework of research performed by the group led by Dr. Cervantes (2004 www.re-urban.com) for León City Council. The methodology of the survey is based on that of the UFZ-Centre for Environmental Research Leipzig-Halle. Over a total population of 5,364 living in 2,443 homes in the historical centre in 2004, 287 homes were contacted and 261 questionnaires were distributed, of which 206 were analysed, representing 8.4% of the homes and 3.8% of the population.

The information is completed from archives and other graphical sources, which allow us to corroborate data, along with bibliographical sources for purposes of comparison.

2. The situation and the problems of historical centres of medium-sized cities

Although the problems of historical centres are manifest and they are a constant source of preoccupation, they have received less attention than they deserve. Their situation has been tackled from different disciplines and in different countries and contexts. A recent review of the literature shows that the preoccupation aroused has come from different scientific angles and is present in diverse cultural contexts. Thus for example, Rossi (2004), writing on the process of urban change undergone by the centre of Naples over the last decade, examines the role of a number of local institutions, such as the judiciary, the new urban political elites, institutionalized civil society and urban social movements. They are the instigators of the dynamics of urban change "from above", and later of dynamics "from below". Another case from the same country is offered by Lo Piccolo (1996), who examines the different attempts to preserve and restore the historical centre of Palermo, a historical city with a unique cultural heritage, typical of the southern Mediterranean cities, for its setting, the wealth and variety of its architecture and its historical buildings, its characteristic habitational problems, and the different preservation plans implemented. More recently, Sancassiani (2005), from an optimistic standpoint, gives data about Italy in general, though with specific reference to a survey on the implantation of Local Agenda 21. Maiques (2003) shows the configuration of the townscape of Valencia, Spain, another European Mediterranean city, from a historical perspective. The creation of an intellectual and political discourse based on biological and medical metaphors helps to create an image of urban society at a definite historical moment. Another case is the one offered by Williams (2002) on the "who", "why" and "to what purpose" of the Historic Cairo Restoration Program, after many years of neglect. A different cultural context, but not for that less interesting, is offered by Middleton (2003), who says that recuperating a city through tourism is having some repercussions in Latin America, for example in Quito, the object of his study. The solutions offered may be similar to European ones, although he says that there could be some disadvantages to them if any type of social conflict arises to prevent their development.

But if we concentrate on the historical centres of many southern European cities, we shall see that they are characterized by a marked development during the second half of the twentieth century, more so in the last quarter. Reviewing these studies and the various institutional forums offering information on the progress and achievements within the development of European programmes and projects, and continuing previous studies (Fernández, 2000), I shall outline the situation of many historical centres similar to those of Bologna in Italy, León in Spain, Leipzig in Germany and Ljubljana in Slovenia, all participants in the Re Urban Mobil Programme; and a number of historical centres of medium-sized cities of southern Europe involved in Local Agenda 21, which offers strategies for changing many situations, including that of historical centres (http://www.un.org/esa/sustev/documents/agenda21/index.htm) plus other cities that are signatories to the Aalborg Charter (http://www.sustainable-cities.org).

It could be argued that the analysis of the period 1960-1990s shows the typical development of many of these cities, as well as in León: the resident population is mainly aging, generally working class, with housing units occupied by a single person paying a low rent, a situation which would seem to have prevented investment in renewal. In the most degraded parts of old districts, where the buildings are on the verge of ruin, the people largely belong to low-income groups and even socially excluded ones. There are often small shops and craftsmen's workshops recalling the economic activities of bygone days, generally the Middle Ages. Aging is therefore accompanied by a functional residualness, a step towards the socio-economic degradation of a neighbourhood, with the planning and social consequences of a social and functional vacuum. The progressive aging and social exclusion of the residual population are factors that have contributed to the decay of neighbourhoods, brought on by the low purchasing power of the population and progressive impoverishment of economic and social relationships. The tendency of the last few years has generally meant a progressive loss of population, which has given rise to the replacing of the original social group of the area. Poverty and the decay of accommodation pushed all those who could afford it to other parts of the city, which meant that those remaining belonged to low-income groups.

Another common phenomenon in historical centres has been the inability to bring about a social unification of all the residents of the neighbourhood. Furthermore, the arrival of new ones implies a structural division within it into old and new residents. The latter are on the road to exclusion, favoured by the fact that the unity of the group, if indeed it can be called that, mainly made up of the elderly, is based on tradition and common symbols, and on the knowledge afforded by years of living in the area. In the case of Spain, for the period in question, there were half a million people living in historical centres, with characteristics like those already mentioned, that is an elderly population with major shortcomings in services and accommodation, with a mean unemployment rate of 27.6% and 21.6% of the population with only elementary education. They are the only areas with a major amount of rented accommodation (44.5%) and a high proportion of empty dwellings (21.3%) (http://habitat.aq.upm.es/doc.html). In general, over half of the residents were born outside the municipal boundaries and are of low social classes. A quarter of the houses were lying empty and of those occupied, only 25% were owner-occupied. About 40% were severely dilapidated.

An analysis of references showed us that the areas in question had suffered deterioration in building and planning quality as well as falling behind socially and environmentally, some of them affected by social exclusion, an increase in the elderly population and people with low resources, neglect of buildings by well-off, etc. this brought about a crisis of economic activity, with the loss of traditional activities, the occupation by marginalized groups of buildings with inadequate sanitation that made them difficult to live in, and a consequent proliferation of tenement slums. The social image of these areas fell, and this deterioration gradually spread to neighbouring districts.

To this situation of depopulation and neglect there must be added the difficult problem arising from the new functionality being acquired by these neighbourhoods, now leisure areas, with an increasing number of bars and restaurants, which in principle met the demand of a cultural tourist sector offered by the area through the historical buildings and museums normally situated in such areas. This saturation of bars and the like, and their impact on night life and their concentration in certain parts of historical centres has discouraged new residents from entering the areas, along with other commercial functions and services. What we have seen is that the rehabilitation of these areas has become a pressing question, for they needed a very careful reformation in order not to affect their historical atmosphere and character and increase their vitality, and even their picturesqueness (various authors, Places to Live, 2003).

A similar process in these southern European centres, according to another study about León, (Fernández, 2000) has consisted, as the catastrophic tendency has waned, in new population groups coming into these districts, and not only the socially excluded, ethnic minorities or foreigners in precarious employment situations. Attracted by the gradual processes of restoration, young professional people are moving in, choosing these areas for their accessibility and central position. Moreover, the rehabilitation plans carried out have generally formed part of public programmes with European funding, mainly through the European Regional Development Fund (ERDF), such as those mentioned above (Urban, Re Urban, etc.) They have the advantage of setting up networks of cities with similar problems, affording the possibility of information exchanges and synergies (http://europa.eu.int/ comm/regional policy/intro/regions1 en.htm).

The effects have been varied and, at times, limited, mainly in small and medium-sized cities. The positive effects have served to establish guidelines for renovation, and grime, filth and damp have gradually given way to colour. The preservation of historical centres has become an aesthetic task, whereby the historical and artistic city is rediscovered for citizens and tourists alike. Carefully chosen light colours for façades have become a visible sign of this recuperation, which we trust is not just a facelift. Another characteristic tendency has been pedestrianization, with non-residents' cars being denied entry, and space thus being liberated for citizens to walk around and enjoy the historical buildings.

On the other hand, however, there has always been a noticeable division of the subgroups living in historical centres, which means that they fail to achieve the group identification necessary for solving problems facing all of them, while bringing out their socio-economic diversity and multiculturalness.

3. The historical centre of the city of León

After this brief overview, we shall now concentrate on the case of León, where, after a period of lethargy, the strategies applied are fairly typical of those outlined above.

In León, three factors come together which, according to J. Borja (1997, 2003), are necessary for the success of urban transformation projects: the sensation of acute crisis brought on by the awareness of the globalization of the economy, the harmonization of public and private urban key players together with the generation of local leadership (both political and civic), and the joint will and consensus of the citizens for the city to take a step forward, both physically and socio-economically and culturally.

Thus, León, with the brokerage of the council and the impulse of the global macroeconomic current imposed and facilitated by the European Union, and under the auspices of the Regional Development Fund, in the 1990s, sought the revitalization of its old quarter but with criteria that in some way were imposed on it from outside. Thus arose the Urban pilot project "Building León: a Development Proposal for the Old City 1995-1999", which, while also potentially applicable to other historical cities centres in the European Union, sought to bring new life to the historical centre of León. It was felt necessary to involve young people and revitalize the economy of the area by attracting small and medium-sized firms, and to improve the quality of life of the residents. In this way, the historical centre would not merely be a collection of historical buildings for tourists and visitors but the living district that it historically used to be, a place with its own identity and, in short, the definite and dynamic urban reality which, also historically, it has sought through the consensus of physical, social and economic structures. Furthermore, these projects include the idea of competitiveness at different levels, both internal, within the city, and external, with other similar cities, both Spanish and European. And this idea of competitiveness has made itself very much felt in the territorial debate, according to Díaz Orueta (1997). Because of it, and in line with globalizing and subsidy-based policies, there has been an on-going struggle to attract investment. Much effort has been made to modernize infrastructures, and much money has been spent on defining strategies to determine the place León should take in the hierarchies of cities.

The work of social and political groups functioning in the district, which has had partial and isolated successes in line with Rossi's ideas (2004), is to be interpreted as directed at organizing the group and establishing an infrastructure for social dynamics leading to co-existence, solidarity and an improvement of the quality of life. A significant case was the

"León Típico" Residents' Association, which had a major social and political impact (Fernández, 1997), originally based on opposition to the previous political regime in its struggle for the artistic and cultural heritage of the district. But now, because of the absolute lack of stimuli, and the loss of interest of the citizenry, it has largely given up solidarity to concentrate on improving the quality of life. This means peace, being able to enjoy a walk, no noise at night and, if necessary, the possibility of bringing charges against annoying bars or other premises causing any kind of nuisance (noise, drugs, and so on). But this piecemeal work, though positive, must include a search for alternatives to the stagnation of the problems of the resident population.

The Urban pilot project had praiseworthy aims, for it sought the recuperation of the historical centre as an active and integral part of the city, while giving it a live role, in order to avoid it becoming a merely monumental centre, and improved the appearance and sanitation of the district with a pneumatic system of refuse collection. It did not, however, solve the structural problems concerning the aging population, the situation of its housing, or the functionality of the area as one for night life, which quite definitely affected the quality of life through noise pollution. These are the main problems facing the old quarter of León and those which the residents consider the most worrying.

Specifically, the residents mention such problems as the deficient state of their homes, the loss of crafts and commercial activities, an aging population and the abandonment of the area as one of residence, in relation to its night life, which engenders high levels of pollution, especially at weekends. Although the old quarter is set in the centre of the city of León, major functional changes have taken place in it and in the city as a whole: as a place of residence it is in decline, as 1900 residents have left in the last ten years; there has been a loss of activity as a result of the changes in the location of administrative services; religious and educational functions have been run down, though not as regards space, owing to the drop in the number of students and in the numbers of residents of religious homes an convents. On the other hand, art, culture and tourism have grown in importance. Trade has been strengthened in some of the better-known streets, which have been renovated, but to the detriment of other parts of the district, where it has disappeared altogether. The southern part of the old quarter, known as the *Barrio Húmedo*, or "Damp Quarter" retains its major role as a night spot. Bars, restaurants and discothèques take up most of the business premises, almost to the exclusion of other shops or services, causing much nuisance to residents through the great amount of noise generated at night.

As has already been said, the population of the historical centre was 5,364 in 2004. Over 60s account for a third of the total, at 1,832. Another significant fact that we observed is that 16% of the homes are occupied by people living alone. Although the population is mainly an aging one, young professional people are moving in, as there are signs of the rehabilitation of the area. Although most of the population comprises Spanish nationals, at given moments, foreign immigrants and ethnic minorities have set up in this area owing to the availability of cheap housing, usually with deficient sanitation. At present, foreigners account for only 3% of the population.

As for the socio-economic structure, over 85% of the population has at least received primary education, 34.6% having been to university. Although a variety of occupations are represented, 27.8% of the population is retired.

The district's central position means that the normal way of getting about within the city is walking, 70.4% of the population using this means, with journey times of 15-20 minutes, although half of the homes have one car and 18% have two while a third have none. In turn, of those who have cars, half have access to garage space or pay for a parking place, while the other half leave their vehicles parked in the streets, which causes thoroughfare problems because of the recent pedestrianization of a number of streets.

Fieldwork shows that one of the means of creating cohesion in the quarter is through the mutual personal acquaintance of residents. The certainty of meeting the same people every day creates affinity, which is borne out by statistics: more than a quarter of the residents of the old city (25.7%) have lived there all their lives, and the remainder have been there fore varying periods, ranging from 10-25 years (18%), 6-10 years (11.2%) to under 5 years (10.2%). Of those who have moved into the area, over half (54.3%) said that the accommodation was what they wanted, that the centre was near, and a third said that their places of work or study were near. Regarding social life in the quarter, most residents consider themselves fairly or very involved in it, which contrasts with the fact that most of them find out what is happening in the area through the local press or television and radio, while 10.3% find out from the residents' association, 4.6% from the parish church and a few (4.1%) because they see what happens personally. Just over half (54.4%) do not feel adequately informed about local issues. A similar percentage (54.6%) say that they know of some association at work in the area like the residents' association, a social or community grouping working for the district or for some sector such as tradesmen or caterers, or for the local heritage or NGOs. Nevertheless, the real percentage of active participation in the life of the district is only 17.9%, as opposed to 82.1% who say that they do not participate.

Relationships between residents may be said to be good, and over half of them (53.3%) say that they have relatives or friends in the area, and see them often, perhaps daily and at least weekly. Half of the residents of the old town would recommend others to move into the area, for its central position and convenience, although they would not recommend it from the noise point of view.

Most of the residents are owner-occupiers (61.2%). Given the age of the buildings, few upper flats have lifts, only 15.5%, and the same may be said for central heating (12.1%), and that in a city where the weather is cold for 8 months of the year. 68.4% of residents tend to use diverse types of gas, diesel, coal and electric fires and heaters.

The level of satisfaction of residents of the area with their housing is very variable. The main complaints concern the condition of the roofs and ceilings, heating and the quality of the windows, all of which are connected with the external environment (climate, noise pollution, age of the buildings, etc.).

Another source of complaints is the lack of facilities for children, of play areas, and the excess of noise at night, together with the lack of services for the elderly and of cultural activities and facilities. The points of greatest satisfaction are the numbers of restaurants and cafés, street cleaning and the novel system of pneumatic selective refuse disposal, installed under the auspices of the Urban pilot project.

But the general feeling among the residents is one of improvement, especially over the last few years, at least as far as outward appearances are concerned, as in this regard, the changes have been far-reaching. This has been helped by the policy of pedestrianization, as the removal of motor traffic has improved the quality of the environment, while the renovation of streets, squares and façades has improved the general appearance. But the main problem, nocturnal noise pollution, has yet to be solved. It may also be the source of another problem, vandalism directed against renovated street furniture.

The perception of the district's future for most residents (60.4%) is positive, while a minority of 19% consider it negative. Despite this, answers about the future of the quarter are equally divided into those who think that it will be restored, and that it will have nothing but bars and restaurants, and those who think that it will depend on the will of the council and of the politicians in power at any given moment. It is significant that 18.5% of the people would move to another district if they could, because of the problem of noise at night, especially at weekends.

In short, we have found some positive feelings among the population of the historical centre regarding a gradual recuperation of the population (which is false, as the statistics show), the accessibility of the city centre, improvements brought about by the pedestrianization of certain areas, improvements to façades and streets and the area's becoming attractive as a residential one because of its architecture and historical setting. Also mentioned are the social atmosphere, the air quality, the cleanliness of the streets and the system of refuse disposal, together with good relations among residents, the number of bars and restaurants, a favourable opinion concerning the overall situation of the last few years and positive feelings regarding the district's future as a residential area. Negative feelings recorded concentrate on the excessively aged population and the number of homes with over sixties living alone and who depend on their old-age pensions, which are usually low. There is also some unrest concerning the state of housing, usually with regard to its age, the conditions of the streets and pavements, the levels of noise at night, the lack of green areas and trees, of children's play areas, of cultural and leisure facilities and of public social centres, activities for the young, sports fields, centres for the aged and public social centres in general, together with the physical state of buildings, poor access to information, and so on.

4. Towards the sustainability of historical quarters

It has now become quite necessary to apply the criterion of sustainability to city planning and development. At the world level, one of the initiatives of the Earth Summit held in Rio de Janeiro in 1992 was to foment local initiatives to support sustainable development. In Europe, the Treaty of Amsterdam of 1999 confirmed in articles 2 and 6 that balanced and sustainable development, together with the protection and improvement of the environment, were basic aims of the European Union. The criterion of sustainability, however, has not been felt excessively in historical cities, which are in the process of change and development.

Brugmann (1992) and Tjallingii (1995) proposed considering the city as an ecosystem and using ecological concepts to understand the problems of urban sustainability and find solutions for them. It is certainly true that advances are being made in the construction of the city, approaching the aims of sustainability, of a modern and dynamic city developing in an attractive setting, and one that is healthy from the environmental point of view. Through a strategy of integration of the environmental elements of different sectorial policies, a kind of progress is being sought that satisfies the needs of the present without jeopardizing the ability of future generations to satisfy their own needs.

One of the typical sustainability proposals that we have seen applied in many historical centres has been the freeing of spaces from traffic. As motor traffic is the main source of urban atmospheric pollution, action must be taken to reduce its impact. As an example, Handy and Clifton (2001) have evaluated the possibility that providing local shopping opportunities will help to reduce automobile dependence. Although local shopping does not show great promise as a strategy for reducing automobile use, but it does show promise as a strategy for enhancing quality of life in neighbourhoods, at least partly by making driving once again a matter of choice. The topic has a complex solution, because, as Black, William R., and Peter Nijkamp (2002) say, "solutions will succeed or fail on the basis of social response". As emissions cannot be stopped altogether, for it is impossible to prohibit all traffic, an effort must be made to rationalize and improve the quality of air through innovative traffic policies, the use of noise-reducing surfaces and public transport policies directed at the improvement of the environment, although, as we have seen, where traffic noise has been eliminated, other forms of noise pollution have taken its place.

Nevertheless, I believe that the role of sustainability in the study of historical centres of this type need not be limited to environmental questions. It would seem necessary to bear in mind other kinds of indicator which in principal could be considered as emerging, such as society and culture, principally in the historical areas that we are dealing with, where taking the old town as a consolidated fact has caused other aspects just as basic to the lives of the citizens to be forgotten. We shall thus be able to bring out the value of the city as a cultural heritage, as a living place, one of co-existence, of commercial and cultural exchange, of the exchange of knowledge, a seat of institutions and a place of leisure. This must also be translated into actions capable of recuperating the normal aspects and activities of the historical city, which are the basis of its formation and development, and which are at risk of disappearing as the result of the burgeoning development of suburbia, with alternatives for coexistence totally detached from the traditional urban nucleus.

In this regard, one of the aspects to consider is the concentration of the cultural heritage characterizing historical cities, especially their old quarters. This fact determines their close association with a growing type of tourism: cultural tourism, or, more specifically, "heritage tourism". This is the case that Ian Strange (1999: 302) explores. He presents the argument that in some places attempts to reconcile the potentially conflicting and incompatible demands of urban competitiveness and urban sustainability are being pursued through the application of sustainable development policies to the management of local environmental and historical assets. He shows analyses the varying ways in which policy makers in historic cities are engaging in action to regulate localised patterns of economic and physical growth. Indeed, a major dilemma for many small- to medium-sized historic cities revolves around the simultaneous need to manage the conservation of the physical fabric of the city and accommodate the pressures associated with an expanding range of economic development and tourist-related activities and functions. Nevertheless, not all urban heritage is a resource for tourism, and not all spaces attract visitors or offer any significant use for tourism. There are other spaces in the historical quarters which are also lived in, and which have value for the collective memory of the residents. Evidence tells us that the rehabilitation of certain spaces with a monumental or historical value has been to the detriment and neglect of areas without such value. It cannot be denied that the association of culture, tourism and historical cities has obvious advantages for communities and the places that are home to them, such as income from visitors, the physical and functional renovation of the areas in question, and so on, but it also has a number of negative effects of growing importance. As far as space is concerned, these effects are especially felt in a small part of the old city defined as a "historical city of interest to tourists". Together with the advocacy of urban sustainability, there arises the need for models of tourism development able to maximize the benefits and minimize the costs of attracting tourists, especially in the more fragile and vulnerable sectors. Likewise, a better flow of information concerning the projects and programmes in any way affecting the neighbourhood and its inhabitants will contribute to social sustainability, while also stimulating the participation of the citizens in the face of the tendency for today's society to become less and less solidarity-minded, as we have seen in the countryside. Of course, it would be desirable to enhance local management with integrated decision making involving public and private institutions and the community at large. In many Local Agenda 21 cities, an opportunity to renew, innovate and increase the processes of participation is arising in this way. The community's involvement in these processes of gradual and strategic change is a fundamental condition for their success. In this regard, Agenda 21 offers a forum and processes that could strengthen the community. Some Spanish cities, such as Barcelona and Granada, or even, Swedish (Adolfsson Jörby, 2002) or Austrian (Astleithner and Hamedinger, 2003, and Narodoslawsky, 2001) or Italian ones, as borne out by Sancassiani's recent study (2005) have approached them as governance experiences, where different key players take part in the processes and take on their share of the joint responsibility regarding the shared goals. These are, then, participative processes directed at establishing a different relationship between the local authorities and the various social actors at work.

Another point to consider is the increased value of the built-up city and the improvement of its infrastructures, fomenting their recycling and preserving the architectural heritage. This would enrich cultural identity, while improving the aesthetic quality of the townscape. Generating urban diversity and complexity (in residence, economic activity, culture and services) would help to reduce the need for mobility, which would make the city more attractive as a place to live, work and offer services. The substantial improvement of living conditions and social cohesion, and of quality of life (housing, education, work, health, culture, leisure, and so on) would help us to regain the idea of a city as a common project of its citizens.

Likewise, there arises the need to stimulate economic dynamism, which would be helped by the setting up of workshops and schools of traditional activities linked with the area, such as restoration, plastic arts, crafts, etc., where education and training would be a first step towards economic consolidation. In this concern, Summers, M., Childs, A., Corney, G. (2005) show the case of education for sustainable development (ESD) in initial teacher training. They find that schools are not yet well developed as sites for student teacher learning in the domain; student teachers generally have greater understanding of sustainable development than their mentors; geography mentors perceive themselves to be better prepared for mentoring in this area than their science counterparts (who feel ill-prepared); for both students and mentors, there are significant gaps in understanding of education for sustainable development. The transmission to lower levels may be significant also. But certainly, it is not possible to base sustainability only on souvenir shops in the daytime and bars and restaurants at night. While on the subject of education, I would suggest that there should be education in ethics for the young, together with alternatives to alcohol-based nocturnal leisure in bars and discothèques that are open well into the small hours of the morning. Such measures could be evening sports activities and a management of culture allowing for a more affordable culture for young people. The noise pollution caused by existing leisure activities is a new source of annoyance in these cities, and, as we have seen, many historical centres that have managed to rid themselves of the noise generated by traffic are suffering from the noise produced by the aggressive spread of nocturnal leisure. It is certainly true that southern European cultures, especially Mediterranean ones, and noticeably in Spain, are characterized by celebrations and street life, and therefore by rowdy open-air partying. But it is just as true that in our cultural environment, noise is becoming more and more a source of urban annoyance. There is a problem of the proliferation of fixed points of noise pollution such as discothèques, bars, open-air terraces and other leisure-time activities. The conflict between the right to rest and the right to leisure and social relationships is generating new urban unrest, in some cases of a serious nature, which is making it necessary to set up mechanisms of dialogue, shared responsibility and arbitration. However, the socio-cultural life of such places is also an asset that makes them attractive and contributes to local vitality and vibrancy in a way that buildings and artefacts do not, (and is not easily separated into what is either critical or tradable capital). In this particular case, communication channels must be opened up between those who run bars and shops, youth groups and residents to find a solution based on the satisfactory compromise of rights. In this regard, in the local setting, campaigns of awareness, education and environmental information have an effective background for generating social change, for it would seem that they get feedback from processes of community participation and are more permeable to the influence of ecological and environmental organizations. Therefore, cities play a basic role in changing their citizens' habits and values regarding the new paradigm of sustainable development.

Most of these problems are typical of the historical centres of our cities. But we might consider that the solution to many of them depends on transforming these historical centres into places of convergence where there are activities with an obvious function within the framework of the city as a whole. Considering and studying them in a piecemeal way and in isolation will give us a sectorial view, with an obvious dysfunctionalness leading to degradation and destruction through their separation from the rest of the city. They will have to be considered within global strategies, at the same level as the rest of the city and its zones of influence, as part of a cultural legacy belonging to the whole community and forming part of a dynamic process in which the old city has a number of attributes that can neither be repeated nor regained. When the elderly people of the area speak to us of bygone days, of time immemorial, what is really essential is the message: the proclamation of the prestige and age of the city, the idea of a past with a presence in the present. But the most interesting thing is how they cling on to them emotionally. This is all the clearer when they tell their personal experiences, however disastrous. As long ago as 1946, Annoni pointed out that the historical centres of cities collaborated in the development of a modern town, insofar as they offer culture, rest and leisure. This obliges us to find a concept of Historical Centre within the context of the city, not just to preserve its history, but also to look after its social life.

Now may be the time to consider again the city as a joint project, as a space for human relationships, as a place where there will inevitably be conflicts, which can and must be resolved via consensus and not by the imposition of a dominating element, by applying to problems created over a long period of time solutions that are not quantified in units of time measured by political legislatures. But above all, it is fundamental that we should seek a project of a city for the future from a more comprehensive point of view than today's, transcending the interests of political parties and economic groups, with our sights always on the people. In short, what is sought is to maintain, or even regain the original urban structure of the city as an area that is lived in, and stressing the value of the historical centre and its importance in the development of the city.

5. Conclusion

The historical centres of cities still have their character as multifunctional areas but the pace of renovation of facilities is quite slow and this impedes competition with other more dynamic neighbourhoods.

In the object of our study, the old city of León, through its historical buildings, it is a symbol of the whole city and the seat of many of its urban, social and cultural values. Moreover, it is a dynamic urban reality which throughout history has striven to strike a balance between physical, social and economic structures. The old city is living through a critical situation where there are obviously many difficulties when it comes to finding an operational model that allows it to get away from the cycle of physical, social, cultural and functional deterioration.

The residents of the historical area have lived there for a long time and many of the old buildings are inadequate and occupied by single people living alone. There is a low percentage of young people, and hardly any in middle age. Although over half the population feel that the district has improved in general over the last few years, people would like to see green spaces, public, social and cultural services, shops, more trade in general and anything that will give more daytime life to the area.

In this regard, we have drawn up a proposal of sustainability for areas of this type which, through criteria of culture and society, seeks to contribute to revitalizing the commercial and leisure activities normal in a consolidated city, by means of specific actions like the maintaining of the resident population, encouraging the refurbishment of old homes, along with the preservation of their topological characteristics and their adaptation to the demands of today's life, and attracting new generations so that they will, together with the present population, guarantee the survival of activities and urban co-existence. Furthermore, one of the greatest problems detected is the danger that these historical centres will become mere monumental areas with an exclusively museum-like character, which leads us to conclude that it is important to integrate historical centres in the daily life of whole cities.

Our cities require sustainability, and to achieve it, we need changes in our concept of what a city should be, and to reconsider the way decisions are made regarding certain policies and new outlooks on how to get tomorrow's problems into a political debate focussed on today. This is an urgent task which requires specific commitments and a great capacity to learn.

Many of the challenges and problems mentioned will only be sorted out (legal, fiscal and regulatory policies aside) with a new social consensus allowing for a change in attitudes and daily behaviour patterns of most of the people. The historical city is the background to the processes of collective identification, of belonging, which creates community, and it is in that community environment that new alliances of sustainability must be created. The positive role of local government and of the city as a whole are, owing to their proximity and permeability, generally recognized, but there is still a long way to go. And the way is not without obstacles, for as humans we have to recognize that we are all somewhat contradictory and we want a lot of things at the same time. But we must remain aware of the demands that the discourse on urban sustainability creates if we want more than well-meaning rhetoric.

The cultural standpoint, and that of Europe's historical heritage, should regain the taste for human projects, advocating the city of Erasmus (quoted by A. Clayton and N. Radcliffe, 1994), who, far better than modern town-planners and humanists, recognized that the ultimate goal of any discipline had to be the improvement of the quality of life of the human being in a development compatible with Nature and the environment.

6. References

Adolfsson Jörby, S. (2002) Local Agenda 21 in Four Swedish Municipalities: a Tool Towards Sustainability?, *Journal of Environmental Planning and Management*, 45 (2): 219–244.

Annioni, A. (1946) *Scienza ed Arte del Restauro Architettonico. Idee ed Esempi*, Artistiche Framar. Milano

Ashworth, G. & Tunbridge, L. E. (1990) *The Tourist-Historic City*, Belhaven, London.

Ashworth, G. J., Tunbridge, J. E. (1999) "Old cities, new pasts: Heritage Planning in selected cities of Central Europe", *GeoJournal*, 49 (1): 105-116.

Astleithner, F., Hamedinger, A. (2003)"Urban Sustainability as a New Form of Governance: Obstacles and Potentials in the Case of Vienna" *Innovation: The European Journal of Social Sciences*, 16 (1): 51-75.

Baker, S., Kousis, M., Richardson, D. & Young, S. (1997) Introduction: the theory and practice of sustainable development in EU perspective, in: S. Baker, M. Kousis, D. Richardson & S. Young (Eds) *The Politics of Sustainable Development*, pp. 1-42. Routledge, London

Black, W. R., and Nijkamp, P. (eds.) (2002) *Social Change and* Sustainable Transport, Indiana University Press, Bloomington.

Blowers, A. & Evans, B. (Eds) (1997) *Town Planning into the 21st Century*. London: Routledge.

Blowers, A. (1997) Society and sustainability: the context of change for planning, in: A. Blowers & B. Evans (Eds) (1997) *Town Planning into the 21st Century*, pp. 153-168, Routledge, London.

Borja, J. (1997) Las ciudades como actores políticos, *América Latina*, Hoy 15: 15-19.

Borja, J. (2003) *La ciudad conquistada*, Alianza, Madrid.

Borja, J., & Castells, M. (1996) *Local y Global: la gestión de las ciudades en la era de la información*, Taurus, Madrid.

Boyko, C. T., Cooper, R., Davey, C., (2005) Sustainability and the urban design process, *Proceedings of the Institution of Civil Engineers: Bridge Engineering*, 158 (3) 119.

Brugmann, J. (1992) *Managing Human Ecosystems: Principles for Ecological Municipal Management*, ICLEI, Toronto.

Castells, M. (2000) Urban sustainability in the information age, *City*, 4 (1) 118.

Cheshire, P. (1979) Inner Areas and Spatial Labour Markets: A Critique of the Inner Areas Studies, *Urban Studies*, 16: 29-43.

Clayton, A. and Radcliffe, N. (1994) *Sustainability: a Systems Approach*. London: Institute for Policy Analysis and Development for the World Wide Fund for Nature.

Commission of the European Communities (2001) *Consultation Paper for the Preparation of a European Union Strategy for Sustainable Development*, SEC, 517, Brussels.

Committee for Economic Development (CED) (1995) *Rebuilding Inner City Communities: A New Approach to the Nations Urban Crisis*, CED, New York.

Costantini, V., Monni, S. (2005) Sustainable Human Development for European Countries, *Journal of Human Development*, 6 (3): 329-351.

Cowell, R. & Owens, S. (1997) Sustainability: the new challenge, in: A. Blowers & B. Evans (Eds) *Town Planning into the 21st Century*, pp. 15-32, Routledge, London.

Dekay, M., O'Brien, M. (2001) Gray City, Green City. *Forum for Applied Research & Public Policy*, 16 (2) 19-28.

Díaz Orueta, F. (1997) Las ciudades en América Latina: entre la globalización y la crisis, *América Latina*, Hoy, 15: 5-13.

European Commission (1994). *Local Development Strategies in Economically Disintegrated Areas: A Proactive Strategy Against Poverty in the European Community*. European Commission Social Papers, 5.

European Commission (1996). *European Sustainable Cities. EU Expert Group on the Urban Environment*. http://europa.eu.int/comm/environment/urban/home_en.htm

Fernández, O. (1997) *Aproximación Antropológica a la ciudad de León: El Centro Histórico*, University of León, León.

Fernández, O.   (2000) Renewal of historic-city areas, between Globalization and Localization, *Metropolitan Ethnic Cultures: Maintenance and Interaction. The relationship between traditional culture and modernization in cities*. Beijing Inter-Congress, IUAES: (A2): 1-12.

Fernández, O. (2007) Towards the Sustainability of Historical centres: a Case-study of León, Spain. *European Urban and Regional Studies*, 14 (2): 181-187.

Finco A., Nijkamp, P. (2001) Pathways to urban sustainability, *Journal of Environmental Policy & Planning*, 3 (4) 289.

Fung M, Kennedy CA. (2005) An Integrated Macroeconomic Model for Assessing Urban Sustainability, *Environment and Planning B-Planning & Design*, 32 (5): 639-656.

Garde, AM. (2004) New urbanism as sustainable growth? A supply side story and its implications for public policy, *Journal of Planning Education and Research*, 24 (2): 154-170.

Gibbs, D. (1997) Urban sustainability and economic development in the United Kingdom, *Cities*, 14 (4) 203-209.

Glasson, L., Godfrey, K. & Goody, B., Absalom, H. & Van der Borg, J. (1995) *Towards Visitor Impact Management. Visitor Impacts, Carrying Capacity and Management Responses in Europe's Historic Towns and Cities*, Avebury, Aldershot.

Handy, SL; Clifton, KJ (2001) Local shopping as a strategy for reducing automobile travel, *Transportation*, 28 (4): 317-346.

Hannerz, U., (1986) *Exploración de la ciudad. Hacia una antropología urbana*, FCE, México.

Harris, J. M. (2003) *Rethinking Sustainability: Power, Knowledge, and Institutions.*: University of Michigan Press, Michigan.

Hudson, R. (2000) *Production, Places and Environment. Changing Perspectives in Economic Geography*, Prentice Hall, Harlow.

Hudson, R. (2005) Towards Sustainable Economic Practices, Flows and Spaces: Or is the Necessary Impossible and the Impossible Necessary? *Sustainable Development*, 13 (4): 239-252.

ICLEI (International Council for Local Environment Initiatives) (1995) *European Local Agenda 21 Planning Guide. How to Engage in Long-term Environmental Action Planning Towards Sustainability*, ICLEI, Fribourg.

Jackson P. 2002. Commercial cultures: transcending the cultural and the economic. *Progress in Human Geography* 26: 3-18.

Kelly, R. and Moles, R. (2000) Towards Sustainable Development in the Mid-west Region of Ireland, *Environmental Management and Health*, 11 (5): 422–432.

Lake,A. (1996) The City in 2050: How Sustainable?, *World Transport Policy and Practice* 2/1 (2): 39–45.

Lewis, G., Brabec, E. (2005) Regional land pattern assessment: development of a resource efficiency measurement method, *Landscape and Urban Planning*, 72 (4) 281.

Lo Piccolo, F. (1996) Urban Renewal in the Historic Centre of Palermo, *Planning Practice and Research*, 11 (2): 217-227.

Low, S. M. (ed.) (1999) *Theorizing the city: the new* urban anthropology *reader*. Rutgers, New Brunswick.

Lynch, K. (1960) *The Image of the City*, MIT Press, Cambridge, MA.

Maiques, JVB, (2003) Science, politics and image in Valencia: a review of urban discourse in the Spanish City, *Cities*, 20 (6): 413-419.

Marat-Mendes T, Scoffham E. (2005) Urban Sustainability And The Ground Rules That Govern Urban Space, *Urban Morphology*, 9 (1): 45-46.

Marvin, S. & Guy, S. (1997) Creating myths rather than sustainability: the transition fallicies of the new localism, *Local Environment*, 2 (3): 311-318.

McMahon, S.K. (2002) The Development of Quality of Life Indicators – a Case Study from the City of Bristol UK, *Ecological Indicators*, 2 (1–2): 177–185.

Meethan, K. (1997) York: managing the tourist city, *Cities*, 14 (6): 333-342.

Middleton, A. (2003) Informal traders and planners in the regeneration of historic city centres: the case of Quito, Ecuador, *Progress in Planning*, 59: 71-123.

Narodoslawsky, M. (2001) A Regional Approach to Sustainability in Austria, *International Journal of Sustainability in Higher Education*, 2 (3): 226–237.

National Research Council Board on Sustainable Development (2002) *Our Common Journey*, National Academies Press, Washington.

Newman, P. (1998) From Symbolic Gesture to the Mainstream: next steps in local urban sustainability, *Local Environment*, 3 (2): 299-311.

Newman, P. (2005) Story and Sustainability: Planning, Practice and Possibility for American Cities, *International Journal of Urban and Regional Research*, 29 (2): 468-469.

Pareja Eastaway, M., Støa, E. (2004) Dimensions of housing and urban sustainability, *Journal of Housing and the Built Environment*, 19 (1) 1.

Pendlebury, J. (1999) The conservation of Historic areas in the UK. A Case study of "Graninger Town", Newcastle-upon-Tyne, *Cities*, 16 (6): 423-433.

Rees, W. E.; Devuyst, D., Luc H., and Walter de Lannoy, (eds.) (2001) *How green is the city? Sustainability assessment and the management of urban environments*. Columbia University Press, New York.

Richardson, D. (1997) The politics of sustainable development, in: S. Baker, M. Kousis, D. Richardson & S. Young (Eds) *The Politics of Sustainable Development*, pp. 43-60, Routledge, London.

Rossi, U. (2004) The multiplex city - The process of urban change in the historic centre of Naples, *European Urban and Regional Studies*, 11 (2): 156-169.

Rydin, Y. (1997) Policy networks, local discourses and the implementation of sustainable development, in: S. Baker, M. Kousis, D., Richardson & S. Young (Eds) *The Politics of Sustainable Development*, pp. 152-174, Routledge, London.

Sancassiani, W. (2005) "Local agenda 21 in Italy: an effective governance tool for facilitating local communities' participation and promoting capacity building for sustainability". Local Environment, 10 (2): 189-200.

Santos, M. A. (2005) Environmental stability and sustainable development, *Sustainable Development*, 13 (5): 326-336.

Singh, R. B., (ed.) (2001) *Urban Sustainability* in the Context of Global Change: *Towards Promoting Healthy and Green Cities*, Science Publishers, Inc. Enfield.

Stephen M. W., and Timothy B. (2004) *The Sustainable Urban Development Reader*, Routledge, London.

Steven B. S.B. Kraines; D.R. Wallace (2003) Urban sustainability technology evaluation in a distributed object-based modelling environment, *Computers, Environment and Urban Systems*, 27 (2) 143.

Strange, I. (1996) Local politics, new agendas and strategies for change in English historic cities, *Cities*, 13 (6): 431-438.

Strange, I. (1997) Planning for change, conserving the past: towards sustainable development policy in historic cities?, *Cities*, 14 (4): 227-234.

Strange, I. (1999) Urban Sustainability, Globalisation and the Pursuit of the Heritage Aesthetic, *Planning Practice & Research*, 14 (3) 301-311.

Summers, M., Childs, A., Corney, G. (2005) Education for sustainable development in initial teacher training: issues for interdisciplinary collaboration, *Environmental Education Research*, 11 (5): 623-647.

Tjallingii, S. (1995) Ecopolis. *Strategies for ecologically sound urban development.* Leiden: Backhuys publishers.

Tomé, S. (1982), Memoria urbana y crisis de los barrios históricos a través del ejemplo de la ciudad de León, *Tierras de León*, 4: 23-36.

Urry, J. (1995) *Consuming Places*. London: Routledge.

Valentin,A., and Spangenberg, J.H. (2000) A Guide to Community Sustainability Indicators, *Environmental Impact Assessment Review*, 20 (3): 381–392.

Van Dijk Mp, Zhang Ms (2005) Sustainability indices as a tool for urban managers, Evidence from four medium-sized Chinese cities, *Environmental Impact Assessment Review*, 25 (6): 667-688.

VVAA (2003) *Places to Live. Case Studies in Urban Regeneration*, CECODHAS.

Walton, J. S., M. El-Haram, N. H. Castillo, R. M. W. Horner, A. D. F. Price, C. Hardcastle (2005) Integrated assessment of urban sustainability, *Proceedings of the Institution of Civil Engineers: Engineering Sustainability*, 158 (2) 57.

Wen Zg, Zhang Km, Huang L, et al. (2005) Genuine saving rate: An integrated indicator to measure urban sustainable development towards an ecocity, *International Journal of Sustainable Development and World Ecology*, 12 (2): 184-196.

While, A; Jonas, A. & Gibbs, D. (2004) The environment and the entrepreneurial city: searching for the urban 'sustainability fix' in Manchester and Leeds, *International Journal of Urban and Regional Research*, 28 (3): 549-569.

William M. Lafferty (Ed.) (2001) *Sustainable Communities in Europe*. London: Earthscan.

Williams, C. (2002). Transforming the old: Cairo's new medieval city. *Middle East Journal*, 56 (3): 457-475.

2

Sustainability Challenges: Changing Attitudes and a Demand for Better Management of the Tourism Industry in Malaysia

Janie Liew-Tsonis and Sharon Cheuk
School of Business and Economics Universiti
Malaysia Sabah

1. Introduction

1.1 The acceptance of the sustainable development principles in reference to the brundtland report

Before discussing how tourism can be an essential tool for sustainable development, it is necessary to present the importance of sustainable development as a background for understanding its concept and principles. The notion about sustainable development started to come together in 1983, when the United Nations General Assembly (UNGA) established the World Commission on Environment and Development (WCED). One of the main objectives of the WCED was to prepare long-term planning on environmental concerns towards the year 2000 and beyond. The WCED also focused on heightening cooperation among developing countries, including countries at different stages of economic and social development. This was with the intention of creating mutually supportive objectives which took into account the interrelationships between natural resources and economic development.

The term, sustainable development, was popularised in *Our Common Future*, a report published by the WCED in 1987. Also known as the Brundtland Report, *Our Common Future* included the now accepted definition of sustainable development as development which meets the needs of the present without compromising the ability of future generations to meet their own needs. Acceptance of the report by the UNGA gave the term political salience and in 1992 leaders set out the principles of sustainable development at the United Nations Conference on Environment and Development in Rio de Janeiro, Brazil. This is probably the most crucial conference to date in promoting the concept of sustainable development; the event is now interchangeably referred to as the Rio Earth Summit, Rio Summit or the Earth Summit. The Rio Summit was the largest environmental conference ever held, attracting over 30,000 people including more than 100 heads of states. The objectives of the conference were to build upon the hopes and achievements of the Brundtland Report, in order to respond to pressing global environmental concerns and to agree on major treaties for biodiversity conservation, climate change and forest management. It also focused on environmental development and conceived frameworks for strategies and measures in minimising and controlling the effects of environmental degradation. This had greatly assisted governments in the context of increasing domestic and global efforts in the promotion of sustainable development in their own and UN member countries.

Consequently, the event rapidly contributed directly in shaping the concept of sustainable development which led the United Nations Commission for Sustainable Development (UNCSD) to issuing Agenda 21. This was further affirmed, with a globally accepted political statement, called the Rio Declaration on Environment and Development. Agenda 21 is a framework which provides action plans consisting of forty workable chapters, detailing the future of sustainable development from 1992 into the 21st century. Although Agenda 21 is a nonbinding set of recommended approaches, this had since translated into greater co-operation between countries at different stages of economic and social development in the achievement of global objectives.

It is generally accepted that sustainable development calls for a convergence between the three pillars of achievement: economic development, social equity and environmental protection. Sustainable development is a visionary development paradigm and over the past 20 years, governments and private-sector businesses have accepted it as a guiding principle. This had involved progress on sustainable development metrics, and improved private-sector businesses and NGO participation in the sustainable development process. Yet for many, the concept remains elusive and institutional implementation has proven difficult as unsustainable trends continue and political entry points in making real progress remain generally very limited. As a result, market forces have taken over to become, in all intents and purposes, the understudy for the implementation of the sustainable development agenda. However, frameworks of private sector negotiations are not always appropriate platforms for broader strategic management discussions of sustainable tourism or sustainable development.

Unfortunately, while sustainable development is intended to encompass the three pillars, the general perception is that sustainability is often compartmentalised as an environmental issue. In addition to this, and potentially more limiting for the sustainable development agenda, is the orientation of development growth which is predominantly assessed as, economic growth. This is largely due to traditional economic frameworks used by developed countries in attaining unprecedented levels of wealth; in which rapidly developing countries, including Malaysia, aspires to attain. The dilemma with such an approach is that natural resources are often displaced and / or their quality compromised to an extent which threatens biodiversity and environmental preservation. Due to global changing attitudes and more environmentally knowledgeable societies, the demand for better management of businesses, including that of tourism is increasingly evident.

Although there is increasing affluence and technical capacity in implementing more sustainable policies and measures, the required level of political leadership and the engagement of society in Malaysia is still a long way off. This is compounded by economic growth which follows the resource-intensive model of developed countries. Without a concerted effort in addressing levels of consumerism and resource use, it would be difficult to expect a receptive audience when attempts are still directed towards pure economic development practices. In other words, more sustainable development directions are needed which will require levels of dialogue, cooperation and most importantly, conviction; which are simply not reflected in the practices of multilateral institutions and organisations across tourism and its related businesses.

Even where attempts had been made to turn policies into action, the results have been limited. There is a huge gap between the multilateral processes in which broad goals and strategies had not transpired into national actions, which reflects domestic political and economic realities. Deep structural changes are needed in the addressing the practical side of businesses which allows society at large to manage its economic, social, and environmental affairs. Hard choices may need to be made in translating ideals into workable actions in making tourism more sustainable. However, while fractions argue that Malaysia has been unsuccessful in achieving its sustainable development obligations, it is worthwhile to consider that 20 years is a relatively short time frame to implement the required changes in such a mammoth area. As the country continues to achieve rapid growth, the needed systemic changes will require far-reaching ways in which businesses are managed. This will have impacts on lifestyles and consumption patterns—especially so in a fast developing country, with a rapidly growing middle class. The current global economic and environmental crises and the use of the liberalisation and globalisation models, in the trading of services, could bring renewed receptivity towards a shift in the sustainable development paradigm. The new economic models can demand development directions which could focus on reducing resource use, and integrating economic, environmental, and social issues in policy decision making. The opportunity is certainly there, for progress to go beyond concepts in its move towards actual systemic changes.

2. Malaysia's decision in taking tourism as a sustainable development option and hence, obligation to global agendas

In Malaysia, the tourism industry and its related services have emerged to become the second most important industry over the last twenty years. The industry has remained strong despite several economic slowdowns in Asia and continues to be a key foreign exchange earner, contributing to growth, investment and employment. The allocation of public funds has continued to increase in order to meet the demands of a growing tourist industry. However, global obligations towards Agenda 21 and Local Agenda 21 have not created the planning and management measures necessary to ensure that tourism growth is sustainable. Additionally, new legal measures have not been designed nor implemented, in order for concerns relating to uncontrolled development to be recorded. This challenge is complicated by differing concepts of economic growth and its relevance to sustainable development. The application and approach to sustainable development, including tourism, requires new legal tools which can mitigate and adapt to a knowledgeable traveling market base. Meeting this challenge will involve multi disciplines which are not traditionally deemed to be critical in international negotiations.

The tourism industry has tremendous potential in its contribution to sustainable tourism development, particularly for environmentally and culturally sensitive sites, which can be linked to Local Agenda 21 processes in planning and monitoring. In order to achieve sustainability, the tourism industry is perhaps in the better position in integrating and balancing competing economic, environmental and social interests. These interests are represented as pillars on the sustainable development model. However, the impediments to achieving sustainable development arise not only from an imbalance of the economic, environment, and social sustainable development pillars, but also from political biases. Based on the economic potential tourism can bring to the country, Malaysia has taken this sector as one of its core development options. The government also recognises the vast potential of tourism as a major source of employment which can create a foundation for entrepreneurial resourcefulness. Hence, a great deal of marketing and promotion effort is placed in attracting increase arrivals to a destination which has much to offer in terms of competitive shopping, and natural icons of global significance. This has led to continued year-on-year growth in tourism arrivals, which are grounds for enthusiasm and concern.

There is no doubt that tourism has the potential to generate the needed revenue for the economy, increase awareness for the host environment and culture, in addition to increasing political incentives for the conservation of natural resources. Nevertheless, there are also increasing concerns that poorly managed or uncontrolled tourism can cause more damage to the environment, culture and society in general. Sustainable tourism is supposed to have the potential to meet each of these challenges. However, if sustainable tourism is to fully achieve its potential, well-founded principles and clear guidelines for the active involvement of stakeholders such as planners, developers and private entrepreneurs, are no longer obligatory, but a necessity. The guidance in facilitating site-specific research on socio-economic and environmental impacts of visitors, and the development of appropriate local, national and regional tourism strategies will need to be imposed.

Sustainable Development –the goal universally agreed to at the Rio Summit in 1992 – has become the main challenge, against the background of a rapidly growing tourism industry in Malaysia. Within the concept of the Brundtland Report and the framework of sustainable tourism, it is now necessary to address the key sustainable development issues especially for environmentally and culturally sensitive areas. The perception that sustainable development is a complex process which is too unachievable or simply too difficult cannot be given too much credence. Twenty years after the publication of the Brundtland Report, the concept behind *Our Common Future*, remains true to form and more critical today. The aims and objectives are realistic in that emphasis is focused on its practicality and application. The attention needs to be on manageable steps in which individuals and society can identify which supports sustainability.

There must now be less concentration on the design and promotion on all-embracing theoretical frameworks and idealistic concepts but rather on, applicable practices which leads to sustainable development; concentrating on identifying and applying lessons from practical experience and implementing the principles and commitments that have already been agreed. Consequently, the development of the tourism sector must begin with the goal of achieving a balance of the three pillars, as tourism offers a good starting point for analysing the relationship between sustainable development and progress. Sustainable tourism can be an effective tool in achieving sustainable development because when practiced, it benefits all the three fundamental pillars.

3. The need to address rapid adaption of technological changes, product innovation and new markets

There is no doubt that tourism is a dynamic industry; and with it, continued challenges are expected from its technologically confident market base. With technology developing, tourist experiences are evolving at a rapid pace. Tourists acquire sophisticated and refined tastes and needs in differentiating tourism products and destinations based on a variety of sources. These levels of sophistication has transform into real trends where existing tourism products requires deep innovation; especially as the competitive focus is on the quality in service and comfort, combined with the uniqueness of the destination. In order to achieve continuous product innovation, destinations require a change in organisational policies and practices, at macro and micro levels, as well as, in their corresponding strategies. These changes can impact dramatically on profit margins if the engagement of organisationwide strategies, technology skill development and procedures are not in place in addressing this knowledgeable and technology savvy market base.

The 21st century had brought profound changes in the international and Asia-Pacific markets. Accordingly, the operations within the tourism industry in Malaysia, had faced a series of changes which required new strategies in capturing a bigger share of the global market. Among the changes was the increase in disposable incomes of developing nations, economic crises and natural disasters which had a direct impact on the industry. The adjustment of tourism suppliers in meeting the demand had required modifying the content of tourism products and adapting them to the requirements of global agendas, including the creation of products which meets sustainability concerns.

The changes generated by international trends, which already had an echo among the tourism stakeholders, are obvious. In most cases, the base for the development of new markets and new tourism products is represented by market demand. It is noteworthy to recognise that although in Malaysia, national policies for stimulating mass market tourism is still maintained, there is also a shift towards segmentation, specialisation and diversification of markets especially in the eastern states of Sabah and Sarawak, which focuses on nature and culture, as their primary markets. As travelers are becoming increasingly more sophisticated in terms of destination selection, the industry must adapt to rapid technology changes in order to meet demands. New technology, more experienced travelers, global economic restructuring and environmental limits to growth are only some of the challenges facing the industry. The desire to innovate is increasingly enhanced as it becomes the only way in which a highly competitive industry, such as tourism, can survive and prosper.

In most cases, competition can push destination management organisations to invest in innovation. Unless a competitive advantage is secured through innovations in terms of destination or products offered, there will still be uncertainty, in terms of risks and instability. Consequently, the key to success in meeting market demands will rely on the industry's ability to innovate and take risks. The tourism industry must be aware of and anticipate changes in the global tourism market, or risk losing their share of that very market. In order to increase market share, there is also the need for innovations to be backed up by real strategies in order to secure competitive advantages. The main types of innovation which is likely to trigger interest in the development of tourism is the adaptation to tourist purchasing power and behaviour.

The strategies of tourism operators bear resemblance to those in other fields of activity. In this situation, market size expansion, market share enhancement, cost cutting and product mix adjustments are looked into. As strategies go, they are closely associated with product innovation. For instance, tour operators that offered trips to undiscovered areas, expanding its market share by attracting new customers who had not previously considered such options. Product innovations involve the ability to make one's product stand out from those of the competition. The essence of product differentiation exists in creating a niche as opposed to the competition, either at similar or different approaches, and thus, gains expansion by being a market leader with benefits to match. On the contrary, it is worth mentioning that opening the product market towards mass tourism has been significantly influenced by marketing and advertising playing their due parts, and also by investments in research into new product development and technology.

For instance, the development of holiday packages from Peninsula Malaysia towards the two eastern states of Sabah and Sarawak was heavily influenced by the technological changes in the purchase of plane tickets and the expansion of air accessibility. This has subsequently led to cheaper flights (due to competition), as well as by investments in hotels which further promotes domestic travel. This had underlined the importance of re-discovering domestic attractions whereby the tourism industry took advantage of international trends and adapting them locally. As far as process innovations are concerned, it has been recorded with respect to the length of time it takes to complete travel related transactions and subsequently, bear witness to varied methods of combining decision-making aspects, which led to increased business competitiveness.

Additionally, a combination of marketing innovations (including online retailing of airline tickets or accommodation), selectivity (seat assignment on board the aircraft), and the ability for price comparisons (between airlines) are now available at a touch of a button. Process innovations can be achieved through exclusive technological changes that are linked to the information technology revolution or ones which are specific to the tourism industry. By adapting to e-commerce technologies, the sales process within the tourism industry will have the highest impact. Technology changes the framework of competition regardless of the field of activity, location or size. This has a bigger impact on competitive advantages; as it gives opportunity for the set up of price and product differentiation.

The rapid changes in technology provides the tourism industry with a series of advantages linking it to unprecedented access to markets in terms of distribution channels, pricing policy, cost in shipping, and increase effectiveness. And, technology improvements also offer the travelling markets virtual tourism; even if costs generated by this have displayed a decreased trend in time, innovations constantly need substantial capital investments, either capital-related or pertaining to work force training. Therefore, it can be concluded that certain risks can be associated to technology innovations, despite technological evolutions. Information technology not only changes the volume of information transferred but it also influences the long-term relation within the distribution chain and gives rise to new forms of competition.

This issue has contributed to the metamorphosis of tourism; from a standardised, rigid mass phenomenon to a more flexible, customer-oriented industry which is more sensitive to the latter's needs and expectations. Information technology is certainly deemed as one of the factors which facilitated change. At the same time, it also triggers changes in the traditional organisation of production and can decrease the level of dependence on tourism agencies, as customers can purchase airline tickets or holidays directly via the internet.

Additionally, advancement in technology has contributed in some cases to the disappearance of customer relationships. From the perspective of information technology use, it is also noted that there are differentiations between tourism service providers, as not necessarily all markets enjoy equal rights to accessibility and has the required technical skills. Competitive capacity is still reliant not only on the development and implementation of new technologies but also on the capacity of the tourism industry in learning and adapting itself to changes. Further, the competition is not limited to the way in which the destination is positioned but also, can be due to its performance in the market. The cooperation within increasing levels of competition can be seen particularly in airline and hotel alliances.

4. The adoption of knowledge management strategies and governance in managing tourism development

The tourism industry has become one of the most dynamic industries globally; and, rapid adaptation to technological changes, product innovation and new markets must be explored. The United Nations World Tourism Organisation (UNWTO), which is responsible for overseeing global tourism growth, states that "… tourism is firmly established as the number one industry in many countries and the fastest-growing economic sector in terms of foreign earnings and job creation." The UNWTO also shows that growth of the tourism industry is indeed remarkable, with the number of international arrivals documenting an evolution from a mere 165 million international arrivals in 1970 to over 846 million in 2006. Moreover, the UNWTO forecasts an increase in the arrival number of international tourists to 1.6 billion in 2020. The critical relationship between the tourism industry and sustainable development is an examination of global trends and the challenges they raise. As environmental and societal decisions can be irreversible, individuals who hold key positions in determining policies which affect society must identify and solve unstructured problems which require the use of multiple information sources.

In Malaysia, the tourism industry's ability to maintain global arrivals in the future will depend largely on solid research, in order to better understand and accept new trends and concepts as they appear. The wellspring of future tourism growth in Malaysia is a commitment to good and structured frameworks as determined by differing policies in governance. The Government whether national, state or local, the private sector and non-government organisations (NGOs) is working cohesively in making this a reality. It realises that managing tourism growth depends on forward-looking policies and sound management philosophies which include a harmonious relationship amongst the public sector, private sector, non-government agencies and society at large.

If the assumption that policy decisions are made based on the delivery of rapid development which serves only to address tourist needs, this will undoubtedly cause setbacks to vital questions about the future of the destination, appropriate scale and type of development, and residents' quality of life. This is largely because this said assumption is questionable at times of rapid social change, especially where tourists' knowledge of sustainability, is not taken into consideration. The rise in tourism and international arrivals can be explained by many factors, including population growth, increased tourism segmentation, the development of information technology, and marketing. The internet has transformed the tourism industry, providing a medium for marketing through websites, email, and website pop ups in addition to a dissemination of advice on where and how to get to the best destination, and a means for reserving airline tickets, car rentals, and hotels.

Consistent with the growth of information technology is the increase in the use of digital cameras and social networks such as Twitter and Facebook, which allow tourists and marketers alike to immediately share images of exotic destinations all over the world via email, blogs and websites. Such exchanges are increasingly influencing travel decisions. Furthermore, increasing efficiency in the transportation sector, such as aviation, increases the range, and capacity of travel. Despite discouraging factors, such as acts of terrorism (Bali bombings, 9/11), the economic crisis (USA, UK and Europe), and natural disasters (earthquake and tsunami in Japan), which sometimes slow the growing rate of tourism. However, the overall long-term growth rate of tourism will continue to increase.

In exploring current solutions, the management of tourism development offers insights for policy makers in seeking better solutions when confronted with issues on sustainability. The continuing importance of market-based strategies for balancing the development of the tourism sector demands a change in the way sustainability is viewed. As the tourism industry has become one of the fastest and largest growing economic sector in Malaysia, it can make an important contribution to sustainable development. Nonetheless, the implementation and development of policies, plans and strategies regarding sustainable tourism is a challenge for policy makers in balancing economic growth with sustainability.

Indeed, many methodologies and assessment tools for sustainable tourism management have been developed by researchers, which recognise the multiple facets of sustainability. Amongst them, Spenceley (2003) developed the Sustainable Nature-based Tourism Assessment Toolkit (SUNTAT) which provided a mechanism to measure sustainable tourism at two levels; strategic (for policymakers and planners) and enterprises (for tourism enterprises and developers), and took into account policy and planning, economics and tourism management, environmental and conservation management and social and cultural issues. Cernat and Gourdon (2007) developed a Sustainable Tourism Benchmarking Tool (STBT) in order to detect sustainability problems in a tourism destination. The tool, using benchmarks and policy-relevant indicators, was also aimed at enabling policymakers to make informed decisions and improve the prospects for sustainable tourism development in their respective countries. The STBT encompassed the key dimensions of economic sustainability (tourism assets, tourism activity, linkages, and leakages), socio-ecological sustainability, infrastructure sustainability and destination attractiveness.

Although there is much discussion that the tourism industry will be affected with the global economic crises, it is clearly evidenced that there will always be a market for tourism, due to an individual's wish to travel and discover new and foreign places. It is without doubt that the tourism industry has proven to be resilient and will continue to expand. As sustainable tourism is not only about protecting the environment, and tourism initiatives do bring benefits to some people and costs to others, what level of development and where to develop has become not only an academic field of study, but a highly political one. In any case, progress towards more sustainable forms of tourism will depend far more on the activities the industry and the attitudes of tourists, rather than solely the actions of public sector bodies and policies. In the highly competitive tourism market, success will come from understanding the target markets, and focus will be on the most profitable prospects in terms of motivation to visit, economic yield and appreciation of what Malaysia, as a destination, to ensure its long term security as a profitable industry.

There is also continuing efforts to promote domestic tourism for future growth, within the thirteen (13) states of Malaysia. The lack of direct international access to Sabah and Sarawak, the two eastern states of Malaysia, will be the key constraint to these areas. There is clear emphasis that an expanded direct international service and more domestic links are important initiatives for the respective State governments. However, any major changes to the current scenario will not be achieved until open air accessibility is further developed to stimulate commercial demand for access. As international interest grows in authentic destinations, these two eastern states are positioning itself as nature and cultural hubs as specialty destinations, based on natural and cultural experiences.

5. The growth of tourism in Malaysia

The main government machinery behind tourism development is the Ministry of Tourism. This was first established as the Ministry of Culture and Tourism (MCT) in 1987, and further designated in 1990, as the

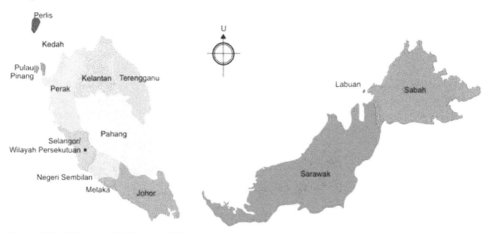

Fig. 1. The thirteen (13) States of Malaysia

Ministry of Culture, Arts and Tourism (MOCAT). In 2004, it reaffirmed itself as The Ministry of Tourism (MOTOUR) in order to fully reflect the responsibility of coordinating and implementing government policies and strategies pertaining to tourism development in the country. MOTOUR's vision is to develop Malaysia into a leading tourism nation and its mission is in the implementation of the National Tourism Policy and building the industry to be the nation's main source of income for socio-economic development. Currently, tourism and its related services is the second largest contributor to the nation's Gross Domestic Product (GDP) and contribution to employment is 1.7 million jobs.

The Malaysian Tourism Promotion Board, or commonly referred to as Tourism Malaysia, is a statutory body established under the Malaysia Tourism Promotion Board Act 1992. Tourism Malaysia's ultimate aim is to increase the number of foreign tourists to Malaysia, extend the average length of their stay and increase Malaysia's tourism revenue. Its activities centre mainly on promotion and the increase of arrivals for both international and domestic tourism. It also coordinates all tourism related marketing and promotional activities conducted by any organisation; government, non-government or private sector. Tourism Malaysia also offers recommendations for the adoption of appropriate methods, measures and programmes, in order to facilitate or stimulate the development and promotion of the tourism industry within the country.

Malaysia is made up of thirteen (13) states; each has its own government and autonomy in its decision for tourism development. Eleven of the states are located in the Malaysian Peninsula (West Malaysia) and two in the Island of Borneo (East Malaysia). In addition to the thirteen states, Kuala Lumpur, Labuan and Putrajaya are called Federal Territories. At the national level, the direction for tourism development is led by the Ministry of Tourism. At state levels, such as Sabah and Sarawak, this is led by the Ministry of Tourism, Culture and Environment and Ministry of Tourism and Heritage, respectively. Not every state is represented by a designated Ministry for tourism development; of importance, as Sabah and Sarawak relies on nature and culture as the foundation of their tourism attraction, there is more emphasis for both these East Malaysian states to focus on sustainability in the development of tourism.

Tourism development is continuing to be an important economic activity for Malaysia. In the 1960's, tourism was virtually unheard of as an economic option, let alone as a sustainable development option. With complete autonomy to decide on how tourism is developed within the country, each State Government is not only in competition with another state, but also in competition with similar destinations regionally and around the world. This is evidenced by the year-on-year increase of funding allocation towards tourism development and promotion within the country. From the 1970s, the Malaysian government's priority was the provision of basic tourism infrastructure, like highways, airports and upgrading of attractions in each state. This was fast tracked in the 1980's, as a means to meet several development objectives. Tourism was actively promoted, and the focus was in terms of marketing and the improvement of services.

It was anticipated that tourism will increase foreign exchange earnings, lead to an increase in employment, and raise the standard of living of the population; in addition to fostering regional development. However, in spite of two decades of rigorous public sector intervention, tourism was only identified and taken seriously as an industry that had the potential to stimulate the socio-economic development of the country in the late 1980s to mid-1990s, when economic recession hit the region. In order to boost private sector investments, the Malaysian government concentrated on promoting tourism by providing incentives to develop accommodation, visitor centre facilities and actively encourage Bumiputera participation. Bumiputera or Bumiputra is a term widely used in Malaysia. This category of the populace embraces ethnic Malays and other indigenous ethic groups in the eastern Malaysian states of Sabah and Sarawak.

The sustainable development approach is particularly important in tourism because this sector depends almost entirely on attractions and activities that are related to the built and natural environment; including any historic and cultural features these extends to in society. The tourism industry is both dynamic and diverse. Services and tourism go hand in hand

and to implement a tourism strategy, or make informed decisions involving major tourism projects and events, there is a critical need to have the right information and decision criteria. The concern for the impact of tourism is not confined to developed countries; it is part of a growing concern in regard to the sustainability of tourism in Malaysia, as well. Although meeting the needs of travelers by providing tourism related goods and services have proven to be an attractive form of economic development, unplanned tourism growth can also lead to problems. Attempts to encourage the development and growth of tourism activities are often desirable because tourism creates jobs and offer much needed foreign exchange into the country. However, year-on-year visitor growth into areas which hold the major tourism attractions is likely to have impacts on the environment on site and society, as a whole.

The complex interplay of balancing the sustainable development pillars includes economical, environmental and socio-cultural dimensions. This requires a broad understanding and acceptance from different stakeholders, working directly and / or indirectly in the tourism sector. Malaysia is ranked 16th in terms of global inbound tourism receipts, capturing approximately 2% of global market share in 2008. Of increasing relevance, the tourism industry provides 1.7 million jobs or approximately 16% of total employment in 2008. From its 11th position in 2008, with 21.3 million international arrivals, Malaysia entered the UNWTO *Top 10 Major Tourism Destination Ranking* in the 9th position in 2009. From 2006-2009, revenue from the tourism industry increased 67.1% to MYR 53.4 billion and tourist arrivals increased 43.6% to 23.6 million. For 2011, the Ministry of Tourism is targeting an increase in arrivals to 25 million. National planning and budget allocation in Malaysia is projected every five years. For the latest, 10th Malaysia Plan period of 2010-2015, the target is to improve Malaysia's position to be within the Top 10 in global ranking, in terms of tourism *receipts* and increase the sector's contribution by 2.1 times. It is also projected that this will contribute MYR 115 billion in receipts and provides 2 million jobs for the industry by 2015.

However, the projections for Malaysia from the World Travel and Tourism Council (WTTC), is more conservative. WTTC anticipate that direct contribution of the travel and tourism to Malaysia's GDP for 2011 is expected to be MYR 56.9 billion, which represents 7.2% of the total GDP), and expected to rise by 5.1% per annum to MYR 93.6 billion (7.7%) in 2021. This is expected to support 768,000 jobs directly; representing 6.7% of total employment. Despite these achievements, several issues need to be addressed, including the need to develop vibrant and iconic tourism products, improve maintenance of existing tourism sites and adoption of more focused tourism promotions.

It is without doubt that the key trend of the tourism industry in Malaysia is that, it is on its way up. This trend points to the increasing role and significant contribution of the industry to the Malaysian economy, particularly in the next ten years. From virtually a zero base in the 1960's, there are now clear indications and a recognition that the tourism industry will continue to play a significant role in the Malaysian economy. In order for tourism to be sustainable, it is vital that effective policies and planning take place *today*. Hence, policy makers, planning officials and stakeholders must identify emerging trends in tourism and orchestrate new measures that will lead to orderly growth (and production of quality products) which benefits both tourists and society. While domestic tourism is more difficult to measure, the figure is often estimated to be up to ten times the number of international visitors. The statistics demonstrate that tourism can create economic development, investment, and income growth within the country. However, the rapid growth of tourist arrivals can also increase pressure on the natural, cultural and socio-economic environments.

Tourism development in Malaysia, like most parts of the world, is growing at a rapid pace and no doubt, is an essential part in the trade in services and economic development. With continued enhanced spending, bringing with it widespread development, an understanding of the complexities and relationships which co-exist in tourism and the roles played within the concept of sustainable development remains unclear. With the expected rise in tourist arrivals, comes the complex interplay of the social, cultural and environmental dimensions of which sustainable tourism commits to. Therefore, a broad understanding, acceptance and commitment of the different stakeholders, either working directly and indirectly in the industry, must be in place. To capitalize on the emerging trends and opportunities, the tourism industry and the public sector will need to improve on the markets' knowledge of Malaysia and what it has to offer.

In determining the direction of the government's aspirations and policies, lie the challenges and interplay of global declarations, such as Agenda 21, and the creation of sustainable policies at national and local levels. These will cumulatively, lead to sustainable development of the tourism sector amongst its thirteen states. In making the decision to take tourism as a sustainable development option, tourism must be planned and managed so that its influences on environmental, socio-cultural and economic benefits are spread widely throughout society. The question is – can Malaysia remain competitive if the global concern for sustainability is not addressed today?

6. The potential and pitfalls in tourism development

Different approaches to the valuation of goods and services can lead to an inconsistent reporting of outcomes. The dilemma lies in the different approaches to the valuation of goods and services by the three fundamental stakeholders within the tourism industry; policy makers (Government direction and guidelines), the private sector tourism industry (in the business of making money) and local residents (the victims or the beneficiaries?). Sustainable tourism represents a value point of referencing in which the management of tourism impacts takes precedence over market economics – although tension between the two is ever present. Further, tourism impacts rarely take precedence over market economics in practice, even though the term *sustainable tourism* is often used. In tourism ventures within natural environments, economic success and environmental impacts are often negatively associated; therefore, policy makers and the private sector must find a balance between the interest *for* nature and the *impact* of outside influence to these natural areas. As tourism is essentially dependent on the unspoilt nature of a destination's attractions (natural or built), it follows that tourism has the responsibility for, and a need to invest in, the maintenance of the natural environment.

Tourism's relationship with the environment is complex. It involves many activities which can have adverse environmental effects. Tourism, as an industry can no longer claim its success based on economics alone. Many of these negative impacts are linked to the construction of general infrastructure such as roads and airports, and of tourism facilities.

Water, as a natural resource, is often exploited where tourism facilities are concerned. Infrastructure built specifically to generate tourism arrivals such as hotels and resorts, swimming pools and golf courses, generally overuse water which can result in water shortages and degradation of water supplies, as well as generating a greater volume of waste water. A common example is that an average 18-hole golf course built in a tropical country, such as Malaysia, needs a minimum of 1500kg of chemical fertilisers, pesticides and herbicides per year and uses as much water for a minimum of 60,000 adults. Without doubt, the quality of the environment, both natural and built, is essential to tourism. What needs to be recognised is that the negative impacts of tourism development can gradually destroy the natural resources on which it depends, if infrastructure is indiscriminately developed for tourism.

The negative impact of tourism usually occurs when the level of visitor use is greater than the environment's ability to cope with this use, within the acceptable limits of natural changes. Uncontrolled tourism development generally poses enormous pressures on any natural area, and the quality of the environment, is unquestionably essential to the success of tourism. In spite of the large amount of literature available documenting the pressures which unplanned and uncontrolled tourism development can have on natural resources, especially in cases of consumption increases in areas where resources are already scarce, the success of tourism is still being qualified by its economic contributions.

Although tourism can cause the same forms of pollution as any other industry like air emissions, noise, solid and liquid waste, littering, sewage, oil and chemical releases, the increase in transport by air, road, rail or sea amplifies an eroding quality of air. Other negative impacts such as land degradation and pollution are also not taken into consideration when economic benefits are recorded. Tourism can create pressures on local resources such as energy consumption, food supply and other raw materials that may already be in short supply. Greater extraction and transport of these resources intensify the physical aspects associated with their exploitation. Due to the seasonal nature of the industry, during peak season, transient residents such as visitors can be as high as ten times that of the low season for some areas. The higher demands and expectations of tourists places pressures that surpasses the natural development process caused by the increased construction of tourism and recreational facilities, including land clearing and extraction of resources for the use of building materials.

As tourist arrivals continuously increase, year-on-year, tourism is now responsible for an important share of carbon emissions. It is estimated that a single transatlantic return flight emits almost half the carbon dioxide emissions produced by everyday usage and consumption of sources, such as lighting and air-conditioning in vehicles, consumed by an average person yearly. Transport emissions and emissions from energy production and use are linked to acid rain, global warming and air pollution. Some of the impacts are quite specifically caused by tourism activities. For example, tour buses often leave their engines running for hours while waiting for arrivals at airports or while waiting to go out on excursions and to, during and from excursions.

However, it has also been maintained that if properly managed, tourism has a huge potential in creating positive effects on the environment by contributing to environmental awareness, protection and conservation. An example made popular by the travelling public is that tourism, unquestionably, has facilitated a greater awareness of environmental values. Tourism has also successfully served as a tool to finance the protection of natural areas and thus, led to an increase in their economic importance. Therefore, there is an obvious need to balance the use of tourism as a driver of economic development and the management of its public resource consumption. With the continuous increase in tourism arrivals, there is a need to find a balance between the maximisation of income from tourism and that of exploiting resources beyond its extent of being sustainable. The latter would be equivalent to killing the familiar goose which lays the golden egg, and does not make economic sense for either the public or the private sector.

The indication is that sustainability or conservation can mean very different things in different environments and social circumstances, which makes the practicing of the concept even more challenging. If Malaysia wants its tourism industry to succeed, it needs suitable plans and policies to enhance the development of tourism. A large part of tourism is about travel and the role of the public and private sectors must come together in addressing common goals. In Malaysia, it is largely due to the improvement of such private-public relationship which had contributed to the expansion of tourism. Inevitably, in order for all stakeholders to benefit from tourism, attention needs to be given towards the perceptions and understanding of the participants, including the tourists. This need to take into account the level of involvement each brings to the table and impacts of such involvement. Nevertheless, it is also critical to accept the roles played by the different stakeholders and their inter-relationships. In order for tourism to be sustainable, it is vital that a win-win situation be identified and its implications for sustainable development which benefits economic, social and environmental carrying capacity. Whichever way you look at it, sustainability of the natural and cultural environment is an integral part of the tourism industry.

7. The development of sustainable tourism

The tourism industry is important to Malaysia, as it assists in fulfilling global agendas, whilst serving as a source for economic development. Tourism, like many other sectors of the economy, uses resources, generates wastes and creates environmental, cultural and social costs. The processes laid out in Agenda 21 (and, Local Agenda 21) revitalized the commitment on goals and objectives set out at the Rio Summit which addressed environmental conservation and socio-cultural interactions. The main issues for integration under Agenda 21 are within the areas of social and economic involvement of major groups such as women, children, youth, non-governmental organisations and local authorities; especially in the promotion of education, public awareness and training.

The Ministry of Tourism has advanced the development of tourism by concentrating on the development of policies, strategies and master plans for sustainable tourism. This had required continuous cooperation and consultation among all stakeholders, including the private sector, academic institutions, local communities, and relevant non-government organisations. It also called for capacity building across sectors and public participation to include and involve rural communities. Policies had included financial support and incentives for tour operators and accommodation investors to play a continuing role in developing the tourism sector; however, these policies had not included programmes nor indicated tools and instruments to cover appropriate institutional, legal, economic, social and environmental monitoring frameworks, nor do they include voluntary initiatives and agreements such as a commitment to corporate social responsibilities (CSR), amongst the stakeholders.

A lot of sustainability debates in Malaysia are about process failures; failures to engage and listen to the right people and organisations in the consideration of long-term, as well as immediate impacts. To take into account wider and less obvious upstream or downstream effects, consultative decision-making can narrow gaps between differing motivations to development and quite rightly, there has been much attention in addressing these deficiencies. It has to be noted that progress has been made in many areas with a commitment to open and transparent participative decision-making. This is accepted as an essential foundation for sustainable solutions across the national Government and its agencies.

To further enhance the sustainability agenda, the extent and depth of discussions has revealed many gaps in the lack of analytical tools in assisting comprehension and acceptance. There has also been a growth in the number of conceptual approaches such as life-cycle analysis, which has been vital in providing a structure for analysis, especially in the identification of priorities and the monitoring of progress in fractions of government agencies. But it is important to keep in mind that such techniques are tools to help decision-making and they do not make the decision itself. All they can practically do is provide direction to ensure that all the economic, social and environmental factors are identified. But assessing the trade-offs between, say, jobs, social cohesion and environmental damage will always be a political decision, and not an academic equation; thus, clear and defined net cost/benefit conclusions are not always the main consideration. This broadly considers how sustainable development has developed in Malaysia and continues to be managed in recent years.

8. Tourism as a tool for sustainable development

Tourism is considered to be a trade in services and differs from the trade in goods. Tourists travel to consume at the host destination, and in the case of international travel this is often, referred as cross-border consumption. With this, the development of tourism has direct environmental, economic and socio-cultural impacts on the consumption patterns within a host country. In terms of economic development, the appeal of tourism is that it can create jobs and stimulate business opportunities. It also generates foreign exchange earnings, injects capital and new money into the development of the local economy. Tourism contributes to government revenue generation through taxes and levies, either directly and/or indirectly. It can also stimulate regional development and the development of infrastructure such as roads, airports, and improve telecommunication links. Therefore, the quality and standard of life for local residents can be enhanced by economic diversification through tourism. Bearing in mind that tourism development can have a positive or a negative impact, or both, it is considered to be a powerful tool for sustainable development.

Where environmental and cultural conservation can lead to an increase of economic opportunities through the development of tourism, this can also be an incentive to protect natural resources; rather than allow for further degradation. Tourism can create financial resources which can be used for overall conservation programs and activities, such as improved park ranger salaries, park maintenance, and the establishment of more national parks and protected areas. Tourism can also significantly contribute to environmental protection, conservation and the restoration of biological diversity and the sustainable use of natural resources. By nature of tourism's demand of aesthetic standards, pristine sites and natural areas are considered valuable for it to be an attraction; and the need to keep the attraction appealing can lead to the creation of additional natural areas to be protected as public parks. These protected areas can then further contribute to sustainability, in ways that have not yet been fully explored; such as providing a base for new medical treatments or new industries or just by serving as carbon sinks. On the other hand, negative consequences from tourism arise when the level and type of visitor use is greater than the environment's ability to cope with this use within the acceptable limits of natural changes.

For example, everyone drinks water and generates waste. Uncontrolled conventional tourism poses potential threats to natural areas if visitors are not managed properly. It can also put a strain on water resources, which can lead to local populations competing for the use of critical resources and increased costs for purification. Uncontrolled tourism development can also put enormous pressures on an area and lead to land degradation, increased pollution and discharges of solid and liquid waste, habitat loss, and heightened the vulnerability of environmentally sensitive areas such as marine and terrestrial habitats. The negative physical impacts of tourism development include construction activities, infrastructure development, deforestation and unsustainable use of the land. The unplanned rapid development of tourism can also create significant social disruptions and increase environmental and ecological pressures.

The government's focus on the economic pillar alone can lead to deficiencies affecting the environmental and social pillars. Tourism development is often compared to a two-edged sword. On the one hand, it can be a tool for sustainable development. On the other, if not managed adequately, tourism can significantly impede sustainable development. Of central importance, national policies need to reflect a high level of commitment to environmental management, which include strategies to effectively limit social and environmental impacts, in the short and long term, which ensures equitable sharing of benefits.

9. Malaysia's response to the principles of sustainable tourism development

Today, sustainable development is a core issue implicating every step of development. Sustainable development aims to allocate the limited natural resources not only for this generation, but also for future generations. It aims to balance development and the environment, in addition to, maintenance of an appropriate balance between economic and environmental development. As the travel and tourism industry continues to grow in Malaysia, the industry must face up to serious and difficult choices about its future. The decisions made now will, for years, affect the way of life, standards of living, and economic prospects of residents in the country. Based on the attractions which Malaysia is promoting in order to differentiate itself regionally and internationally, many of these decisions may be irreversible. Once Malaysia loses its character which makes it distinctive and attractive to tourists, it will also lose the opportunities that go with a tourist-based economy which is increasingly competitive.

In ensuring that development is going in the right direction, a number of political actions have been determined. Sustainable development has become a principle practice and many of the recommendations of Agenda 21 have guided the new pathway of policies in addressing the issues. In the recently released 10th Malaysia Plan (2011-2015), the recognition that the quality of the environment, both natural and man-made, is essential to tourism is duly acknowledged. However, tourism's relationship with the environment is complex. It involves many activities that can have adverse environmental effects. Many of these impacts are linked with the construction of general infrastructure such as roads and airports, and of tourism facilities, including resorts, hotels, restaurants, shops, golf courses and marinas. The guidelines and details of how negative impacts of tourism development, which can gradually destroy natural resources on which it depends, is however, lacking.

The conflict of interest between public and private sectors in tourism development has been well-documented. If private enterprises remain, to a great extent, unregulated, it is expected that short-term profit maximisation will be sought over and above the interests of sustainability. The growing attention towards tourism development in Malaysia has been centrally driven by the potential economic benefits the industry can bring. This traditional view is now complicated not only by trends towards more socially and environmentally responsible travel but also in the intangible nature of the industry whereby stakeholders are seeking direct involvement in the development of tourism, especially where natural and socio-cultural integrities are concern.

To achieve the 2015 targets set under the 10th Malaysia Plan, the focus will be on attracting a larger share of high spending travelers; and capturing a higher share of high growth segments. The target segments are Russia, India, China and the Middle East, in addition to increasing the overall total number of tourist arrivals from current markets. For this purpose, select key strategies are promoted. The marketing approaches for unique and distinctive travel patterns and needs for visitors are defined. The aim is to attract the markets seeking nature and adventure (including ecotourism), cultural diversity, family fun, affordable luxury, and Meetings, Incentives, Conferences and Exhibitions (MICE).

There is also emphasis for the improvement of existing tourism products through the creation of focused tourism clusters that will leverage on existing and new iconic tourism products, which supports sustainable tourism; such as natural icons within the thirteen states in Malaysia; e.g. the Geopark and Pulau Payar Marine Park in Langkawi, Georgetown (a UNESCO World Heritage Site in Penang), Sipadan Island and Kinabalu Park, (a UNESCO World Heritage Site in Sabah) and the Sarawak Cultural Village and Gunung Mulu National Park (a UNESCO World Heritage Site in Sarawak). Subsequently, considerable budgets are allocated for the improvement and maintenance of tourist sites through multiple approaches. The framework for funding mechanisms is anticipated to be via Government linked companies and corporate sponsorship. It is also expected that stronger enforcement and imposition of entrance fees, particularly in environmentally sensitive and heritage sites, will further enhance the importance in conserving these sites.

It has also been recognised that by realigning promotional and advertising activities with the physical presence of Tourism Malaysia offices overseas, this will in turn enhance Malaysia's presence with the identified core market segments. To further affirm the government's commitment to sustainable tourism development, the beginning of a progressive certification of tourism products and activities will be implemented to ensure that quality, sustainability and safety are accentuated. With this in mind, particularly where national guidelines exist, the concept of sustainable development had influenced legislation, and policies are continuously building on recommendations from the Brundtland Report and processes within Agenda 21.

10. Conclusion

The debate over environmental conservation and protection is often about the balance between leaving the area in its natural state and / or exploiting it for economic development. This choice is often fraught with pressures from stakeholders with differing motivations. Thus, an atmosphere of mutual cooperation is necessary to ensure that tourism economic opportunities are translated to social and environmental benefits for all and not limited to segments of society. One of many ways to achieve this is by forming strategic alliances with partners from all sectors in creating and establishing links between sustainable economic development and the government's social and environmental obligations.

The call from the UNCD since the Brundtland Report remains as valid and urgent as ever today. The description of similar processes identified in the Brundtland Report has been applied to organisations, including tourism businesses, and these have to some extent encouraged real progress towards the vision for sustainable development. Although it had been necessary to make changes in achieving the Brundtland Report's goals for sustainable development, it is easy to overlook that progress had been made. The continuous effort now is to concentrate on replicating and dispersion of that progress. The twenty years since the Brundtland Report has seen all the processes associated with globalisation developed at a bewildering pace. This, in turn, has helped encourage much greater awareness and understanding of the economic, environmental and social challenges across the world.

There is a strong recognition that effective sustainable development can only be achieved with the engagement of society as a whole. Sustainable development thinking needs to be integrated and ingrained in policy and decision-making at all levels. This had required the development of processes and procedures from the United Nations through all levels of government and individual businesses and organisations. Over the next twenty years, the aim is for sustainable development processes to improve to an extent whereby, sustainability thinking is the norm which does not require special units, procedures or written instructions as it does now.

The same can be said of that which has been applied to Corporate Social Responsibility (CSR). The concept of CSR has become so ingrained in some business practices that it is a self promoting and sustaining end in itself. This has since provided a broad framework for businesses in generating profits which maximises positive contribution to society. Although tourism as an industry will never be completely sustainable, as every industry has its impacts, it can work towards being more sustainable in many ways. Key issues identified include the need for responsible planning and management, where a balance must be found between limits and usage so that changes are monitored. In Malaysia, there remains a strong sentiment that environmental management is the responsibility of the public sector, as effective implementation of sustainable development practices and environmental monitoring, is seen to be for the long-term. This has proven to be a difficult task as methods of information gathering had not expanded, and reinforced basic communication, intellectual, and interpersonal values, are not embedded into private-sector management practices.

Tourism development has helped with environmental protection by creating awareness on its values and in generating mass opinions for conservation; it has also contributed to the promotion of intercultural understanding and acceptance within the country itself. As Malaysia has a multiethnic population, tourism has narrowed the gap between the different ethnic groups. As tourism undergoes fundamental changes globally, from the experiences sought, to setting demands on regulations and an increase in budget for environmental protection, signs of these shifts have been progressively evident. This has varied from statements on natural and cultural values to stipulation of conservation fees. In spite of these, the challenges deriving from differing policy approaches indicates that the interests of all parties are not safeguarded from commercial exploitation and tourism benefits are not transcribed to society and tourists alike.

More attention can be paid to applying existing techniques, concepts and tools where they can add most value. There is certainly the need to build on and exploit the core competencies of different stakeholders. The Malaysian Government has set the broad policy frameworks and through its fiscal, regulatory, incentives and disincentives, catalysed and inspired actions which support sustainability. The Brundtland Report has set out to do, and has done well, in providing a platform and focus for exchanging experience and learning between governments, society and businesses. This had encouraged the networking and scaling up of successes in practical progress whereby sustainable development can work. However, there is a need for a more cohesive interaction amongst implementing stakeholders in embracing tourism within the sustainable development option; taking into consideration the varied depth of knowledge, working practices and priorities of the different stakeholders. The linkage to sustainability and key issues including the need for responsible planning and management is found, between limits and usage, and changes monitored. This requires long-term management and recognition that change is often cumulative, gradual and irreversible.

Hence, in order to address the sustainability of tourism, the economic, social and environmental aspects of sustainable development must include the collective interests of all stakeholders. The tourism industry in Malaysia has to face some serious and difficult choices about its future. The decisions made now will, for years after, affect the lifestyles and economic opportunity of the country. For any tourism development to have the desired effect, it is necessary to position tourism as an improvement to the quality of life for society at large and not just to tourists and visitors. The pitfall of a bottom-up or top-down approach in tourism intervention is largely influenced by the values, rights and responsibilities of the implementing stakeholders. The ability to use data, exercise judgment, evaluate risks, and solve genuine and emerging concerns will determine options for continuing on the road of the past, or address emerging concerns about the rapid development of the industry now to ensure that its future remain just as profitable. The effects on the quality of life of host residents can no longer be determined by past performance.

Tourists, visitors and residents are increasingly demanding that the industry pursue sustainability and care of the environment; as opposed to unconstrained economic growth. It can be argued that the GDP or per capita income is incomplete measures of well-being. These measures not only do not accurately portray the distribution of economic benefits among local people nor do they realistically reflect on important quality of life factors, social distribution of existing and anticipated costs and benefits of resource use. Although by its very nature, the concept of sustainability makes it difficult to coordinate and monitor, the implementation of sustainable development in the formulation of effective policies and, the deciding factors will have to be how policy assessments are based. Globally accepted principles can contribute to accelerated and effective implementation of sustainable development, even where the prevailing institutional approaches compartmentalise ecological, social, economic and cultural issues, as separate factors. In conclusion, there is certainly a need to systematically explore the linkages that exist, whether recognised or not, between tourism, the environment and sustainability as the Brundtland Report recommended more than twenty years ago.

11. Acknowledgement

The empirical findings of this chapter are part of a research project on "Redefining Tourism Management: Identifying Critical Success Factors on Tourism Sustainability and Corporate Social Responsibility". This was funded by the Ministry of Science and Technology under its Science Fund Project No: 06-01-10 SF0134 from 2009–2010. The authors hereby express their sincere gratitude to the Ministry of Science and Technology Malaysia, for the opportunity of this research.

12. References

Abdul Aziz, W, Hani, N, Musa, Z., (2007). Public-Private Partnerships Approach: A Success Story in Achieving Democracy in the Home Ownership for Urban Inhabitants in Kuala Lumpur, Malaysia.

Ap. J., (1992), Residents Perceptions on Tourism Impacts, *Annals of Tourism Research,* 19(4), pp 665-690.

Ap. J, Crompton, J., (1993), Residents Strategies for Responding to Tourism Impacts, *Journal of Travel Research,* 32(1), pp 47-50

Ayala, H., (1996). Resort Ecotourism: A Paradigm for the 21st Century, *Cornell Hotel and Restaurant Administrative Quarterly,* 52 (16), 256-269.

Baloglu, S, McClearly, K., (1999). A Model of Destination Image Formation, *Annals of Tourism Research,* 26 (9), 868 - 897.

Bernama, (2009). New Economic Model Expected in Second Half 2009, retrieved March 11, 2011 from http://www.bernama.com/bernama/v5/newsindex.php?id=414374

Biehl, D. (1991), The Role of Infrastructure in Regional Development. (pp. 9-35). *In Vickerman, R.W. (eds). Infrastructure and Regional Development. European Research in Regional Science.* I. Pion. London

Biederman,P.,(2008), *Travel and Tourism: An Industry Primer,* Pearson Education, New Jersey

Baloglu, S and Mangaloglu, M. (2001). Tourism Destination Images of Turkey, Egypt, Greece and Italy as Perceived by US based Tour Operators and Travel Agents. *Tourism Management.* 22:1-9.

Baloglu, S and McClearly, K. (1999). A Model of Destination Image Formation. *Annals of Tourism Research.* 26:868-897.

Bond, P. & O'Flynn, D. (2005, August 2). London Olympics 2012: corporate greed and privatization, *World Socialist Web Site.* Retrieved April 6, 2009, from http://www.wsws.org/articles/2005/aug2005/olym-a02.shtml

Butler, R., (1980). The Concept of a Tourist Area Cycle of Evolution: Implications for Management of Resources, *Canadian Geographer*, 24(1), 5-12.

Butler, R., (1991). Tourism, Environment, and Sustainable Development, *Environmental Conservation*, 18(3), 201-209.

Cater, E. and Lowman, G., (1994). *Ecotourism: A Sustainable Option?*, John Wiley, Great Britain

Ceballos-Lascurain, H. (2001). *Integrating Biodiversity into the Tourism Sector: Best Practice Guidelines*, UNEP/UNDP.

Cernat, L. and Gourdon, J. (2007). *Is the concept of sustainable tourism sustainable? Developing the Sustainable Tourism Benchmarking Tool (STBT)*. New York and Geneva: United Nations. Retrieved September 5, 2011 from http://www.unctad.org/en/docs/ditctncd20065_en.pdf.

Cheuk, S., Liew-Tsonis, J., Phang Ing, G., Razli, I., (2009). An Establishment of the Role of Private and Public Sector Interests in the Context of Tourism Transport Planning and Development: The Case of Malaysia, *Proceedings from EABR Conference, Prague, Czech Republic*, June 8-11, 2009.

Cook, R, Yale, L, Marqua,J, (2010). *Tourism: The Business of Travel*, 4th Edition, Pearson.

Cook, R, Yale, L, Marqua, J., (2006). Tourism: *The Business of Travel*, 3rd Edition, Pearson International Edition

Chon, K., (1991). Tourism Destination Image Modification Process: Marketing Implications.

Creswell, J.W. (2003). *Research Design: Qualitative, Quantitative and Mixed Method Approaches*, 2nd Ed.. US: Sage Publications.

Creswell, J.W. (2003). *Research Design: Qualitative, Quantitative and Mixed Method Approaches* (2nd Ed.). US: Sage Publications.

Crouch, G, Ritchie, J. R.B., (1999), Tourism, Competitiveness, and Social Prosperity, *Journal of Business Research*, 44(3), 137-152.

Davis, D, Allen, J, Cosenza, R., (1988), Segmenting Local Residents by their Attitudes, Interests, and Opinions Toward Tourism, *Journal of Travel Research*, 27(20), pp 2-8.

Dogan, H.Z., (1989), Forms of Adjustments: Sociocultural Impacts of Tourism, *Annals of Tourism Research*, 16(2), 216-136.

Doxey, G., (1975), *A Causation Theory of Visitor-Resident Irritants: Methodology and Research References*, Travel and Tourism Research Associations Sixth Annual Conference Proceedings, TTRA, San Diego.

Dyer, P, Gursoy, D, Sharma, B, Carter, J., (2007), Structural Modeling of Resident Perceptions of Tourism and Associated Development on the Sunshine Coast, Australia, *Tourism Management*, 28(2), 409-422.

Dredge, D. & Jenkins, J. (2007). *Tourism planning and policy*. Queensland, Australia: John Wiley & Sons. Economic Planning Unit Malaysia. (2006). *Ninth Malaysia Plan, Realising Tourism Potential*. Retrieved April 5, 2009, from http://www.epu.gov.my/rm9/english/Chapter8.pdf

Echtner, C., Ritchie, J., (1992). The Meaning and Measurement of Destination Image. *Journal of Tourism Studies*. 2(2):2-12.

Fakeye, P. and Crompton, J. (1991). Image Difference between Prospective, First Time and Repeat Visitors to the Lower Rio Grande Valley. *Journal of Travel Research*. 30(2):10-16.

Frias, D. M., Rodriguez, M.A. and Castaneda, J.A. (2007). Internet vs Travel Agencies on previsit Destination Image Formation: An Information Processing View. *Tourism Management*. 29:163-179.

Forester, J. (1989). *Planning in the Face of Power*. Berkeley, CA: University of California Press.

Forester, J. (1993). *Critical Theory, Public Policy and Planning Practice: Toward a Critical Pragmatism*. Albany, NY: State University of New York Press.

Gartner, W. (1993). Image Formation Process. *Journal of Travel and Tourism Marketing*.2:191-216.

Gartner, W. and Bachri, T. (1994). Tour Operators' Role in the Tourism Distribution: An Indonesian Case Study. *Journal of International Consumer Marketing*. 6(3/4): 161-179.

Getz, D., (1983). Capacity to Absorb Tourism: Concepts and Applications for Strategic Planning, *Annals of Tourism Research*, 10(2):239-263

Getz, D., (1994), Residents Attitudes Towards Tourism: A Longitudinal Study in Spey Valley, Scotland, *Tourism Management*, 15(4), 247-258.

Goeldner, C.R., Ritchie, J.R.B and McIntosh, R.W. (2000). *Tourism Principles, Practices, Philosophies*. 8th Ed. John Wiley & Sons Inc. New York.

Gold, J.R. and Ward, S.V. (1994). *Place Promotion*, John Wiley & Sons, Chichester.

Godfrey, K, Clarke, J., (2000). *The Tourism Handbook: A Practical Approach to Planning and Marketing*, Continuum, London

Goodrich, J. (1978). A New Approach to Image Analysis through Multidimensional Scaling. *Journal of Travel Research*. 16(3):3-7.

Hall, M, Lew, A., (1998). *Sustainable Tourism: A Geographical Perspective*, Longman

Hunt, J. (1975). Image as a Factor in Tourism Development. *Journal of Travel Research*.13 (3):1-7.

Healey, P. (1997). *Collaborative Planning: Shaping Places in Fragmented Societies*. London: Macmillan.

Holloway, J.C. & Taylor, N. (2006). *The Business of Tourism* (7th ed.). England: Pearson Education Limited.

Inskeep, E. (1987). Environmental Planning for Tourism, *Annals of Tourism*, Vol.14: 118-135.

Innes, J.E. (1996). Planning through consensus building: A new view of the comprehensive planning ideal. *Journal of the American Planning Association*, 62(4), 460-473. Introduction to the Ministry of Unity, Culture, Arts and Heritage. Retrieved March 25, 2009, from http://www.heritage.gov.my/about/pengenalan/?c=6

Khalifah, Z. & Tahir, S. (1997). Malaysia: Tourism in Perspective, *Tourism and Economic Development in Asia and Australasia* (pp. 176-196). London: Cassel.

Lerner, M. and Haber, S. (2000). Performance Factors of Small Tourism Ventures: the Interface of Tourism, Entrepreneurship and the Environment. *Journal of Business Venturing*. 16:77-100.

Liew-Tsonis, J, (2008). Ecotourism Development: Government, Industry and Community Linkages, *Proceedings from the Asia Pacific Tourism Association (APTA) 2008 Conference, 9-12 July 2008, Bangkok, Thailand*.

Liew-Tsonis, J. (2007). Ecotourism as a Tool for Conservation: An Analysis of the Role of Implementation Stakeholders, *Proceedings from the 5th Tourism Educators' Conference on Tourism and Hospitality, 2007, 1-4 August 2007, Penang, Malaysia*. Malaysia Tourism Promotion Board Act 1992. Retrieved September 12, 2009 from http://www.agc.gov.my/agc/oth/Akta/Vol.%2010/Act%20481.pdf

Mayo, E. (1973). Regional Images and Regional Travel Behaviour. *Proceedings Travel Research Association Fourth Annual Meeting*. Sun Valley, ID. pp. 211-218.

Mayo, E. and Jarvis, L. (1981). *The Psychology of Leisure Travel*. Boston: CBI.

Mehta, M. (2009, March 15). Tourism Malaysia Optimistic About Tourist Arrivals. *Bernama (Malaysia)*, Retrieved April 5, 2009, from http://www.bernama.com/bernama/v5/newsgeneral.php?id=396348

M. Schrenk, V. V. Popovich & J. Benedikt (Eds.), *REAL CORP 007: To Plan Is Not Enough: Strategies, Plans, Concepts, Projects and their successful implementation in Urban, Regional and Real Estate Development* (pp. 159 – 165). Wien/Schwechat: CORP. Ministry of Tourism Malaysia Corporate Website, *Profile Section*. (2009). Retrieved August 6, 2009 from http://www.motour.gov.my/index.php/kem_profil.html Ministry of Transport (2009): http://www.mot.gov.my/index.php

Musa, G. (2000). Tourism in Malaysia. In C.M. Hall & S. Page (Eds.) *Tourism in South and South-East Asia: Issues and Cases* (pp. 144-156). Oxford, England: Butterworth Heinemann.

Page, S.J. (1999). *Transport and tourism*. England: Pearson Education Ltd.

Parsons, W. (1995). *Public Policy: An Introduction to the Theory and Practice of Policy Analysis*. Aldershot: Edward Elgar.

Pearce, P. (1982). Perceived Changes in Holiday Destinations, *Annals of Tourism Research*. 9:145-164.

Pierce, P, Moscardo, G, Ross, G., (1996), *Tourism Community Relationships*, Elsevier Science Ltd, Oxford.

Phang Ing, G., Liew-Tsonis, J., Cheuk, S., Razli, I. (2009). An Examination of the Challenges Involved in Distributing a Strong and Consistent Destination Image in the Marketing of Tourism in Malaysia, *Proceedings from EABR Conference, Prague, Czech Republic*, June 8-11, 2009.

Phelps, A. (1986). Holiday Destination Image: The Problem of Assessment, *Tourism Management*. 7: 168-180.

Pike, S. (2005). Tourism Destination Branding Complexity, *Journal of Product & Brand Management*. 4(4):258-259.

Purdue, R, Long, P, Allen, L. (1990). Residents Support for Tourism Development, *Annals of Tourism Research*, 17(4), 586-599.

Simon, D., (1996), *Transport and Development in the Third World*, Routledge, London

Spenceley, A. (2003). *Managing Sustainable Nature-Based Tourism in Southern Africa: A Practical Assessment Tool*. Unpublished PhD thesis, University of Greenwich, United Kingdom. Retrieved September 6, 2011, from http://anna.spenceley.co.uk/files/ManagingSustainableNBTToolkitSpenceley.pdf

Taylor, G, Stanley, D. (1992). Tourism, Sustainable Development and the Environment: An Agenda for Research, *Journal of Travel Research*, 31(1), 66-67.

Taylor, M. (1995). *Environmental Change: Industry, Power and Policy*, Avebury, Aldershot.

Thompson, D., Wilson, M. (1994). Environmental Auditing: Theory and Applications, *Environmental Management*, 18(4), 605-615. Tourism Concern, 2009. Retrieved 26 March 2011, from www.tourismconcern.org.uk Tourism Malaysia Corporate Website, *About Us Section*. Retrieved March 25, 2009, from http://www.tourism.gov.my/corporate/aboutus.asp Tourism Malaysia Corporate Website, *Research Section*. Retrieved April 5, 2009, from http://www.tourism.gov.my/corporate/research.asp United Nations. (1997). *Public-private partnerships: The enabling environment for development* ST/SG/AC.6/1997/L/6). Retrieved April 5, 2009 from http://unpan1.un.org/intradoc/groups/public/documents/UN/UNPAN000727. pdf

Wall, G., Mathieson, A., (2006). *Tourism: Changes, Impacts and Opportunities.* England: Pearson Education Ltd.

Wikipedia, (2009). http://en.wikipedia.org/wiki/Bumiputra#Definition

Williams, P., Gill, A., (1991). *Carrying Capacity Management in Tourism Settings: A Tourism Growth Management Process,* Simon Fraser University, BC

Williams, J, Lawson, R, (2001), Community Issues and Resident Opinions of Tourism, *Annals of Tourism Research,* 28(2), 269-290.

Woodside, A, Lysonski, S. (1989). A General Model of Traveler Destination Choice, *Journal of Travel Research.* 27(4):8-14. World Future Council. Retrieved March 1, 2011 from http://www.worldfuturecouncil.org. World Tourism Organisation. (1997). *Tourism 2020 Vision,* Madrid: TO UN Conference on Environment and Development; Report UN: New York, NY, USA, 13 June 1992; Volume I, Doc A/CONF.151/26 31 ILM 874. *Rio Declaration on Environment and Development (Annex 2);* UN World Commission on Environment and Development; Report UN: *Our Common Future;* Report, 4 August 1987; UN Documents: A/42/427; Retrieved 16 April 2011 from http://www.un-documents.net/ocf-10.htm UN Conference on Environment and Development; *Agenda 21 (Annex 2);* 13 June 1992; Volume I, Doc A/CONF.151/26. UN Documents: *Background Paper for the UNCSD;* UNCSD: New York, NY, USA, 18 April-3 May 1996; Retrieved 8 March 2011 from http://www.un.org/documents/ecosoc/cn17/1996/background/ecn171996-bp3.htm UN Documents: *Documents from the World Summit for Sustainable Development;* WSSD Documents; World Summit for Sustainable Development (WSSD): Mumbai, Maharashtra, India, August 2002; retrieved 01 March 2011 from http://www.un.org/jsummit/html/documents/summit_docs.html UN Documents: *Johannesburg Declaration on Sustainable Development and Johannesburg Plan of Implementation;* Report of the World Summit on Sustainable Development; UN: New York, NY, USA, 4 September 2002; retrieved 01 March 2011 from http://www.un.org/jsummit/html/documents/summit_docs.html UN Documents: *United Nations Millennium Development Goals;* Retrieved 01 March 2011 from www.un.org/millenniumgoals.

3

Sustainable Development Global Simulation: Analysis of Quality and Security of Human Life

Michael Zgurovsky

National Technical University of Ukraine "Kyiv Polytechnic Institute", Kyiv, Ukraine

1. Introduction

This research is based on the concept of "sustainable development" being the further development of studies of V. Vernadskij about noosphere (Vernadskij, 1944). It has been theoretically and practically proved that on the edge of the centuries studies about the noosphere appeared to be a necessary platform for the development of three-dimension concept of ecological, social and economic sustainable development (Summit Planet Earth, 1992) and (Johannesburg Summit, 2002).

Economic approach is based on the optimal usage of limited resources and application of natural-, power- and material saving technologies for creation of the gross income flow which would at least provide the preservation (not reduction) of the gross capital (physical, natural or human), with the use of which the gross income is created.

From the ecological point of view the sustainable development is aimed at provision of the integrity of both biological and physical natural systems as well as their viability that influences the global stability of the whole biosphere. The ability of such systems to renovate and adapt to the various changes instead of maintenance of the biological variety in the certain static state, its degradation and loss is becoming extremely important.

Social constituent is aimed at human development, the preservation of stability of social and cultural systems, as well as the decrease in the number of conflicts in the society. A human being shall become not the object but the subject of the development participating in the processes of his/her vital activity formation, decision-making and implementation of the decisions, in the control over their implementation. To meet such requirements it is important to fairly distribute the wealth between the people, to observe pluralism of thoughts and tolerate human relationships, to preserve cultural capital and its variety, including first of all, the heritage of non-dominant cultures.

Systemic coordination and balance of these three components is an extremely difficult task. In particular, the interconnection of social and ecological constituents causes the necessity to preserve equal rights of present and future generations to use natural resources. The interaction of social and economic constituents requires the achievement of equal and fair distribution of material wealth between people and help provision to the poor. And finally, the correlation of environmental and economic components requires the cost estimation of anthropogenic influences on environment. The solution of these tasks is the main challenge of the present time for the national governments, influential international organizations and all progressive people of the world.

In this research a Sustainable Development Gauging Matrix (SDGM) (Zgurovsky, 2007) within three abovementioned components is proposed and these processes are globally modeled in terms of quality and security of the human life. With the help of this Matrix the sustainable development processes have been globally modeled for a large group of world countries in terms of quality and security of the human life.

2. The methodology of sustainable development evaluation in terms of quality and security of the human life

2.1 Sustainable development as the quaternary functional of quality and security of the human life

The important issue in the process of implementation of the concept of sustainable development is the formation of the measurement system (Matrix) for the quantitative and qualitative assessment of this extremely complicated procedure.

The process of sustainable development will be characterized according to two main components: security (C_{sl}) and quality (C_{ql}) of the human life as it is shown in fig.1.

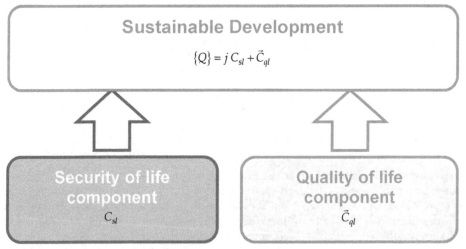

Fig. 1. Quaternary approach to the description of sustainable development process

Under this concept, the generalized measure (index) of sustainable development can be presented by means of the quaternion $\{Q\}$:

$$\{Q\} = j\,C_{sl} + \vec{C}_{ql}(I_{ec}, I_e, I_s). \tag{1}$$

The quaternion $\{Q\}$ includes an imaginary scalar part jC_{sl} which describes the security of human life and a real scalar part as a projection of the norm of vector radius $\vec{C}_{ql}$ to an ideal vector with coordinates (1;1;1) which describes the quality of human life within three dimensions: economic (I_{ec}), ecological (I_e) and socio-institutional (I_s). Under this condition j gains a value of a real unit for a normal regular state of society development at $C_{sl}>0$ and a value of an imaginary unit when a society enters conflict state ($C_{sl}=0$):

$$j = \begin{cases} 1 & , \ for \ C_{sl} > 0; \\ \sqrt{-1}, & for \ C_{sl} = 0 \ \ (conflict). \end{cases}$$

2.1.1 Sustainable development estimation methodology in the context of quality of human life

For every country the Euclidean norm of vector radius of human life quality ($\vec{C}_{ql}$) is given in the following form:

$$\left\| \vec{C}_{ql} \right\| = \sqrt{I_{ec}^2 + I_e^2 + I_s^2} \ . \tag{2}$$

In this case the indicators and policy categories included are calculated as a weighted total:

$$I_i = \sum_{j=1}^{n} w_j x_{i,j}, i = \overline{1,m}, \sum_{j=1}^{n} w_j = 1 \ , \tag{3}$$

where I_i is a value of an indicator or a category of policy for i^{th} country (the number of the countries is m), w_j is weight of the jth component of I index (the number of the components is n), $x_{i,j}$ is a value of the jth component for i^{th} country.

Such representation of integrated indices (indicators and categories of policy) envisages that components of $x_{i,j}$ in the formula (3) must be non-dimensional and vary within the same range.

Considering the fact that all data, indicators and indices included into the model are measured by virtue of different physical values, may be interpreted differently and change within the different ranges, they were aggregated to the standard form in such a way that all their variations would occur within the range from 0 to 1. The following formula was used:

$$l_{i,j} = \left(1 + e^{\frac{\overline{x_j} - x_{i,j}}{\sigma(x_j)}} \right)^{-1} \ , \tag{4}$$

where $x_{i,j}$ and $l_{i,j}$ are respectively the initial and standard j^{th} value for i^{th} region, $\overline{x_j}$ is the average value of x_j at sampling and $\sigma(x_j)$ is the corresponding standard deviation. To calculate a mean value and a standard deviation value the following formulae are used:

$$\overline{x_j} = \frac{\sum_{i=1}^{m} x_{i,j}}{m}, \sigma(x_j) = \sqrt{\frac{\sum_{i=1}^{m} (x_{i,j} - \overline{x_j})^2}{m+1}} \ .$$

Such data setting provides that values of indicators being the worst from the point of view of sustainable development correspond to numerical values near to 0, and the best values approach 1.

This normalization gives the possibility to calculate each of I_{ec},I_e,I_s indices and with the help of them the components with appropriate weighting coefficients. Then the quantitative value of human life quality can be identified as projection of the norm of this vector to an ideal vector with coordinates (1; 1; 1), (Fig.2):

$$C_{ql} = \sqrt{I_{ec}^2 + I_e^2 + I_s^2} \cdot \cos(\alpha). \tag{5}$$

The deviation angle α of the vector's radius C_{ql} from the ideal vector (1,1,1) is estimated on the basis of the values of dimensions I_{ec}, I_e, I_s in the following way:

$$\alpha = \arccos\frac{I_{ec} + I_e + I_s}{\sqrt{3} \cdot \sqrt{I_{ec}^2 + I_e^2 + I_s^2}}, \ 0 \le \alpha \le \arccos\frac{1}{\sqrt{3}}. \tag{6}$$

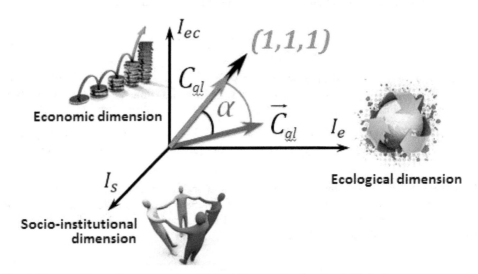

Fig. 2. Human life quality component (C_{ql}) and harmonization level ($G=1-a$)

Thus, the projection of the norm of the vector's radius $\vec{C}_{ql}$ to the ideal vector (1,1,1) characterizes the human life quality and the attitude position of the vector $\vec{C}_{ql}$ in the coordinate system (I_{ec}, I_e, I_s) characterizes the "harmonization" level of sustainable development. We should mention that when the angle α approaches 0, the harmonization level of sustainable development increases, i.e. the equidistance of the vector $\vec{C}_{ql}$ from each of coordinates (I_{ec}, I_e, I_s) will correspond to the highest harmonization value of sustainable development. If this vector approaches one of these coordinates, this will indicate the priority direction of the corresponding dimension development and neglect of two others. Let the value $G=1-a$ be the harmonization level of sustainable development. It will increase when G approaches 1 and decrease when G approaches 0.

As the researches of human life quality and security are conducted with the help of different methods and sets of initial data, it is worth performing them separately in three stages. At the first stage we will analyze the human life quality as one of the components of sustainable development. At the second stage we will investigate the human life security as another component of sustainable development. And at the third stage we will calculate the aggregate value of the Sustainable Development Index using two components and investigate this index.

In order to conduct the research of the life quality component of sustainable development, it is necessary to sample the data with the help of which each of three dimensions of sustainable development will be characterized in the most appropriate way. These data shall conform to the following important requirements: they have to be formed annually on continuing basis by respected and recognized international organizations.

Thus, the life quality component of sustainable development C_{ql} and the harmonization level of sustainable development $G=1-\alpha$ are calculated on the basis of their constituents I_{ec}, I_e, I_s. Considering the requirements to initial data mentioned above the value of every dimension I_{ec}, I_e, I_s will be calculated according to five global indices widely used in the international practice (Tab.1), being annually formed by the recognized international organizations. Let us consider all of them.

Life quality component C_{ql}	Global index	Constituents	Source
Economic (I_{ec})	I_c – Global Competitiveness Index	12 policy categories, 25 indicators	World Economic Forum [www.gcr.weforum.org]
	I_{ef} – Economic Liberty Index	10 indicators	Heritage Foundation &The Wall Street Journal [www.heritage.org/index/]
Ecological (I_e)	EPI – Environmental Performance Index	10 policy categories, 25 indicators	Yale and Columbia universities, USA [www.epi.yale.edu]
Socio-institutional (I_s)	I_{ql} – Life Quality Index	9 indicators	International Living [www.internationalliving.com/]
	I_{hd} – Human Development Index	3 policy categories, 4 indicators	UNDevelopment program [www.hdr.undp.org]

Table 1. Global indices used for calculationC_{ql} and $G=1-a$

The Economic Dimension Index (I_{ec}) will be made of the two following global indices (Table 1.)

1. *The Global Competitiveness Index* (I_c) was created by the organizers of the World Economic Forum. This index is annually estimated for 139 world economics and published in the form of so-called "Global competitiveness report" (World Economic Forum, n.d.). We will use the

data of such report for 2010-2011. The Global Competitiveness Index is formed of the following three groups of indicators: 1 – *the group of indicators of basic requirements (Basic requirements); 2 – the group of indicators of efficiency enhancers (Efficiency enhancers) and 3 – the group of indicators of innovation and sophistication factors (Innovation and sophistication factors).*

The first group includes four complex categories of economic policy: *Institutions; Infrastructure; Macroeconomic stability* and *Health and primary education.* The second one consists of six policy categories: *Higher education and training; Goods market efficiency; Labor market efficiency; Financial market development; Technological readiness* and *Market size.* The third group involves two important complex indicators: *Business sophistication* and *Innovation.*

2. The Index of Economic Freedom (I_{ef}) was created by the Heritage Foundation (The Heritage Foundation, n.d.). This index is formed of the following ten indicators: a level of business freedom; a level of trade freedom; a level of fiscal freedom; a dependence degree of economics on the government; a level of monetary freedom; a level of investment freedom; a level of financial freedom; private property rights; a level of freedom from corruption; a level of labor-market freedom. These ten indicators are calculated according to the expert assessment and usage of different economic, financial, legislative and administrative data.

The Ecological Dimension Index (I_e) will be estimated with the help of EPI (Environmental Performance Index 2010 (Yale Center for Environmental Low& Policy, n.d.)). This index is formed by the **Yale Center** of Environmental Law and Policy together with Columbia University (USA) for 163 countries of the world.

To calculate this index the aggregation method is used according to which EPI 2010 index is formed of two categories of top-level environmental policy (Environmental health, being the sanitary state of environment, and Ecosystem vitality, which is the vital ability of the ecosystem), ten medium-level ecological indicators and 25 low-level indicators.

The presented index and its indicators identify the ability of every country to protect its environment both during a current period of time and also in long-term perspective, on the basis of availability of national environmental system, the ability to resist to environmental impacts and decrease in human dependence on environmental impacts, social and institutional resources of a country to meet the environmental challenges, possibility of global control over the environmental state of the country etc. Moreover, they can be used as a powerful tool for making decisions on the analytical basis including social and economic dimensions of sustainable development of the country.

The Social Dimension index (I_s) will be formed of two global indices:

1. *The Life Quality Index* (I_{ql}) which is created by the international organization International Living (International Living, 2009). This index is formed with the help of nine indicators: human life cost, leisure and culture of people, economic state of the country, environmental state of the country, human freedom, human health, an infrastructure state, life risks and safety, climate conditions.

2. *The Human Development Index* (I_{hd}), which is annually calculated under the UNO program 'United Nations Development Program' (UNDP) for the majority of countries which are members of this organization. It is formed on the basis of the aggregation method according to which three policy categories of human development are used on the top level i.e. health, education and welfare of the population of the country.

These policy categories are formed of four indicators that characterize peculiar features of the education system of a country, nation poverty factors, level of unemployment, human health-care activities, gender conditions in the country and other constituents of human development.

Table 2 shows the groups of policy categories and indicators used for global modeling of sustainable development processes in 2010.

Economic dimension		
1. Global competitiveness index I$_c$		
Object	*Policy category*	*Indicator*
1. Basic requirements	Institutional environment	1. Property right 2. Ethics and corruption 3. Improper influence 4. State inefficiency 5. Safety
	Economic infrastructure	6. Transport infrastructure 7. Power and communication infrastructure
	Macroeconomic stability	8. Macroeconomic stability
	Human health and basic education	9. Population health 10. Basic education
2. Effectiveness increase	Higher education and education system	11. Education quantity 12. Education quality 13. Correspondent education
	Goods market effectiveness	14. Competition 15. Demand condition quality
	Labor market effectiveness	16. Flexibility 17. Talent use effectiveness
	Financial market perfection	18. Effectiveness 19. Reliability and confidentiality
	Technological readiness	20. Technology adaptation 21. ICT usage
	Market scales	22. Domestic market volume 23. Foreign market volume
3. Innovation	Business perfection	24. Business perfection
	Innovations	25. Innovations
2. Economic Freedom index I$_{ef}$		
		1. Business freedom 2. Trade freedom 3. Fiscal freedom 4. Dependence of economics on government 5. Monetary freedom 6. Investment freedom 7. Financial freedom 8. Private property right 9. Freedom from corruption 10. Labor market freedom

Ecological dimension		
Ecological dimension index I_e, (EPI)		
Object	*Policy category*	*Indicator*
1. Ecological health	1.Ecological disease load	1.Ecological disease load
	2. Air pollution (influence on human)	2. Air pollution in facilities
		3. Dust pollution of city atmospheric air
	3. Water (influence on human)	4.Potable water availability
		5. Availability of sanitation means
2. Ecosystem viability	4. Atmospheric air pollution (influence on ecosystems)	6.Sulphur dioxide emissions 7. Nitrogen dioxide emissions 8. Non-methane organic volatiles emission 9. Surface ozone concentration (in ecosystems)
	5. Water (influence on ecosystems)	10. Water quality index 11. Water resources load index 12.Water resources deficiency index
	6. Biodiversity and natural habitat	13. Protected nature territories (biomes protection) 14. Marine protected areas 15. Index of Alliance against complete species extinction
	7. Forestry	16. Growth change of woodland coverage 17. Woodland area change
	8. Fishery	18. Marine trophic index 19. Trawling intensity
	9. Agriculture	20. Intensity of fresh water consumption for agricultural purposes 21. State-subsidizing of agriculture 22. Pesticides usage control
	10. Climate changes	23. Greenhouse gases emission per capita 24. Carbon dioxide emission per unit of generated energy 25. Intensity of industrial greenhouse gases emission

Socio-institutional dimension	
Life quality index I_{ql}	Human development index I_{hd}
Indicators	Category of policy, indicators
1.Life quality	**1. Population health**
Life cost	Life expectancy index
Leisure and culture	**2.Population education**
State of economy	Adults literacy index
State of environment	Education coverage index
Human freedom	**3. Population welfare**
Human health	GDP index
State of infrastructure	
Life risks and safety	
Climate conditions	

Table 2. Policy categories and indicators for global modeling of sustainable development processes in 2010

As it is shown in Table 1 and 2, life quality component of sustainable development C_{ql} and its harmonization degree $G = 1 - \alpha$ in the year 2010 were determined with the usage of twenty two categories of policy and 73 indicators.

On the basis of description of relations between different categories of policy and indicators reduced to common calculating platform, the mathematical SDGM model was developed, the structure of which is presented in Figure 3.

It was taken into account that all data, indicators and indexes included into model (Figure 3) are measured with the help of different physical quantities, may be interpreted differently

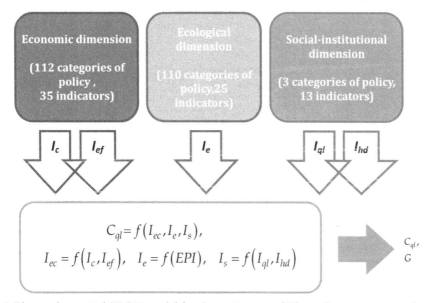

Fig. 3. The mathematical SDGM model for determination of life quality component of sustainable development and its harmonization degree

and change within different ranges. That is why they were normalized for their changes to occur within range from 0 to 1. In this case the worst values of mentioned indicators conform to numeral values close to 1. Such normalization gives the opportunity to calculate every index I_{ec}, I_e, I_s and component C_{ql} through their components with appropriate weight coefficients. In their turn the weight coefficients in the formula of calculation of life quality component of sustainable development C_{ql} are selected in order to give the possibility to provide equal values of economic, ecological and social dimension in the coordinate system (I_{ec}, I_e, I_s).

Therefore, the SDGM model gives the possibility to calculate life quality component of sustainable development C_{ql} and harmonization degree of this development $G = 1 - \alpha$ for every country of the world for which data about global indexes and indicators exist (Table 2).

2.1.2 Methodology of sustainable development assessment in terms of the human life security

Let us consider the global threats to the sustainable development to be those determined in the beginning of the XXI century by such recognized international organizations as UNO, World Health Organization (WHO), international organizations "World Economic Forum, Transparency International", "Global Footprint Network", "International Energy Agency", "World Resources Institute", company "British Petroleum" and others. The analysis of every threat will give the possibility to determine the vulnerability level of different countries of the world to the influence of these aggregated threats. Let us analyze each of the global threats separately.

Threat 1. Global decrease in energy security (ES)

For the first part of the XXI century one of the main critical challenges to the mankind is the rapid decrease in organic fuel resources that are extracted from entrails of the earth, and the increase in consumption of such resources, first of all, by India and China. In the beginning of the 20-ies of the current century, the curves of energy consumption and production of energy from oil will be crossed (AlenkaBurja, n.d.). In other words, the "production-consumption" balance of energy, produced from oil, will change its value from positive to negative (Figure 4). The similar phenomena will occur for "production-consumption" balances of energy, made from gas in the beginning of 30-ies and for the energy generated from uranium-235 in the beginning of 50-ies, accordingly (Figure 4).

Thus, until the mankind invents the energy resources that could fully replace the organic types of fuel and nuclear energy, the energy security of a country in particular and the world in general, will decrease. In order to quantitatively estimate the energy security of different countries of the world let us introduce the energy security index (Energy Security Index, ES) that will be calculated by the formula

$$ES_i = \frac{Exhaustables_i + Renewables_i}{2}, i \in \{countries\},$$

$$Exhaustables_i = \frac{NuclearR_i + CoalR_i + OilR_i + GasR_i}{\max_{\forall j \in \{countries\}} [NuclearR_j + CoalR_j + OilR_j + GasR_j]}, \tag{7}$$

$$Renewables_i = \frac{RenewablesUsed_i}{\max\limits_{\forall j \in \{countries\}} RenewablesUsed_j},$$

where:

- $ES \in [0;1]$, {countries} - set of explored countries,
- *Exhaustables* is the component that characterizes the dynamics of resource deflation;
- *Renewables* is the component that characterizes the volumes of usage of renewable sources in national energetic;
- *NuclearR, CoalR, OilR, GasR* –resources of uranium-235, coal, oil and gas (Nation Master, n.d.);
- *Renewables Used* – part of renewable energy produced and consumed by the country (at the expense of use of the energy of water, sun, wind, geothermal heat, biomass and rubbish burning) in percents from total energy consumption (Human Development Report 2007/2008, n.d.).

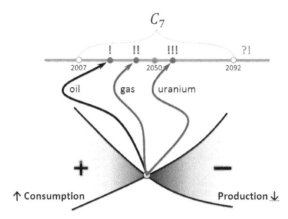

Fig. 4. Change in "production-consumption" balance from positive into negative for energy production from oil, gas and uranium-235, accordingly

Threat 2. The imbalance between biological abilities of the Earth and human needs in biosphere in terms of the change in the world demographic structure (BB)

In February 2011 the population of the planet has exceeded 7 million people living on the total area 510 072 000 km2. Daily growth of population is 211 467 people (GeoHive, n.d.). According to the method of arithmetic extrapolation the Earth population will have been 9,75 billion people by the year 2050. That is why the first threat appears being related to the fact that the Earth will be inhabited by the number of people that will exceed its abilities to sustain on the basis of the present natural resources. The Pentagon experts consider that the real problems for the mankind will have occurred by the year 2020, and will be connected with the catastrophic shortage of water, energy, foodstuff that can cause new conflicts on the Earth (Membrane, n.d.).

Nature can satisfy human requirements for business activity and only while this activity remains within the biosphere renewable capacity on the populated part of the planet. The calculation of ecologically disturbed area (Ecological Footprint) (Global Footprint Network, n.d.) gives the possibility to establish some limit according to which the ecological requirements to the world economics are within or exceed the biosphere abilities to supply the people with goods and services. This limit helps people, organizations and government to create strategies, establish the goals and provide the process according to the requirements of the sustainable development.

Ecologically disturbed territory (Ecological Footprint) determines which its part is necessary to preserve present population according to the present level of consumption, level of technological development and usage efficiency of natural wealth. The unit of measurement of this dimension is average (global on the whole Earth) hectare. The most substantial component of the Ecological Footprint is the territory of the Earth used for foodstuff production, forest area, biofuel amount, ocean (seas) territory, used for fishing and the most important element is the Earth area, necessary to support the life of plants absorbing the emissions of CO_2 as a result of organic fuel burning.

Ecological Footprint envisages that in world economy the people use resources and ecological services from all over the world. Thus, the indicator for a country may exceed its actual biological possibilities. On the basis of it, the essence of Ecological Footprint for a country is the extent of its consumption and global impact on environment.

The same methodology can be used for calculation (in the same values) of biological abilities of the Earth, biological productivity of its territory. In 2011 biological abilities of the Earth were approximately 11.2 billion or 1.8 global hectares per capita (non-human species were not considered). Now the human need in biosphere, i.e. its global Ecological Footprint is 18.1 billion global hectares or 2.7 global hectares per capita. That is why, today global Ecological Footprint exceeds biological abilities of the Earth by 0.9 global hectares per capita or by 50%. This means that vital resources of the planet disappear faster than the nature can renew.

This threat has substantial correlation degree with demographic structure change of the planet population. For example, according to UNO (Human Development Report 2007/2008, n.d.) the biggest growth of population over a period of the following 50 years is expected in the poorest regions of the world: in Africa it will increase in 2 times, In Latin America and Caribbean basin will increase in 1.5 time, at the same time in Europe it will decrease in 0, 8 times. Essential threat is also uncontrolled increase in the urban population in underdeveloped countries. By the year 2050 it will have been doubled approximating to 10 billion people. It will lead to intensification of transport, ecological and social problems, an increase in criminality and other consequencess of chaotic urbanization.

The important tendency of the nearest decades is rapid change in the structure of religious groups of the Earth population. So, from 1980 to 2005 the number of Muslins will increase from 16,5% to 30%, the number of Christians will decrease from 13.3% to 3%, the number of Hindus will decrease from 13.3% to 10%, the number of Buddhists will decrease from 6.3% to 5%. The number of representatives of other religious groups will also decrease from 31.1% to 25% (Science Council of Japan, 2005). These changes will cause the necessity of searching new methods of tolerance coexistence of people on the Earth.

For estimation of increasing threats, connected with imbalance between biological capability of the Earth and human requirements in biosphere, in terms of demographic structure change of the world we will use the indicator which is ecological reserve ("+") or deficit ("-") in global hectares per capita for a country (Global Footprint Network, n.d.).

Threat 3.Growing inequality between people and countries on the Earth (GINI)

According to the World Bank data, in the year 1973 the difference in incomes between the richest and poorest countries were determined by ratio 44:1, and today it is 72:1. The assets of three world's richest people exceed the wealth of 47 countries of the world. Assets of the whole mankind are controlled by 475 richest people. Assets of 50 richest people of Ukraine which amount to 64,4 billion dollars in 2007 exceeded two national budgets of the country, in particular (Donbass Internet Paper. News.dn.ua, n.d.). The correlation between one fifth of the richest and one fifth of the poorest parts of the Earth population has reached 1:75. Wealth of civilization still remains unachievable for the poorest group. Its representatives spend less than two dollars a day; 700 million of them live in Asia, 400 million live in Africa and 150 in Latin America. The gap between the richest and the poorest groups of people of the Earth has risen approximately tenfold according to their living standards in the course of the last 20 years. The threat is considered to be dangerous due to the growing number of the world conflicts, growth of corruption, terrorism and crime, ecology deterioration, a decrease in the level of education and health service support.

In order to estimate the distribution inequality of economical and social boons for each country the SP-index (CIA, n.d.) which identifies these characteristics will be used.

Threat 4.The spread of global diseases (GD)

The World Health Organization considers such diseases as cancer, cardio ischemia, cerebrovascular disease (paralysis), chest troubles, diarrhea, AIDS, tuberculosis, malaria, diabetes to be the most dangerous for mankind as they may not only have bad consequences but also globally spread all over the world.

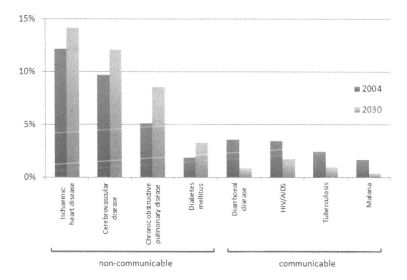

Fig. 5. Comparison of mortality factors, 2004 -2030 (Mathers, 2006)

During the next 20 years the sufficient increase in mortality caused by all non- infectious global diseases and decrease in mortality caused by AIDS, tuberculosis and malaria are expected. Such diseases as cardio ischemia, cerebrovascular disease, lung cancer and diabetes will become main global diseases during this period. At the same time the rate of total mortality from tobacco consumption will increase from 5.8 million people in the year 2009 to 6.4 million in the year 2015 and 8.3 million in 2030. Thus, tobacco is expected to kill by 50% people more than AIDS. Total human mortality on the Earth will be by 10% predetermined by the tobacco consumption.

But for estimation of the level of protection of the countries against quickly-spreading diseases it is reasonable to use the data on infectious diseases. In the further modeling the data on total mortality of the population of the world countries (million per year) caused by the totality of infectious diseases such as diarrhea (the most common mortality factor in underdeveloped countries), AIDS, tuberculosis, malaria and others will be used (Mathers, 2006).

Threat 5. Child mortality (CM)

The child mortality rate or under-5 mortality rate is the number of children who die by the age of five, per thousand live births per year. According to the data of United Nations Children's Fund 11 million children aged less than 5 die every year. Poverty which leads to bad health of mothers, insufficient nutrition and unsatisfactory sanitary is the reason of child mortality. Such factors as infectious diseases, poor health care and conflicts also increase child mortality. Africa, for example, has high rates of child mortality which are connected with AIDS epidemic, poor sanitary conditions and bad nutrition. The increase in child mortality in Iraq and Afghanistan is mostly caused by the conflicts.

According to UNICEF, most child deaths (and 70% in developing countries) result from one the following five causes or a combination thereof: acute respiratory infections, diarrhea, measles, malaria, malnutrition.

There is a significant difference in the indices of child mortality for different countries. In western industrially developed countries from 4 to 7 out of 1000 children die under the age of 5 years. The average rate of child mortality in developing countries is 158. In Sierra Leone, for example, every fourth child dies at infant age. Every tenth child doesn't live to 5 years in Iraq.

The rate of child mortality in the countries of the former Soviet Union in 10-12 times exceeds the rate of child mortality in the countries of Western Europe. It is particularly high in Armenia, Azerbaijan, Georgia, Kazakhstan, Kyrgyzstan, Tajikistan, Turkmenistan, Uzbekistan.

Leaders of the countries took the responsibilities to decrease the rate of death of children aged under 5 years by two thirds by the year 2015. The United Nations Children's Fund now warns that 98 countries of the world will not be able to succeed in the specified task.

One of the UN Millennium Development Goals (MDGs) is to reduce child mortality, and the target is to "Reduce by two thirds, between 1990 and 2015, the under-five mortality rate". According to the UN MDG Report 2010 child deaths are falling, but not quickly enough to reach the target. Revitalizing efforts against pneumonia and diarrhoea, while bolstering nutrition, could save millions of children. Recent success in controlling measles may be short-lived if funding gaps are not bridged.

Such tendencies signify another global threat due to marginalization of social and economic processes, a decrease in ecological and sanitary standards, impoverishment of people in the majority of countries of the world. In the further modeling, the data on the child mortality rate or under-5 mortality rate will be used. This data is collected by World Health Organization (WHO) and published in WHO Annual Reports and Statistical Information System. That data is also accessible at World Data Center for Geoinformatics and Sustainable Development (WDC-Ukraine).

Threat 6. The growth of corruption (CP)

Corruption is the biggest obstacle to the economic and social development of society. It endangers every change. Corruption has become not only one of the main reasons of poverty but also a source which prevents its overcoming. Although corruption had existed for a long time it became more widely spread in the process of globalization at the end of the 20th at the beginning of 2the 1th centuries.

Corruption in one country had negative impact on the development of other countries which means that countries with the high level of corruption are not limited to the Third World. The process of liberalization in the former socialist countries was accompanied by unprecedented position abuses in 90-ies. Thus, Financial Times proclaimed 1995 to be "the year of corruption". The following years were marked with the spread of this phenomenon almost throughout all countries of the world and corruption itself became of global and international character.

Wellbeing did not become the prerequisite of successful elimination of corruption. The analysis of long-term tendencies revealed by the international organization «Transparency International» showed that during last 12 years the level of corruption has decreased in such countries as Estonia, Columbia, Bulgaria. Nevertheless, the growth of corruption occurs in such developed countries as Canada and Ireland. Such factors of risks as opacity of state authorities, excessive influence of separate oligarchic groups, violation in financing of political parties, etc. exist both in poor and rich countries and unfortunately, tendencies in increase of corruption scale are the same.

Usually, the structure of corruption is different in different countries of the world. Figure 6 illustrates countries and segments of society with the highest level of corruption according to (Transparency International, n.d.).

Figure 7 shows average indices of corruption in different segments of society according to (Transparency International, n.d.).

To estimate the influence of corruption on socio-economical and cultural development of different countries of the world we will use "the index of corruption perception" established by the international organization "Transparency International" (Report on the Transparency International Global Corruption Barometer 2007, 2007).

Threat 7. Limited access to drinking-water (WA)

According to the data of the WHO **and** the UNICEF (Corruption Perception Index 2008, 2008) the world is under the threat of increase of limited access to drinking-water and sanitary facilities. The fifth part of all mankind (11 billion people) does not have access to drinking-water and 2,4 billion of people do not have minimal sanitary facilities. That is why

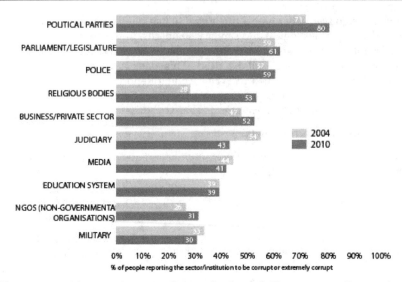

Fig. 6. The structure of corruption according to the data of «Transparency International» (Transparency International, n.d.)

2003 was proclaimed as year of drinking water by the General Assembly of UNO. The period of 2005-2015 starting from the International Day of Water Recourses (22nd of March, 2005) was proclaimed as International decade of actions "Water for life".

The urban regions of underdeveloped countries have complicated situation where due to the rapid increase in the population the problem is exacerbating rapidly. These factors negatively influence the children health. According to the data of the WHO in the year 2005, 1,6 million children aged under 5 (4500 children per day approximately) died as a result of consumption of the dangerous water and inappropriate sanitary facilities.

The more the population of the planet increases, especially in underdeveloped countries, the more struggle for the control of drinking-water recourses will exacerbate another global threat for mankind.

The limited access to the drinking- water will be estimated by the inversed magnitude to the indicator of the access to drinking water (Human Development Report 2007/2008, n.d.).

Threat 8. Global warming (GW)

Global warming is the process of gradual increase in the average annual temperature of the Earth and World Ocean. According to conclusions of the International UNO Expert Group in Climate Control (UNICEF Joint Monitoring Programme for Water Supply and Sanitation, n.d.) and National Academies of Sciences of the Group of Eight, from the end of 19th century the average temperature of the Earth has risen by 1°C and "the major part of warming observed during the last 50 years had been caused by human activities" preliminary by gas emissions which cause green-house effect (carbon dioxide, CO_2) and methane (CH_4).

Estimates obtained with the climate models and cited by the International UNO Expert Group in Climate Control show that the average temperature of the Earth can increase from one to several °C (in different regions of the world or in the Earth in average) in 1990-2080 years. The warming is expected to cause other climate changes such as an increase in the level of Word Ocean by 0,1-5 m. (probably, in 30-40 years), the appearance of new viruses and also the change of atmospheric condensation and their distribution. This may result in an increase in such natural disasters as floods, draughts, hurricanes etc; a decrease in harvests of agricultural crops, the emergence of new epidemic diseases and the extinction of many biological species. As a result of the control over decreasing natural resources the struggle not only between countries but also between separate groups of population can exacerbate. This process will cause new global conflicts. The influence of carbon dioxide emissions on the global warming is much higher than the corresponding influence of methane. That is why the danger of global warming will be estimated by the amount of carbon dioxide emissions CO_2 in metric tons per capita.

Data about emissions is obtained by WDC-Ukraine from Carbon Dioxide Information Analysis Center (CDIAC). It can be obtained with data extraction tool (http://wdc.org.ua/en/data). Original data is only the amount of Carbon (C) and calculation has been done to convert Carbon into Carbon Dioxide (CO2): values were multiplied by according coefficient (12+16*2)/12. Per capita emission data is based on calculations: CO2 emission / population for each country correspondingly.

Threat 9. The state fragility (SF)

After the end of Cold War and Soviet Union collapse (1991) the world has entered the era of new dramatic geopolitical processes. The following 18 years were marked with the blistering growth of globalization. Technical revolution in the field of information-communication technologies has made the world policy more transparent and led to an increase in changes influence which occurred in one region and affected the other parts of the planet. Due to these new qualities of the globalized world it became clear that new geopolitical system is full of unstable, unsuccessful and weak countries. The weakening of retaining mechanisms peculiar to bipolar world and conflict exacerbation between fundamental values of different countries caused a new wave of oppositions, terrorism, violence, territorial claims and irregular development.

Uncontrolled spread of nuclear, chemical and biological weapon, rebuilding of nuclear energetics in such unstable, unbalanced world significantly increases the threat to sustainable development and global security of mankind.

Under such conditions the stabilization of world development becomes possible due to the international cooperation, investments and support to the weak countries and planet regions by the progress of new paradigm of "tolerant, peaceful world". In order to accomplish such global, stabilizing policy the recognized international organizations and scientific centers began to develop analytical instruments for the estimation of new developing tendencies of the world since the beginning of this century. The first attempt to control the tendencies of the global development was a series of reports "The world and the conflict" which were published in the University of Maryland State (USA) in 2001. Reports devoted to the global tendencies of world development were also published in many countries such as Spain, Canada, and Germany etc.

The final aim of the development of new analytical instruments was the attempt to estimate the ability of different countries to act in such important dimensions as conflict, state administration, economic and social development. Among all these instruments "The index of ability of the peaceful society development" that belongs to the series of reports "The world and conflict", "Indicators of the world management" developed by the World Bank and "Index of unsuccessfulness of the countries" developed by The Fund of Peace can be mentioned.

For the quantitative estimation of the sustainable development threat in our research the State Fragility Index will be used (The Intergovernmental Panel of Climate Change, n.d.). This index is calculated as average arithmetic value between political and economical instability of the country. Data concerning these values are given in the paper (Marshall, 2008).

Threat 10. Natural Disasters (ND)

Natural disasters are the threat which is not so directly dependent on the human activity comparing to the other threats mentioned above. But, taking into account last reports of the international organizations on climate changes (World Economic Forum, 2010) *we cannot state that a human being is beside the point of the dynamics of the natural disasters.* For the quantitative estimation of the degree of vulnerability of the world countries to the natural disasters the index of vulnerability to natural cataclysms was developed. The data of the International Disasters Database (Kotlyakov, 2001) and the Centre for Research on the Epidemiology of Disasters (CRED) of the World Health Organization (WHO) are used for its calculation.

Experts of UNO and WDC-Ukraine determined 6 major natural disasters (in the order of danger decrease): draughts, floods, hurricanes, extreme temperatures, earthquakes and tsunami (UNDP, n.d.; Aivazian, 1983).

Index is calculated as follows:

1. The summarized total of people suffered from the natural cataclysms in a year in a country is calculated:

$$DisastersAffected_{year,\,state} = DroughtAffected_{year,\,state} + FloodAffected_{year,\,state} +$$

$$+ StormAffected_{year,\,state} + ExtremeTemperatureAffected_{year,\,state} + EarthquakeAffected_{year,\,state} +$$

$$+ TsunamiAffected_{year,\,state},\ \forall year, state.$$

2. Then the summarized total of people affected DisastersAffected is divided by the amount of population in the country and in the given year:

$$DisastersAffected'_{year,\,state} = \frac{DisastersAffected_{year,\,state}}{Population_{year,\,state}},\ \forall year, state.$$

3. After that the obtained data are normalized by the logistic norm:

$$\left\| DisastersAffected'_{year,\,state} \right\| = \left[1 + e^{\frac{DisastersAffected'_{year,\,state} - M[DisastersAffected']_{year}}{s[DisastersAffected']_{year}}} \right]^{-1},$$

where M[.], s[.] – are approximate average and standard deviation values respectively per year in all countries.

As consequences of the natural disasters usually make a long-term influence on the country, gradually disappearing only with time, the final value of vulnerability index on the natural disasters will be defined as Exponential Weighted Moving Average, EWMA, which has the potential smoothing factor $\alpha = 0,25$

$$ND_{year,state} = 1 - \alpha \cdot \sum_{1 \leq t \leq T_{max}} (1-\alpha)^{t-1} \cdot \left\| DisastersAffected'_{year-t,state} \right\| . \tag{8}$$

The value of the coefficient α was chosen by the experts on the basis of the estimation of the average time and level of the impact of disasters on the country. For convenience of calculations only the last significant $Tmax = 25$ years will be considered. At the same time the significance of time series will amount to $\varepsilon = e^{T_{max} \cdot \ln(1-\alpha)} = 0,0007525 \leq 10E - 3$

The values of vulnerability index for the countries to the natural disasters during 1995-2010 were calculated according to the given methodology.

2.1.3 Determination of the aggregate impact of the total global threats on different countries and their groups

The total impact of the total global treats to different countries and their groups will be determined by the component of human security C_{sl} *being the part of index of sustainable development in formula* (1).

Let us formalize this in the following way. Let every **j** country corresponds to the vector In correspondence with each country **j** a vector

$$\vec{Tr}_j = \left(ES, BB, GINI, GD, CM, CP, WA, GW, SF, ND \right) \tag{9}$$

the coordinates which characterize the degree of the development of the relevant threats, where:

ES is a global decrease in energy security (determined by the index of energy security calculated by the formula 7);
BB is misbalanced biological capacity of the Earth and needs of the mankind in the biosphere in terms of changing world's demography (measured in global hectares per person);
GINI *is growing inequality between people and countries of the Earth (measured by* Gini-index which changes within the range from 1 to 100; where 0 is a minimum inequality, 100 is maximum inequality);
GD is the spread of global infectious diseases (measured by the total quantity of the people [millions per year] died from diarrhea diseases, AIDS, tuberculosis and malaria);
CM is child mortality (measured by the number of children who died under 5 per 100 newborn)
CP is the growth of corruption (measured by the index of corruption perception varying within the range from 0 to 10; where 0 is a maximum corruption level and 10 – minimum corruption level);
WA I s the *limited access to drinking-water* (the percentage of the population which has no access to drinking-water);
GW *is global warming* (measured by the quantity of carbon dioxide emissions in metric tones);

SF is state fragility (measured by State Fragility Index (The Intergovernmental Panel of Climate Change, n.d.), which changes in the range from 0 to 23, where 0 - minimum fragility; 23 – maximum fragility);

ND is index of vulnerability to natural disasters (calculated by the formula (8)).

The source data for each danger are normalized by the formula (4) and in the case of necessity converted for the maximum threat to correspond to 0 and minimum threat to correspond to 1. Thus, after normalization the more each threat approaches its zero value it becomes the most "likely to occur" in each specific country. But the more its value approaches 1 it becomes more 'unlikely to happen' in that country.

After the normalization for all global threats, the normalized vector is obtained:

$$\vec{Tr}_j^0 = \left(ES^0, BB^0, GINI^0, GD^0, CM^0, CP^0, WA^0, GW^0, SF^0, ND^0\right), \tag{10}$$

Let us calculate the value for each component of life security C_{sb}, *which is norm of Minskoski, which* is formed of normalized threats according to $P = 3$, $n = 10$:

$$C_{sl} = \left\|\vec{Tr}_j\right\| = \sqrt[3]{\sum_{l=1}^{n}\left(\vec{Tr}_{jl}^0\right)^3}. \tag{11}$$

It should be mentioned that in practice the parameter P is mostly chosen to be equal 2. An increase in this parameter increases the model sensitivity for each part of the vector and vice versa its decrease smoothes (reduces) this sensitivity. That is why on the basis of the data analysis of the mentioned threats it is advisable to enlarge parameter P from the value 2 to 3, to increase sensitivity of the models to the threats being insignificant by their quantitative values if compared to the other models but being important by their substantial values.

Let us also introduce the value of vulnerability of the country to the total of the global threats which is the inverse value to the component of the life securityy C_{sl}:

$$I_{vul} = \sqrt[3]{10} - C_{sl}. \tag{12}$$

Thus, the SDGM model (1-12) combines a lot of indicators and indexes included in it by mathematical correlations making their algebraic convolution. This model combines the data of different nature i.e. economic, ecological and socio-institutional one. Thus, it shows the reverse connection and balance between three integral spheres of society development. With the help of this model it is possible to obtain the numerical value for every dimension of the quality of life and also its single matrix that considers all three dimensions together.

3. The mathematical simulation of sustainable development processes

3.1 Computation for general simulation

The mathematical simulation of sustainable development processes can be performed in three stages. At the first stage we will perform the estimation of life quality dimension C_{ql} as the component of sustainable development index in the formula (1) using Sustainable Development Gauging Matrix (SDGM) (chapter 2.1.). At the second stage we will calculate the total impact of global threats totality on different countries and world countries groups in the form of human life security component C_{sl} as the component of sustainable development index in the formula (1) (chapters 2.1.2, 2.1.3). At the final third stage we will calculate the value of quaternion {Q} according to the formula (1) as the quantitative dimension of sustainable development which considers the human life security and quality of life.

3.1.1 The estimation of human life quality as index of the sustainable development

Calculation of the life quality component C_{ql} of sustainable development and the level of its harmonization $G = 1-a$ will be performed with the use of the mathematical model SDGM (chapter 2.1) and global indices (tables 1 and 2).The initial data for the SDGM model will be taken from the annual reports of such international organizations as UNO, Heritage Foundation, World Economic Forum, International Living, Environmental Law and Policy Center of Yale University, the University of Columbia (USA).

In order to perform comparative global analysis of the life quality component of the sustainable development let us choose five countries of the world: Countries leading by the quality of life component; group of Eight (G8); the Group of giant rapidly developing countries including Brazil, Russia, India, China (BRIC countries); the group of post-socialistic countries; the countries of Africa.

It should be mentioned that owing to its geographical position and economic status Russia enters the 2nd, 3rd and 4th group simultaneously, while Germany, France and Great Britain belong to the 1st and 2nd   groups.

1. **Ten leading countries** in the year 2010 by the life quality component of sustainable development are presented in table 3. This group includes 9 European countries and 1 country of Oceania. Considering the results of the research it can be seen that countries which in 2005-2010 were 5 world leaders by the index of their sustainable development were not superpowers with dominating ideologies and economies. Basic industries of such

Rate Cql	ISO	Country	Life quality component Cql	Economic dimension Iec	Ecological dimension Ie	Socio-institutional dimensionIs	Harmoni-zation degree G
			CLUSTER 1("VERY HIGH")				
1	CHE	Switzerland	1,498	0,872	0,917	0,806	0,947
2	SWE	Sweden	1,398	0,796	0,895	0,730	0,917
3	NOR	Norway	1,379	0,731	0,847	0,810	0,939
4	NZL	New Zealand	1,365	0,816	0,739	0,810	0,956
5	ISL	Iceland	1,357	0,730	0,942	0,678	0,855
6	AUT	Austria	1,343	0,751	0,810	0,765	0,967
7	FIN	Finland	1,342	0,804	0,761	0,760	0,974
8	DEU	Germany	1,338	0,770	0,736	0,812	0,960
9	FRA	France	1,320	0,664	0,812	0,810	0,909
10	GBR	Great Britain	1,319	0,803	0,753	0,729	0,960

Table 3. Ten leading countries according to the life quality component of sustainable development, 2010

countries are not oriented towards the usage of significant natural recourses and cheap workforce. The characteristic feature of these countries is domination of intellectual and highly-technological labor in the additional cost of their economies. All these countries are the world leaders by the ecological dimension of the world. Their innovative activity is of high level; over 4% of their GNP is spent for research and development.

Since the beginning of 1990-s they have been actively working in order to implement the model of the 'environmental economy' and knowledge-based economy. They started large-scale production of new knowledge, 'ecosystem' products and services and in the course of the following few years they included social assets into their strategy as another productive factor of the development. That is why now these counties are the countries with well-harmonized life quality components of the sustainable development i.e. economic, ecological and social ones. These countries have become the closest to the model of the 'smart' society which is the highest form of the developed, knowledge-based society.

2. The Group of Eight countries (table 4), in the year 2010 takes from 8th to 24th positions in the list by the quality of life component in sustainable development (except Russia).

Rate Cql	ISO	Country	Life quality component Cql	Economic dimension Iec	Ecological dimension Ie	Social-institutional dimensionIs	Harmoni-zation degree G
CLUSTER 1 ("VERY HIGH")							
8	DEU	Germany	1,338	0,770	0,736	0,812	0,960
9	FRA	France	1,320	0,664	0,812	0,810	0,909
10	GBR	Great Britain	1,319	0,803	0,753	0,729	0,960
13	CAN	Canada	1,293	0,845	0,608	0,786	0,866
14	JPN	Japan	1,290	0,789	0,725	0,719	0,957
16	USA	The USA	1,268	0,851	0,546	0,801	0,819
CLUSTER 2 ("HIGH")							
24	ITA	Italy	1,169	0,525	0,734	0,767	0,843
CLUSTER 3 ("AVERAGE")							
69	RUS	Russian Federation	0,740	0,358	0,497	0,427	0,868

Table 4. The Group of Eight according to the component of the life quality of sustainable development, 2010

Although they have leading GNP indices in the world they are still on 20-30 places in the world list by quality characteristics of their economic, renewable environmental resources and development of their social assets.

The only exception in this group is Russia (69th position) which being formally included into the Group of Eight is at the same time "excluded" frotm it by the qualitative characteristics. Dependence of Russian economy on the energy sector is extremely high. This field provides the country with almost 25% of GDP and 50% of national export that makes Russia rather sensitive to and dependent on global market conditions. These results in narrowing the diversification of economic interests of Russia, which in its turn, provides aggressive state-monopoly foreign policy of the country in energy field.

3. BRIC-country group (Brazil, Russia, India and China) is characterized by rapid increase in their economies development that annually reaches 8-12 %. This is provided both due to the growth of innovational, highly-technological components of the development of these countries and by intensive use of their own natural and environmental resources, involvement of cheap labor, giant consumption of organic types of fuel (oil, gas and coil).

In spite of the rapid economic growth these countries hold from the 48th (Brazil) to 85th (India) positions in the rating table by the life quality component of sustainable development (Table 5).

This can be explained by the low level of harmonization of sustainable development for this group of countries at the expense of prior economic development and at the same time substantial backlogs in environmental and social spheres. The countries of this group are characterized by the decrease in ecological results, increase in inequality between people, high corruption levels that tend to increase. These and other factors of ecological and social character restrain harmonized sustainable development of the group of BRIC-countries.

Rate Cql	ISO	Country	Life quality componentC ql	Economic dimension Iec	Ecological dimension Ie	Socio-institutional dimensionIs	Harmoni-zation level G
CLUSTER 3 ("AVERAGE")							
48	BRA	Brazil	0,902	0,424	0,544	0,594	0,864
69	RUS	Russian federation	0,740	0,358	0,497	0,427	0,868
CLUSTER 4 ("LOW")							
79	CHN	China	0,647	0,459	0,255	0,406	0,773
85	IND	India	0,572	0,418	0,245	0,328	0,789

Table 5. Group of BRIC countries according to the life quality component of sustainable development, 2010

4. Post-socialist countries (Table 6) turned out "scattered" from the 29th to 99th positions of the rating table by the life quality component in 2010. The leaders in this group were the countries of the Central Europe and Baltic, which outstripped the countries of the East Europe and Middle Asia.

For the countries of this group it is not current position by the life quality component of sustainable development that is of great importance but the dynamics of the qualitative changes and differentiation scale that have been observed for the last 15-20 years. From the approximately equal initial conditions in the late 80-ies of the last century, the countries of this group have passed through very different political, economic and mental changes for historically short period of time. The best examples of successful development were shown by the countries of the Baltic, Central and Eastern Europe, and the worst ones were shown by the countries of the Central Asia and North-Caucasian countries of the former USSR.

5. African countries listed by the life quality component of sustainable development are shown in Table 7. Except for South Africa, Tunis and Algeria, they belong to the poorest countries in the world, the GDP per person of which is lower than 5000 dollars.

According to the data of the International Organization "Transparency International", these countries have the highest levels of corruption, and according to the World Health Organization they have the highest levels of spreading global diseases, such as AIDS, tuberculosis and malaria. In 2010 the characteristics of these countries (except Tunis) greatly decreased in comparison with the previous years, not only by the life quality component in general, but also by all three dimensions of this component. The positive tendency of the sustainable development of Tunis can be explained by significant improvement of innovation climate especially in the sphere of information technologies after the UNO World Summit on Information Society was held in this country in 2005.

Rate Cql	ISO	Country	Life quality component Cql	Economic dimension Iec	Ecological dimension Ie	Socio-institutional dimension Is	Harmonization level G
CLUSTER 2 ("HIGH")							
21	CZE	Czech Republic	1,214	0,669	0,709	0,725	0,967
23	SVK	Slovakia	1,176	0,611	0,757	0,669	0,912
26	LTU	Lithuania	1,125	0,615	0,646	0,686	0,955
27	EST	Estonia	1,121	0,703	0,553	0,686	0,896
29	HUN	Hungary	1,112	0,553	0,662	0,711	0,898
30	LVA	Latvia	1,095	0,526	0,724	0,646	0,872
32	SVN	Slovenia	1,083	0,591	0,577	0,707	0,907
37	POL	Poland	1,009	0,535	0,538	0,675	0,888
38	HRV	Croatia	1,000	0,435	0,653	0,645	0,827
43	ALB	Albania	0,984	0,470	0,705	0,529	0,826
CLUSTER 3 ("AVERAGE")							
40	ROU	Rumania	0,992	0,510	0,620	0,589	0,920
47	BGR	Bulgaria	0,932	0,472	0,525	0,617	0,890
56	ARM	Armenia	0,817	0,506	0,480	0,430	0,933
65	AZE	Azerbaijan	0,761	0,474	0,451	0,394	0,923
69	RUS	Russian Federation	0,740	0,358	0,497	0,427	0,868
CLUSTER 4 ("LOW")							
72	KAZ	Kazakhstan	0,720	0,464	0,413	0,370	0,907
73	UKR	Ukraine	0,714	0,294	0,432	0,511	0,786
74	BIH	Bosnia and Herzegovina	0,707	0,318	0,383	0,523	0,794
78	KGZ	Kyrgyzstan	0,653	0,359	0,463	0,308	0,830
83	MDA	Moldova	0,619	0,146	0,445	0,481	0,602
CLUSTER 5 ("VERY LOW")							
92	TJK	Tajikistan	0,493	0,264	0,295	0,296	0,948
99	UZB	Uzbekistan	0,411	0,247	0,160	0,305	0,755

Table 6. Post-socialist countries ranked by the quality-of-life component of sustainable development, 2010

On the whole, comparing the group of African countries (table 7) with the leading countries by the life quality component of the sustainable development (table 3) and the Group of

Eight (table 4) it is possible to state that in the year 2010 as compared to the year 2006 the gap between the developed countries of the world and the countries of Africa increases both by standard of living (GDP per capita) and by the life quality component of the sustainable development. This is an alarming symptom due to the increase in inequality in the world, spreading of global diseases, a growing number of global and regional conflicts, the growth of corruption and crime.

Rate Cql	ISO	Country	Life quality component Cql	Economic dimension Iec	Ecological dimension Ie	Social-institutional dimensionIs	Harmoni-zation degree G
CLUSTER 3 ("AVERAGE")							
55	TUN	Tunis	0,835	0,509	0,483	0,455	0,954
57	DZA	Algeria	0,796	0,393	0,628	0,358	0,745
60	NAM	Namibia	0,792	0,472	0,455	0,445	0,975
64	MAR	Morocco	0,774	0,434	0,591	0,315	0,753
CLUSTER 4("LOW")							
68	ZAF	Southern Africa	0,746	0,532	0,286	0,474	0,760
71	EGY	Egypt	0,734	0,433	0,514	0,324	0,818
76	BWA	Botswana	0,668	0,579	0,150	0,429	0,568
CLUSTER 5 ("VERY LOW")							
89	MDG	Madagascar	0,508	0,391	0,258	0,231	0,767
90	KEN	Kenya	0,508	0,354	0,296	0,229	0,828
91	UGA	Uganda	0,496	0,393	0,268	0,198	0,726
93	GMB	Gambia	0,473	0,372	0,278	0,170	0,706
94	MWI	Malawi	0,462	0,281	0,298	0,221	0,878
95	ZMB	Zambia	0,453	0,335	0,224	0,225	0,803
96	TZA	Tanzania	0,450	0,353	0,237	0,189	0,742
98	MOZ	Mozambique	0,414	0,276	0,293	0,147	0,732
100	SEN	Senegal	0,411	0,339	0,161	0,212	0,693
103	BEN	Benin	0,380	0,315	0,132	0,213	0,672
104	NGA	Nigeria	0,375	0,343	0,138	0,168	0,604
105	CMR	Cameroun	0,371	0,274	0,190	0,179	0,804
106	ETH	Ethiopia	0,323	0,253	0,171	0,135	0,743
107	ZWE	Zimbabwe	0,227	0,073	0,236	0,084	0,482

Table 7. The countries of Africa ranked by the life quality component of sustainable development, 2010

3.1.2 The estimation of human life security as the component of sustainable development index

Using the method of estimation of the total impact of the global threats totality on different countries and world countries groups represented in chapter 2.1.2. (formulae 7-12) let us calculate the life security component C_{sl} for every country considered in this research. On the basis of the calculation of the standard value of Minkovski threats vector $C_{sl}= \left\| \vec{T} r_j \right\|$ let us introduce for every j country the correlation between the clusters of the countries:

$$K_k \prec K_j \Leftrightarrow \left\| \vec{T} r_k \right\| \leq \left\| \vec{T} r_j \right\|. \tag{13}$$

The calculations will be performed for the 5 groups of countries mentioned above. Table 8 rpresents the list of ten leading countries by the life security component of sustainable development in 2010.

Rate Csl	ISO	Country	Life security component, Csl	Biological balance, BB	Child mortality, CM	Corruption perception, CP	Energy safety, ES	Global diseases, GD	Inequalities between countries and people, GINI	Global warming, GW	Natural disasters, ND	State instability, SI	Limited access to potable water, WA
CLUSTER 1("VERY HIGH")													
1	AUS	Australia	1,549	0,916	0,666	0,874	0,931	0,642	0,562	0,143	0,564	0,624	0,670
2	ISL	Iceland	1,527	0,678	0,682	0,874	0,785	0,644	0,958	0,437	0,576	0,358	0,670
3	NZL	New Zealand	1,483	0,858	0,667	0,905	0,478	0,646	0,543	0,663	0,574	0,640	0,670
4	FIN	Finland	1,480	0,872	0,679	0,884	0,412	0,642	0,717	0,268	0,576	0,708	0,670
5	CAN	Canada	1,478	0,916	0,663	0,874	0,627	0,642	0,615	0,178	0,575	0,635	0,670
6	SWE	Sweden	1,473	0,766	0,681	0,897	0,466	0,642	0,748	0,498	0,576	0,669	0,670
7	NOR	Norway	1,451	0,511	0,679	0,869	0,621	0,642	0,735	0,661	0,576	0,640	0,670
8	LUX	Luxemburg	1,434	0,347	0,683	0,847	0,278	0,634	0,958	0,071	0,576	0,689	0,670
10	DNK	Denmark	1,397	0,284	0,674	0,901	0,377	0,642	0,752	0,353	0,576	0,722	0,670
CLUSTER 3("AVERAGE")													
9	PRY	Paraguay	1,398	0,918	0,537	0,258	0,975	0,586	0,227	0,644	0,546	0,515	0,425

Table 8. Ten leading countries by the life security component of sustainable development, 2010

All leading countries, except Paraguay, are in the cluster with very high values of life security index of sustainable development (table 8). It should be noted that Canada is the only representative of G8 group included in the list of ten leading countries.

Among G8 countries (Table 9) Italy has the worst values (43rd place). It should be mentioned that Russia in spite of rather low values of separate indices ("Corruption perception", "People inequality", "Global Warming") is on the 16th place which is due, first of all, by a large amount of natural resources.

In the group of BRIC countries (Table 10) we can see that Brazil and Russia have the significantly better results by human life security component while China and India the

Rate Csl	ISO	Country	Life security component Csl	Biological balance, BB	Child mortality, CM	Corruption perception, CP	Energy safety, ES	Global diseases, GD	Inequalities between countries and people, GINI	Global warming, GW	Natural disasters, ND	State instability, SI	Limited access to potable water, WA
CLUSTER 1 ("VERY HIGH")													
5	CAN	Canada	1,478	0,916	0,663	0,874	0,627	0,642	0,615	0,178	0,575	0,635	0,670
13	USA	The USA	1,368	0,244	0,656	0,801	0,908	0,634	0,448	0,128	0,505	0,619	0,654
20	DEU	Germany	1,315	0,296	0,674	0,835	0,328	0,642	0,693	0,357	0,575	0,569	0,670
21	FRA	France	1,312	0,374	0,676	0,754	0,304	0,639	0,611	0,476	0,571	0,701	0,670
23	JPN	Japan	1,281	0,244	0,679	0,815	0,282	0,632	0,750	0,345	0,570	0,146	0,670
30	GBR	Great Britain	1,246	0,272	0,667	0,815	0,282	0,633	0,547	0,370	0,566	0,455	0,670
CLUSTER 2 ("HIGH")													
43	ITA	Italy	1,210	0,255	0,678	0,485	0,306	0,644	0,545	0,411	0,575	0,671	0,670
CLUSTER 3 ("AVERAGE")													
16	RUS	Russian Federation	1,353	0,611	0,625	0,267	0,977	0,614	0,391	0,320	0,570	0,679	0,603

Table 9. The G8 countries ranked by the life security component of sustainable development, 2010

Rate Csl	ISO	Country	Life security component Csl	Biological balance, BB	Child mortality, CM	Corruption perception, CP	Energy safety, ES	Global diseases, GD	Inequalities between countries and people, GINI	Global warming, GW	Natural disasters, ND	State instability, SI	Limited access to potable water, WA
CLUSTER 3 ("AVERAGE")													
15	BRA	Brazil	1,353	0,865	0,576	0,418	0,695	0,574	0,202	0,628	0,549	0,720	0,621
16	RUS	Russian Federation	1,353	0,611	0,625	0,267	0,977	0,614	0,391	0,320	0,570	0,679	0,603
CLUSTER 4 ("LOW")													
79	CHN	China	1,115	0,431	0,584	0,407	0,713	0,605	0,433	0,533	0,145	0,472	0,478
83	IND	India	1,100	0,489	0,306	0,385	0,646	0,430	0,530	0,644	0,408	0,577	0,460

Table 10. BRIC countries group ranked by the life security component of sustainable development, 2010

rates of people life security practically coincide with the positions of these countries in the life quality rating of sustainable development.

For the group of post-socialistic countries (table 11) the main feature is the growth of difference by the value of human life security component. Thus, in 2010 the positions for this group vary from 16 (Russian) to 102 (Uzbekistan).

Rate Csl	ISO	Country	Life security component Csl	Biological balance, BB	Child mortality, CM	Corruption perception, CP	Energy security, ES
CLUSTER 2 ("HIGH")							
24	SVN	Slovenia	1,278	0,328	0,677	0,728	0,324
28	EST	Estonia	1,271	0,597	0,667	0,728	0,326
29	HRV	Croatia	1,267	0,431	0,666	0,463	0,306
33	SVK	Slovakia	1,232	0,416	0,655	0,508	0,296
36	LVA	Latvia	1,230	0,625	0,649	0,508	0,460
38	POL	Poland	1,226	0,361	0,661	0,564	0,312
39	LTU	Lithuania	1,225	0,496	0,658	0,553	0,314
40	HUN	Hungary	1,216	0,460	0,663	0,575	0,300
42	CZE	Czech Republic	1,211	0,309	0,677	0,553	0,301
53	ALB	Albania	1,179	0,445	0,620	0,364	0,395
CLUSTER 3 ("MIDDLE")							
16	RUS	Russian Federation	1,353	0,611	0,625	0,267	0,977
41	ARM	Armenia	1,212	0,438	0,570	0,313	0,295
44	AZE	Azerbaijan	1,199	0,438	0,490	0,276	0,292
46	BGR	Bulgaria	1,197	0,374	0,639	0,429	0,296
84	ROU	Rumania	1,099	0,467	0,622	0,429	0,345
CLUSTER 4 ("LOW")							
50	KAZ	Kazakhstan	1,187	0,481	0,527	0,313	0,651
65	UKR	Ukraine	1,152	0,438	0,613	0,267	0,388
77	KGZ	Kyrgyzstan	1,121	0,525	0,480	0,241	0,540
81	BIH	Bosnia and Herzegovina	1,109	0,438	0,618	0,343	0,318
87	MDA	Moldova	1,094	0,467	0,602	0,375	0,276
94	TJK	Tajikistan	1,069	0,489	0,330	0,249	0,514
CLUSTER 5 ("VERY LOW")							
102	UZB	Uzbekistan	1,038	0,460	0,477	0,225	0,319

Table 11. Post-socialistic countries ranked by the life safety component of sustainable development, 2010

Rate Csl	ISO	Country	Life security component Csl	Biological balance, BB	Child mortality, CM	Corruption perception, CP	Energy safety, ES	Global diseases, GD	Inequalities between countries and people, GINI	Global warming, GW	Natural disasters, ND	State instability, SI	Limited access to potable water, WA
						CLUSTER 3 ('AVERAGE")							
32	NAM	Namibia	1,242	0,839	0,459	0,508	0,495	0,202	0,049	0,684	0,505	0,710	0,533
49	TUN	Tunis	1,189	0,452	0,580	0,474	0,347	0,540	0,448	0,615	0,573	0,727	0,568
64	MAR	Morocco	1,155	0,474	0,489	0,375	0,286	0,586	0,446	0,684	0,574	0,736	0,339
98	DZA	Algeria	1,057	0,445	0,461	0,323	0,320	0,526	0,560	0,555	0,564	0,586	0,373
						CLUSTER 4 ("LOW")							
37	EGY	Egypt	1,227	0,438	0,569	0,323	0,313	0,579	0,623	0,620	0,576	0,713	0,654
56	BWA	Botswana	1,171	0,597	0,522	0,630	0,409	0,062	0,135	0,593	0,569	0,689	0,586
104	ZAF	Southern Africa	1,009	0,431	0,035	0,530	0,494	0,153	0,168	0,370	0,527	0,691	0,514
						CLUSTER 5("VERY LOW")							
45	TZA	Tanzania	1,198	0,503	0,161	0,304	0,826	0,133	0,574	0,683	0,542	0,689	0,068
47	ETH	Ethiopia	1,192	0,489	0,145	0,313	0,833	0,179	0,667	0,685	0,394	0,653	0,022
57	CMR	Cameroon	1,169	0,583	0,090	0,267	0,743	0,180	0,374	0,681	0,573	0,731	0,236
59	GMB	Gambia	1,167	0,354	0,154	0,333	0,764	0,326	0,323	0,681	0,568	0,694	0,533
62	MOZ	Mozam-bique	1,158	0,597	0,093	0,294	0,848	0,156	0,326	0,644	0,372	0,701	0,042
66	ZMB	Zambia	1,150	0,618	0,062	0,343	0,825	0,037	0,265	0,681	0,396	0,659	0,101
69	MWI	Malowi	1,138	0,518	0,174	0,375	0,764	0,059	0,484	0,685	0,399	0,666	0,323
71	BEN	Benin	1,136	0,489	0,113	0,333	0,677	0,256	0,493	0,675	0,544	0,724	0,250
76	UGA	Uganda	1,122	0,467	0,083	0,294	0,764	0,123	0,412	0,684	0,482	0,691	0,157
92	KEN	Kenya	1,076	0,481	0,097	0,267	0,784	0,138	0,316	0,676	0,375	0,616	0,095
93	SEN	Senegal	1,071	0,525	0,146	0,343	0,577	0,301	0,481	0,676	0,562	0,600	0,177
95	NGA	Nigeria	1,069	0,496	0,026	0,294	0,853	0,153	0,405	0,392	0,574	0,509	0,089
96	MDG	Madagascar	1,064	0,611	0,153	0,343	0,764	0,339	0,324	0,683	0,454	0,121	0,028
106	ZWE	Zimbabwe	0,991	0,489	0,188	0,267	0,728	0,003	0,275	0,659	0,443	0,147	0,356

Table 12. Countries of Africa ranked by the life security component of sustainable development, 2010

For the countries of Africa (Table 12) we have the average (Namibia, Morocco, Tunis, Algeria), low (Egypt, Botswana, South Africa) and very low values of life security component of sustainable development. This results in permanent political and military conflicts in this region.

Analyzing Ukraine by its vulnerability to the global threats we see that in comparison with 2009 the rate of its national security has become slightly better, but still remains significantly low (by the human life security index Ukraine has reached the 65th position from 78th position). For Ukraine the worst threats still are the following: level of spreading of global diseases, especially AIDS and tuberculosis, which is one of the highest in the world; very high level of corruption; low level of energy security; high child mortality; high level of state fragility.

3.1.3 The estimation of sustainable development index as quarter functional of human life security and quality

Having obtained the values of life quality component of sustainable development C_{ql} (tables 3-7) and component of human life security Csl (table 8-12), let us calculate the value of sustainable development index I_{sd}, as a quarter functional by the formula (1) according to the SDGM methodology. The results of calculations for 5 groups of countries are shown in Tables 13-17 accordingly. All countries have been distributed into 5 clusters by the sustainable development index: "Very high"", "High", "Average", "Low" and "Very low".

According to table 13, ten countries with the highest values of sustainable development index include 7 European countries (Iceland, Sweden, Norway, Switzerland, Finland, Denmark and Luxemburg), one country of Northern America (Canada) and the countries of Oceania (Australia and New Zealand). They are characterized by low level of vulnerability to the global threats (high level of national security), high indices of human life quality in the economic, ecological and social dimensions, high harmonization level of sustainable development (figure 8).

Cluster 1 ("Very low") contains the group of the most "successful" countries of the world, including the G8 countries, except Russia; they have the highest rates of life quality and lowest rate of vulnerability to the impact of global threats totality according to Table 13, 14.

On the contrary cluster 5 ("Very low") contains the countries with low values of life quality component of sustainable development and these countries are more vulnerable to the impact of global threats totality. Ukraine together with China, India, South Africa and other countries has been included to cluster 4 ("Low") with low level of sustainable development. Most of these countries have average and low values of life quality and security components of sustainable development. This means that there is the definite correlation between vulnerability to the global threat totality (global saecurity) and life quality component of sustainable development of these countries.

BRIC countries group hold the following rating positions: Brazil – the 35th position, Russia – the 49th position, China – the 78th position, India – the 86th position.

Rate Isd	ISO	Country	Sustainable development index Isd	Life quality component Cql	Life security component Csl
CLUSTER 1 ("VERY HIGH")					
1	ISL	Iceland	2,883	1,357	1,527
2	SWE	Sweden	2,870	1,398	1,473
3	AUS	Australia	2,859	1,310	1,549
4	NZL	New Zealand	2,848	1,365	1,483
5	NOR	Norway	2,830	1,379	1,451
6	CHE	Switzerland	2,827	1,498	1,329
7	FIN	Finland	2,823	1,342	1,480
8	CAN	Canada	2,771	1,293	1,478
9	DNK	Denmark	2,707	1,310	1,397
10	LUX	Luxemburg	2,691	1,257	1,434

Table 13. Ten leading countries ranked by sustainable development index, 2010

Rate Isd	ISO	Country	Sustainable development index Isd	Life quality component Cql	Life security component Csl
CLUSTER 1 ("VERY HIGH")					
8	CAN	Canada	2,771	1,293	1,478
12	DEU	Germany	2,654	1,338	1,315
13	USA	The USA	2,636	1,268	1,368
14	FRA	France	2,631	1,320	1,312
16	JPN	Japan	2,571	1,290	1,281
17	GBR	Great Britain	2,565	1,319	1,246
CLUSTER 2 ("HIGH")					
26	ITA	Italy	2,380	1,169	1,210
CLUSTER 3 ("AVERAGE")					
49	RUS	Russia	2,093	0,740	1,353

Table 14. G8 countries ranked by sustainable development index, 2010

G8 countries are "scattered" in the table from the 8th (for Canada) to the 49th position (for Russia) (Table 14).

Rate Isd	ISO	Country	Sustainable development index Isd	Life quality component Cql	Life security component Csl
CLUSTER 3 ("AVERAGE")					
35	BRA	Brazil	2,256	0,902	1,353
49	RUS	Russia	2,093	0,740	1,353
CLUSTER 4 ("LOW")					
78	CHN	China	1,762	0,647	1,115
86	IND	India	1,672	0,572	1,100

Table 15. BRIC countries group ranked by sustainable development index, 2010

Post-socialistic countries also took different positions by sustainable development index (table 16). The clusters with very high and high value of sustainable development index contain Slovenia, Lithuania, Estonia, Slovakia, Croatia, Latvia, Hungary, Poland, Czech Republic, Bulgaria.

Rate Isd	ISO	Country	Sustainable development indexIsd	Life quality componentCql	Life security component Csl
CLUSTER 2 ("HIGH")					
22	CZE	Czech republic	2,425	1,214	1,211
23	SVK	Slovakia	2,408	1,176	1,232
24	EST	Estonia	2,393	1,149	1,244
29	SVN	Slovenia	2,360	1,083	1,278
31	LTU	Lithuania	2,350	1,125	1,225
32	HUN	Hungary	2,327	1,112	1,216
33	LVA	Latvia	2,325	1,095	1,230
34	HRV	Croatia	2,268	1,000	1,267
38	POL	Poland	2,235	1,009	1,226
43	ALB	Albania	2,163	0,984	1,179
CLUSTER 3 ("AVERAGE")					
45	BGR	Bulgaria	2,129	0,932	1,197
49	RUS	Russia	2,093	0,740	1,353
50	ROU	Rumania	2,091	0,992	1,099
54	ARM	Armenia	2,029	0,817	1,212
60	AZE	Azerbaijan	1,961	0,734	1,227
CLUSTER 4 ("LOW")					
64	KAZ	Kazakhstan	1,907	0,720	1,187
68	UKR	Ukraine	1,889	0,854	1,036
73	BIH	Bosnia and Herzegovina	1,816	0,707	1,109
75	KGZ	Kyrgyzstan	1,774	0,653	1,121
CLUSTER 5 ("VERY LOW")					
83	MDA	Moldova	1,713	0,619	1,094
97	TJK	Tajikistan	1,562	0,493	1,069
104	UZB	Uzbekistan	1,450	0,411	1,038

Table 16. Post-socialistic countries ranked by sustainable development index, 2010

Russia, Rumania, Georgia, Moldova, Armenia have been included into the cluster with average values of sustainable development index. The countries with low and very low value of sustainable development index include Ukraine, Azerbaijan, Kyrgyzstan, Tajikistan and Uzbekistan.

All countries of Africa, except for Namibia, Morocco, Tunis and Algeria, are in the clusters with low and very low value of sustainable development index.

Rate Isd	ISO	Country	Sustainable development index Isd	Life quality component Cql	Life security component Csl
CLUSTER 3 ("AVERAGE")					
53	NAM	Namibia	2,034	0,792	1,242
55	TUN	Tunis	2,024	0,835	1,189
62	MAR	Morocco	1,929	0,774	1,155
70	DZA	Algeria	1,859	0,761	1,098
CLUSTER 4 ("LOW")					
61	EGY	Egypt	1,961	0,761	1,199
71	BWA	Botswana	1,853	0,796	1,057
80	ZAF	Southern African Republic	1,755	0,746	1,009
CLUSTER 5("VERY LOW")					
88	TZA	Tanzania	1,648	0,450	1,198
89	UGA	Uganda	1,640	0,473	1,167
90	ZMB	Zambia	1,618	0,496	1,122
92	MWI	Malaya	1,600	0,541	1,059
93	KEN	Kenya	1,600	0,462	1,138
94	GMB	Gambia	1,584	0,508	1,076
95	MDG	Madagascar	1,572	0,508	1,064
96	MOZ	Mozambique	1,571	0,414	1,158
98	CMR	Cameroon	1,540	0,371	1,169
99	BEN	Benin	1,517	0,380	1,136
100	ETH	Ethiopia	1,514	0,323	1,192
101	SEN	Senegal	1,482	0,411	1,071
105	NGA	Nigeria	1,443	0,375	1,069
107	ZWE	Zimbabwe	1,218	0,227	0,991

Table 17. Countries of Africa ranked by sustainable development index, 2010

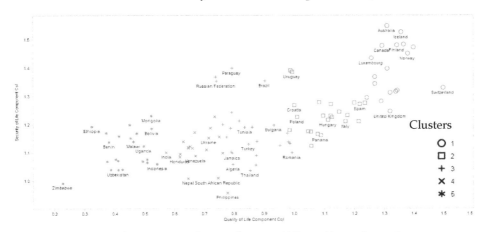

Fig. 6. Clusterization of countries in the coordinates of life quality and security

3.2 Country profiles construction on example of Ukraine

One of the main applications of the Sustainable Development Gauging Matrix (SDGM) is using actual data on indicators and parameters of sustainable development for a given country with the purpose of decision-making at various levels of the country's governance.

Using the country profiles service (http://wdc.org.ua/en/services/country-profiles-visualization) provided by WDC-Ukraine one can easily obtain dashboard for each world country to perform further in-depth analysis.

For 2010 results Ukraine has Isd=1,889, Cql=0,854, Csl=1,036 with rankings #68, #73, #65 correspondingly. Each sustainable development component and its can be displayed in a dimension diagram (Fig. 7).

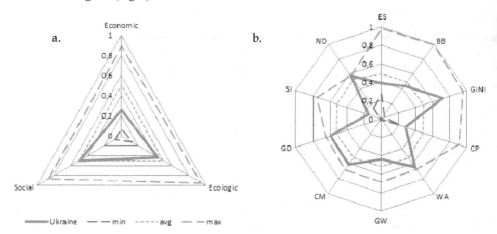

Fig. 7. Dimension diagrams for Ukraine's quality of life(a) and security of life (b).

Given figure gives possibility to handle visual analysis of the strengths and weaknesses of Ukraine through comparison of the values for certain indicators with their extreme and average meanings.

Considering the quality of life diagram one can point out, that Ukraine has better developed social dimension and poorer economic dimension. Analyzing the security of life component we can name as strengths indicators which values are better than average: people inequality (GINI), access to potable water (WA), health (CM, GD) and natural disasters (ND). Accordingly weaknesses are energy security (ES), biological balance (BB), corruption perception (CP), CO_2 emissions (GW) and state instability (SI). The most critical situation is with corruption and state instability that corresponds to the evaluations of experts from many international organizations like World Economic Forum, World Bank, etc. about Ukraine development problems.

4. Conclusion

In this research the system of indexes and indicators has been developed and the gauging matrix for sustainable development processes (SDGM) in three dimensions: economic, ecological and socio-institutional has been offered. Using this matrix and initial data, obtained by the recognized international organizations we have developed the mathematical model that gives the possibility to calculate the components of human life quality and security as the components of sustainable development index and harmonization level of this development for every country. The global modeling of sustainable development processes for the large group of the countries in terms of human life quality and security has been performed. The results of modeling have been explained in details for every dimension of the sustainable development.

5. Acknowledgment

The author expresses his gratitude to the employees of ICSU World Data Center for Geoinformatics and Sustainable Development Kostiantyn Yefremov, Andrew Boldak, Alexis Pasichny, Tetyana Matorina, Olena Poptsova for their assistance in data gathering and computer modeling of the presented results.

6. References

Aivazian S. A. (1983). *Applied statistics: Fundamentals of simulation and primary data processing [Text]* (Reference book), S. A. Aivazian, I S. Yenukov, L. D. Meshalkin, Finances and statistics, Moscow

AlenkaBurja (n.d.). Interview with Hermann Scheer – Energy is a driving force for our civilisation – SOLAR ADVOCATE [Electron. resource]. Access link: http://www.folkecenter.dk/en/articles/hscheer_aburja.htm

CIA (n.d.). The World Factbook [Electron. Resource]. Access link: https://www.cia.gov/library/publications/theworld-factbook/

Corruption Perception Index 2008 [Electron. resource] (2008). Access link: http://www.transparency.org/policy_research/surveys_indices/cpi/2008/cpi_20 08_table

Donbass Internet paper NEWS.dn.ua (n.d.). Parade of moneybags: 50 richest people of Ukraine have amassed $64,4 billion [Electronic resource]. Access link: http://novosti.dn.ua/details/46009/2.html

GeoHive (n.d.). Current World Population [Electron. resource]. Access link: http://xist.org/earth/population1.aspx

Global Footprint Network (n.d.). Ecological Footprint – Ecological Sustainability [Electron. resource]. Access link: http://www.footprintnetwork.org/en/index.php/GFN/

Human Development Report 2007/2008. Fighting climate change: Human solidarity in a divided world [Electron. resource] (n.d.). Access link: http://hdr.undp.org/en/media/HDR_20072008_EN_Complete.pdf

International Living (2009). 2009 Quality of Life Index [Electron. resource]. Access link: http://www.internationalliving.com/index.php/Internal-Components/Further-Resources/qofl2009

Johannesburg Summit (2002). Access mode: http://www.un.org/russian/conferen/wssd/

Kondratiev N. D. (1989) *Problems of economic dynamics [Text]*, Economics, Moscow

Kotliakov V. M. (2001) Global Climate changes: anthropogenic impact or natural variations?, *Ecology and life*, V. M. Kotliakov, №1. Access link: http://www.ecolife.ru/jornal/ecap/2001-1-3.shtml;

Marshall M. G. (2008). Global Report on Conflict, Governance and State Fragility 2008 [Electron. resource], In: *Foreign Policy Bulletin*, M. G. Marshall, B. R. Cole. Access link: http://www.systemicpeace.org/Global Report 2008.pdf

Mathers C. D. (2006). Projections of Global Mortality and Burden of Disease from 2002 to 2030 [Electron. resource], In: *PLoS Medicine*, C.D. Mathers, D. Loncar. Access link: http://www.plosmedicine.org/article/info%3Adoi%2F10.1371%2Fjournal.pmed.0 030442

Membrane (n.d.). Pentagon experts: global warming will ruin the Earth [Electronic resource]. Access link:
http://www.membrana.ru/articles/misinterpretation/2004/03/03/182200.html

Nation Master (n.d.). World Statistics, Country Comparisons [Electron. resource]. Access link: http://www.nationmaster.com/index.php

Report on the Transparency International Global Corruption Barometer 2007 [Electron. resource] (2007). Access link:
http://www.transparency.org/content/download/27256/410704/file/GCB_2007_ report_en_02-12-2007.pdf

Science Council of Japan (2005). Japan Vision 2050. Principles of Strategic Science and Technology Policy Toward 2020 [Text], 30 p.

Statistical Information System of World Health Organization (n.d.). Detailed database search [Electron. resource]. Access link: http://apps.who.int/whosis/data/Search.jsp

Summit «Planet Earth», Rio-de-Janeiro (1992). Access link:
http://www.un.org/ru/development/progareas/global/earthsummit.shtml

The Global Competitiveness Report 2009–2010, World Economic Forum, [Electron. resource] (n.d.). Access link:
http://www.weforum.org/pdf/GCR09/GCR20092010fullreport.pdf

The Heritage Foundation (n.d.). Index of Economic Freedom: Link Between Economic Opportunity and Prosperity [Electron. resource]. Access link:
http://www.heritage.org/index/

The Intergovernmental Panel of Climate Change [Electron. resource] (n.d.). Access link:
http://www.ipcc.ch/index.htm

Transparency International. Visualizing the 2010 Corruption Barrometer. [Електронний ресурс] (n.d.). Режим доступу:
http://www.transparency.org/policy_research/surveys_indices/gcb/2010/intera ctive_2

UNDP (n.d.). Bureau for Crisis Prevention and Recovery – the DRI Analysis Tool: [Electron. resource]. Access link: http://gridca.grid.unep.ch/undp/analysis/result.php;

UNICEF Joint Monitoring Programme for Water Supply and Sanitation (n.d.). Water for life: Making it happen [Electron. resource]. Access link:
http://www.who.int/entity/water_sanitation_health/waterforlife.pdf

UNO Development Program (UNDP) [Electronic resource] (n.d.). Access link:
http://www.un.org/russian/ga/undp/

Vernadsky V. I. (1944). Some words about noo-sphere [Text]. *Success of modern biology*, Vernadsky V. I., № 18, issue 2, p. 113–120

World Economic Forum (n.d). Global Competitiveness Reports [Electron. resource]. Access link:
http://www.weforum.org/en/media/publications/CompetitivenessReports/inde x.htm ;

Yale Center for Environmental Low& Policy (2008). Environmental Performance Index 2008 [Electron. resource]. Access link: http://epi.yale.edu/

Zgurovsky M. Z. (2007) Sustainable development global simulation: Opportunities and threats to the planet [Text]. *Russian Journal of Earth Sciences*, Zgurovsky M. Z., Vol. 9, ES2003, doi:10, 2205/2007ES000273

4

Social Accounting Matrix – Methodological Basis for Sustainable Development Analysis

Sasho Kjosev

University "Ss. Cyril and Methodius", Faculty of Economics
Republic of Macedonia

1. Introduction

The Rio Summit established sustainable development as the guiding vision for the development efforts of all countries. At Rio, and in later commitments, all governments undertook to establish and implement national sustainable development strategies. The strategies for sustainable development, called for at Rio, are foreseen as highly participatory instruments intended "to ensure socially responsible economic development while protecting the resource base and the environment for the benefit of future generations".[1]

In simple terms, sustainable development means **integrating the economic, social and environmental objectives of society**, in order to maximize human well-being in the present without compromising the ability of future generations to meet their needs. This requires seeking mutually supportive approaches whenever possible, and making trade-offs where necessary.

Sustainable development is not an activity that has to be left to the long term. Rather, it constitutes a set of short, medium and long term actions, activities and practices that aim to deal with immediate concerns while at the same time address long-term issues. Achieving sustainable development requires far reaching policy and institutional reforms and the involvement of all sectors at all levels. Sustainable development is not the responsibility of only government or one or two sectors of society.

However, in order to better understand the sustainable development concept, one has to develop an appropriate methodological instrument. In parallel with the introduction of the sustainable development concept, the SNA 1993 introduced the Social Accounting Matrix (where later its extensions, the SESAME and NAMEA approaches, have been developed) as a methodological basis for the sustainable development analysis.

Therefore, the following chapter will firstly present the theoretical foundations of the Social Accounting Matrix (SAM) and its extensions (SESAME and NAMEA approach), and secondly, will present possibilities to implement these approaches in the Republic of Macedonia.

---

[1] OECD, 2001, p. 11

2. Theoretical considerations

The development planning methodology is of highest importance for the unity, complexity and consistency of the sustainable development planning system. It should enable methodological consistency in the process of evaluation of the development conditions, problems and perspectives, perception of interests, objectives and tasks of the relevant stakeholders and their harmonization, the simultaneity of the planning process, as well as the mandatory preparation and execution of plans. Having this in mind, part 2 of this chapter will present the theoretical foundations of the Social Accounting Matrix (SAM) and its extensions (the SESAME and NAMEA approaches).

2.1 Social Accounting Matrix (SAM)

Social accounting matrix (SAM) is a technique related to national income accounting, providing a conceptual basis for examining both growth and distributional issues within a single analytical framework in an economy. It can be seen as means of presenting in a single matrix the interaction between production, income, consumption and capital accumulation.

A SAM is defined as the presentation of System of National Accounts (SNA) accounts in a matrix which elaborates the linkages between a supply and use table and institutional sector accounts. In many instances SAMs have been applied to an analysis of interrelationships between structural features of an economy and the distribution of income and expenditure among household groups. Evidently, SAMs are closely related to national accounts whereby their typical focus on the role of people in the economy may be reflected by, among other things, extra breakdowns of the household sector and a disaggregated representation of labour markets (i.e., distinguishing various categories of employed persons). On the other hand, SAMs usually encompass a somewhat less detailed supply and use table or input-output table.[2]

The SAM is a comprehensive, flexible, and disaggregated framework which elaborates and articulates the generation of income by activities of production and the distribution and redistribution of income between social and institutional groups. A principle objective of compiling a SAM is, therefore, to reflect various interdependencies in the socio-economic system as a whole by recording, as comprehensively as is practicable, the actual and imputed transactions and transfers between various agents in the system. The SAM has, mainly, two basic tasks:

a.   to enable presentation of information about the economic and social structure of the national economy; and
b.   to provide analytical and accounting framework as a basis for construction of macroeconomic models for analyzing the national economy and the effects from the implementation of the macroeconomic and development policy measures.

The increasing interest of recent years in the compilation of this kind of accounting system is intimately related to growing dissatisfaction with the results of growth and development policies. These disappointing results, especially as regards their distributional aspects, gave rise to questions concerning the processes and mechanisms by which the production of

---

[2] European Commission et al., 1993, 20.4

goods and services, and their distribution and redistribution relate to each other. To examine these kind of question on an empirical basis, data were required that would provide a coherent, detailed and consistent picture of all these aspects of the economic process.

The fundamental principle of SAM is balancing within the series of national accounts, whereby revenues and expenditures are presented for each individual account. SAM follows the principles of "single entry" and presents a series of accounts presented in the form of matrix. SAM consists of rows and columns marked with identical titles. The rows and columns present different accounts in the economic system. For each account, and hence, for each pair of row and column, the data in the row shows the revenue (inflow) on the account, and the data in the respective column is an expenditure (outflow) on that same account. Expenditures in one account are revenues in the other and vice versa. In sum, within each economic system, all revenues must correspond to the respective expenditures, i.e. the correspondent rows and columns must be equal. In other words, each component (data) of SAM is revenue for the account in the row, and expenditure for the account in the column (Table 1).

The above presented points to a conclusion that the SAM provides quantitative information on the following aspects of the economic process:

-   the relationship between various types of production activities and various factors of production from the aspect of generation and allocation of income;
-   total income received by various factors of production;
-   allocation of incomes to factors of production between separate institutions via taxes, welfare, transfers;
-   consumption of goods and services among certain institutions;
-   total supply of goods and services from the domestic and foreign markets;
-   savings and capital transfers among institutions, etc.

As a conclusion, one can point out to the following:

1. SAM successfully combines indicators of growth, allocation of income and poverty in one coherent framework. By including elements of input-output table, national accounts and other databases, SAM provides complex quantitative image suitable for macroeconomic analysis and planning;
2. SAM is a useful tool for harmonizing various sources of data and filling the gap in information received from various statistical databases, thus contributing to greater consistency and adequacy;
3. SAM proved its usefulness as an integrated statistical database suitable for preparation of macroeconomic models of the national economy to the end of better understanding and envisaging the interrelationship of the determinants of economic trends in the national economies.

2.2 SESAME approach

SESAME (System of Economic and Social Accounting matrices and Extensions) is a statistical information system in matrix format, from which a set of core economic, environmental and social macro-indicators is derived. The system is driven, to a large

accounts	goods/services	Production	generation of income	allocation of primary income	Allocation of secondary income	use of income	capital	rest of world	total
goods/services		Interme-diate Consu-mption				final consu-mption	gross capital formation	export of goods/services	total demand
production	output and taxes on products less subsidies								total output
generation of income		net domestic product						compensation of employees from ROW	generated income
allocation of primary income			Compensation of employees, taxes on products and import, subsidies, net operating surplus/net mixed income	property income				property income from ROW	total primary income
allocation of secondary income				net national income	current taxes on income, wealth and current transfers			current taxes on income, wealth and current transfers from ROW	total secondary income
use of income					net disposable income	adjustment for change in net equity of households on pension funds		Adjustment for change in net equity of households on pension funds from ROW	disposable income
capital		Deprecia-tion				net saving	capital transfers and acquisitions less disposals of non-produced assets	capital transfers, receivable(=)/payable (-) and acquisitions, less disposals of non-produced asset from ROW	capital transfers
rest of world	import of goods/services		compensation of employees to ROW	property income for the ROW	current taxes on income, wealth and current transfers for ROW	adjustment for change in net equity of house-holds on pension funds for ROW	Net lending (+)/net borrowing (-) of the national economy		total outflow to ROW
total	total supply	total input	Alloca-tion of generated income	Alloca-tion of primary income	allocation of secondary income	Allocation of disposable income	capitalexpenditures	total inflow from ROW	

Table 1. Schematic presentation of a SAM

extent, by the kind of information required for monitoring and policy-making at the macro-level. Although it is impossible to capture socio-economic development in a single indicator, it is equally clear that a prime task of national statistical offices is to comprise the countless numbers they collect to a manageable, "executive" summary. Such a summary typically describes trends in main indicators. At the same time, for analytical purposes a more detailed data framework is required. Obviously, the communication between policy-makers and analysts is optimally served if the core macro-indicators are all derived from an integrated information system such as SESAME.[3]

Keuning, in his paper[4], points out to the fact that, essentially, SESAME integrates economic, social and environmental accounts and indicators, through a conceptual and numerical linkage of related monetary and non-monetary data. It extends the SAM by integrating related information, in non-monetary units. For instance, compensation of employees by industry and labor category in the SAM is broken down into hours worked and an average hourly wage rate. In turn, these hours worked for payment are related to other time use of the employed persons concerned. Subsequently, time use of the employed persons can be combined with the time use of the other members of the same household (group), to arrive at a comprehensive linkage of (social) time use data and (economic) income figures.

Moreover, Keuning states that, in order to achieve a linkage between monetary and non-monetary data, the SAM-values are broken down into price (changes) and volume (changes). The linkages with other data are thus typically established in non-monetary units such as hours, calories, and "volume" changes. In this way, the necessary connections are made without distorting the essentially monetary system of the national accounts A SESAME registers for all variables both the national total value and its distribution among socio-economic household groups, categories of employed persons, etc. As a next step, a range of summary indicators can be derived from such a data set (e.g. Gross Domestic Product, population size, (un)employment, inflation, balance on current account of the balance of payments, income inequality, environmental indicator(s), daily calorie intake of the poorest subgroup, average number of years of schooling). Consistent indices covering distributional aspects can also be derived for all variables included in the system. Whatever set of aggregates is preferred, they would all share two crucial features: first, every indicator is computed from a single, fully consistent statistical system, and secondly, each indicator uses the most suitable measurement unit for the phenomenon it describes.[5]

As a summary, Keuning[6] points out to the following advantages of the SESAME approach:

-   SESAME can serve as a useful extension to present-day national accounts, in two respects. First, the SAM-part of SESAME improves the compilation of national accounts, because it integrates more basic sources at a meso-level. Secondly, SESAME is useful for integrating all kinds of (non-monetary) social and environmental statistics.

---

[3] Keuning, 1998, pp. 353-354
[4] Keuning, 1998, pp. 353-354
[5] Keuning & Timmerman, 1995, pp. 3-5
[6] Keuning, 2000, pp. 289-290

- Just like conventional national accounts, *SESAME provides both core macro-indicators and an underlying information system*. In this way, it simultaneously serves two categories of user: first, the general public, media and policy makers, who want to know the main trends at a glance, and secondly, the analysts, scientists and policy-advisers, who want to disentangle causes and consequences, make forecasts and do policy simulations.
- SESAME promotes the use of uniform units, classifications, concepts, etc. throughout a statistical system; that is, not only in economic statistics, but also in social statistics. Among the advantages of such a harmonization is a much easier matching of results from different surveys. As a consequence, fewer questions per survey and perhaps even smaller samples are needed. It is likely that some groups of specialized users will prefer a different classification or concept for their specific field of interest. However, only an integrated data system can be qualified as a pure public good and therefore the compilation of data according to special purpose specifications might receive a lower priority in official statistics, or be financed to a larger extent by the beneficiaries.
- SESAME is an inherently *flexible* framework. It can readily be adapted to the specific characteristics, needs and capabilities of every country or region. In particular the accounting structure, the classifications and the kind of non-monetary phenomena incorporated can be tailor-made. Because of this modular approach, it is not necessary to include all aspects at once.
- Finally, it should be mentioned that *SESAME essentially aims at a better use, through integration, of existing statistics*. In turn, by integrating information that is already collected, official statistical agencies will increase their own value added.
- Just like present-day national accounts, SESAME is a multi-purpose information system that can be used to test any economic or social theory. It is this property that has made the national accounts the universal language of economics. SESAME may open the door to even richer insights into human welfare.

2.3 NAMEA approach

The economy is a complex system of which extraction of natural resources, production, consumption, technology, investment, imports and exports, and release of wastes (and pollution) are just a few of the many different interrelated dimensions. All these different aspects of the economy may have detrimental or beneficial effects on environmental pressures. Hence, there is a pressing need for promoting integrated economic and environmental information systems as opposed to indicators' lists in order to meet the increasing users' demand for conducting integrated economic and environmental policies. They reiterated that environmental-economic accounting would provide the necessary framework for analysing the impact of economic growth on long-term sustainable development.

The need to account for the environment and the economy in an integrated way arises because of the crucial functions of the environment in economic performance and in the generation of human welfare. These functions include the provision of natural resources to production and consumption activities, waste absorption by environmental media and environmental services of life support and other human amenities.

Having this in mind, in 1989 Statistics Netherlands started to develop a system for describing environmental aspects in conjunction with the national accounts. The system, known as the National Accounting Matrix including Environmental Accounts (NAMEA), creates a link between the national accounts and environmental statistics. NAMEA shows the relationship between a number of important economic indicators (gross domestic product, balance of payments, etc.) and the environment.

The NAMEA has been developed to systematically supplement the national accounts with environmental statistics. Its hybrid accounting structure, i.e. the combined presentation of physical and monetary accounts, indicates that in the NAMEA environmental imputations in the core national accounts framework are avoided.

Therefore, NAMEA has been developed to link environmental and economic statistics. An important characteristic of environmental accounting is that the data are consistent with the National Accounts which mean that the environmental data can be directly compared to well known macro-economic indicators such as GDP, inflation and investment rates, developed in the System of National Accounts (SNA).

The NAMEA system contains no economic assumptions; it is only descriptive. It maintains a strict borderline between the economic and the environmental aspects. It is represented in monetary units, on the one hand, and in physical units, on the other. To get a clear understanding of the interrelationships between the natural environment and the economy, we must use their physical representation. Otherwise, it is impossible to understand these relations. If the NAMEA system would contain monetary values about environmental problems, two problems would occur. Firstly, the environment must be valued in monetary units and secondly it is very complicate to differentiate between price changes and quantity changes.

Therefore, the resulting indicators are measured in physical units. The interrelationship between the economy and the environment has two perspectives, an economic one and an environmental one. The **economic perspective** contains the physical requirements in the economic processes, like energy and material and spatial requirements. The **environmental perspective** puts forward the consequences of these requirements with respect to the availability of the natural environment. Consequently, the optimal allocation of natural resources requires the consideration of both perspectives.

NAMEA is a tool to account for environmental problems combining the data from the environment with the economic data from the core of the SNA. However, no specific economic assumptions are used to compile a NAMEA. Policy–makers are free to decide which kinds of environmental themes and environmental substances should be investigated, and policy–makers should decide how they want to resolve the environmental problems. As a result, the NAMEA does not only produce aggregate indicators in a consistent meso–level information system, it also provides data in the required format for all kinds of analyses.

Two accounts for the environment have been added to the national accounts matrix in the NAMEA: a *substances account* and an *account for environmental themes*. These accounts contain observed environmental data in physical units (emissions and waste in kg and energy use in joules). They show not only emissions (pollution originating from products and consumers), but also immissions (for example, deposition of pollutants) in the environment.

The substances account explains the relationship between the amount of environmental stress attached to current economic transactions and the amount of environmental stress that potentially threatens all properties of resident entities including economic assets, health and the national ecological heritage.

The environmental theme account is denominated in physical units and focus on the consistent presentation of material input of natural resources and output of residuals for the national economy. These inputs and outputs are the environmental requirements of the economy. In the environmental themes account, substances are grouped and aggregated in accordance to their type of environmental stress and subsequently represented in a limited number of aggregated theme indicators. Most themes correspond to national or local environmental problems and the corresponding indicators reflect the net accumulation of pollutants within the country's borders. For global environmental themes, i.e. greenhouse effect and ozone layer depletion, the indicators only review the weighted pollution generated by economic agents that belong to the national economy, representing the national contribution to these global problems. Consequently, for some of the environmental themes it is relevant and possible to determine the total amount of pressure that is put on the national environment in a single accounting period. However, the accounts do not show the actual damages that may result (now or later) from these pressures. By the presentation of the economic accounts in monetary terms and the environmental accounts in the most relevant physical units, the NAMEA maintains a strict borderline between the economic sphere and the natural environment.[7]

De Haan, in his paper[8], points out to the following main characteristics of the NAMEA:

Firstly, the NAMEA maintains a strict borderline between the economic sphere and the natural environment, established by monetary accounts on the one hand and accounts denominated in the most relevant physical units on the other. The non-monetary accounts show the environmental requirements of an economy, which are not subject to market transactions and which are for that reason not included in the core national accounts. Similarly, the physical flows underlying commodity transactions do not enter the accounts for environmental requirements.

Secondly, the NAMEA maintains a clear distinction between physical inputs (extraction of resources) on the one hand and outputs (emission of pollutants) on the other.

Thirdly, most NAMEAs contain environmental themes account in which substances are grouped together and aggregated in accordance to the type of environmental pressure to which they are expected to contribute. In this way, a wide range of substances are represented by only a limited number of aggregated theme indicators on the basis of weighting methods.

---

[7] de Haan & Keuning, 2000, pp. 3-5
[8] de Haan, M, 2004, pp. 84-86

Finally, the NAMEA provides an institutional representation of the economy and its relationship with the environment. This implies that economic activities together with their environmental requirements are defined and subsequently recorded according to statistically observable units, *i.e.* the so-called establishments classified according to the International Standard Industrial Classification (ISIC).

3. Recommendations for the Republic of Macedonia

There is an urgent need for implementation of macroeconomic indicative planning in the Repubic of Macedonia, as a transition country, where the Government should implement managerial activities in the public sector, public finances, etc. A basis for such indicative economic planning is the macroeconomic policy document of the Government, which provides instruments necessary for the realization of the predetermined medium and long-term development goals. The macroeconomic planning document of the Government should be a programme for the Government medium-term economic and social policy, with clearly specified activities for the: public investments, public enterprises, local economic development, social assistance, public revenues and expenditures, etc. This system of indicative economic planning is compatible with those already existing in the market-based economies and enables realization of the Government medium and long-term socio-economic development goals.

Hence, indicative planning methodology has a significant importance on the coherency, complexity and consistency of the national planning system in the Republic of Macedonia. It enables the consistency during the process of obtaining information about the development conditions and problems, adjustment of the policy-makers' tasks and goals, as well as the preparation and realization of the socio-economic development plans and programs.

Therefore, in order to satisfy the abovementioned, it is necessary the effort of the scientists and experts in our country to be focused on preparation of a complex analytical framework, consisting of:

a.   Preparation of a highly disaggregated Social Accounting Matrix (SAM)
b.   Preparation of a SAM extended with social indicators (SESAME approach)
c.   Preparation of a SAM extended with environmental indicators (NAMEA approach)

Consequently, the focus, in the third part of this chapter, will be put on the activities that should be carried out in order to enable implementation of an efficient process of preparation and elaboration of the abovementioned analytical and methodological documents.

3.1 Social accounting matrix (SAM) for the Republic of Macedonia

At the moment, there is no Social Accounting Matrix (SAM) for the Republic of Macedonia. But, there is time series data with a set of sectoral accounts (from the production to the capital account, for five domestic institutional sectors and rest of the world sector), as well as, a high quality time series data for the national accounts

Hence, it is due time the State Statistical Office, together with the relevant ministries in the Government and experts from the scientific and educational institutions in Macedonia, as well as with institutions and experts from the developed market economies to start the preparation and construction of a highly disaggregated SAM for the Republic of Macedonia, based on the positive experiences of the developed market and transition economies. The SAM for the Republic of Macedonia would mainly have two basic tasks:

a. to enable presentation of information about the economic and social structure of the national economy; and
b. to provide analytical and accounting framework as a basis for construction of macroeconomic models for analyzing the national economy and the effects from the implementation of the macroeconomic and development policy measures.

The SAM for the Republic of Macedonia would be a matrix presentation of the transactions in the socio-economic system. The SAM is a comprehensive, flexible, and disaggregated framework which elaborates and articulates the generation of income by activities of production and the distribution and redistribution of income between social and institutional groups. A principle objective of compiling a SAM is, therefore, to reflect various interdependencies in the socioeconomic system as a whole by recording, as comprehensively as it is practicable, the actual and imputed transactions and transfers between various agents in the system. Hence, couple of activities are of significant importance for our country:

-   creation, harmonization and implementation of an integrated analytic-accounting framework as a basis for planning, programming and decision-making for the future socio-economic development of the Republic of Macedonia, based on the United Nations SNA and harmonized with the system and methodology for planning, analyses and decision-making in the developed market economies;
-   affirmation of the role and the importance of the SNA and the SAM for the methodology for preparation, adjustment and implementation of the macroeconomic and development policy and planning documents in the national economy;
-   construction of a SAM for the Republic of Macedonia, based on a comparative analysis of the SAM construction and implementation experiences in the developed market and transition economies.

3.2 SESAME for the Republic of Macedonia

Since 1991, the Republic of Macedonia has been going through a difficult period of transition, from a command to a market economy. This process has resulted in high unemployment rates and increasing levels of poverty. Hence, it is fair to say that unemployment in Macedonia is one of the gravest and most difficult economic, social and political problems. Despite significant progress in macroeconomic stabilization, job creation has been limited, while changes in the sectoral structure of employment and labor reallocation from less to more productive jobs have been modest. This ongoing situation imposes a real necessity for the SESAME approach implementation in the Republic of Macedonia.

The above mentioned is made possible through the main socio-demographic module of the SESAME approach. The main goal of this module is to present the interaction between the economic and demographic changes in the national economy, including the quantitative

National Classification of Activities	Young (0-14)	Potential labor force (15-64)				Pensioners (over 65)	Total
		Fixed-term contracts	Open-ended contracts	Full - time	Part-time		
Agriculture, hunting and forestry							
Fishing							
Minning and quarrying							
Manufacturing							
Electricity, gas and water supply							
Construction							
Wholesale and retail trade; repair of motor vehicles, motorcycles and personal and household goods							
Hotels and restaurants							
Transport, storage and communication							
Financial intermediation							
Real estate, renting and business activities							
Public administration and defense; compulsory social security							
Education							
Health and social work							
Other community, social and personal service activities							
Private households employing domestic staff and undifferentiated production activities of households for own use							

Extra-territorial organizations and bodies								
Social Benefits								
Unemployment subsidies								
Disability and sickness pay								
Social assistance								
Pensions								
Total								
Employment								
Social benefits								
Without income								
Population								

Table 2. Socio-demographic module in a SAM for the Republic of Macedonia

and qualitative changes of the potential labor force, as well as those changes influencing the consumption (ex.: the population age structure). This data can serve for the analysis of the impact of the demographic changes on the income distribution. A proposal for such socio-demographic module is presented in table 2. Republic of Macedonia, in our opinion, can use the socio-demographic module from the SESAME approach, by preparing several tables, using the international standards and classifications, as adopted and implemented by the State Statistical Office of the country (SNA 1993 and 2008, ESA 1995, etc.). The socio-demographic module can reveal numerous socio-economic trends existing in the Republic of Macedonia: population growth rates, average number of household members, the relation between the size of the household and its welfare, the urbanization processes, the process of decreasing size of the agricultural households, number of households depending on income transfers, the participation of women and men in the labor force, the labor force educational level, educational levels differences between the rich and the poor, etc. Hence, it is highly recommendable to use the SESAME approach in the process of macroeconomic and microeconomic policy creation and implementation in the Republic of Macedonia.

The State Statistical Office of the Republic of Macedonia, through its regular annual activities, provides substantial wealth of statistical data sets that can serve for analyzing the different aspects of the socio-economic development, as presented in the SESAME approach. These data include various characteristics and aspects of the population, health protection and social security, education, labor market, different types of revenues and expenditures, consumption and prices, etc. Moreover, substantial data are provided by the already prepared strategic documents in the Republic of Macedonia: National Poverty Reduction and Social Exclusion Strategy, National Sustainable Development Strategy, National Developmnt Plan, etc.

As a conclusion, one must point out to the fact that the SESAME approach is a flexible approach, in which the number of satellite tables and the wealth of presented data will depend on the goals of the research and the analysis of the development management processes in the Republic of Macedonia.

3.3 NAMEA for the Republic of Macedonia

Republic of Macedonia recognizes environmental protection and sustainable development as priorities both in their own right and as an essential part of the process leading to EU accession. Several important policy strategic documents in various environmental sectors have been adopted, defining the countries environmental policy. The existing strategic policy documents in the field of environment are the following: National Strategy for Environmental Approximation, Second National Environmental Action Plan 2006-2011, National Environmental Investment Strategy 2009-2013, Waste Management Strategy 2008-2020 and the National Waste Management Plan 2009-2015.

Other relevant framework strategies, important for the implementation of the above mentioned and tackling environmental performance of the industrial sector, are the following: National Strategy for Sustainable Development 2009-2030, Industrial Policy of Republic of Macedonia 2009-2020, Energy Efficiency Strategy until 2020, Strategy for use of renewable energy resources of Republic of Macedonia until 2020, National Strategy for development of small and medium sized enterprises 2002-2013, National Strategy on organic agricultural production 2008-2011, etc.

Environmental protection is one of the basic and priority values stated in the Constitution of the Republic of Macedonia. The establishment of the Ministry of environment and physical planning contributed to institutional capacity building of the country in the field of environmental policy creation and implementation. Moreover, the Republic of Macedonia follows the modern trends where Ministries of this type are one of the most important in the public administration system in the national economies.

All above mentioned imposes the necessity to implement the NAMEA approach in the country. In addition the above presented strategic documents in the field of environmental protection, the Ministry of environment and physical planning has established the **Macedonian Environmental Information Center**, as its organizational unit. The key functions of the Center are collecting, systematization, analysis, processing and presentation of data and information for the condition, quality and trends in the environment, as well as production of easy to understand and scientifically credible information on the environment. Such information is available to both decision-makers and the general public, thus contributing to enhanced awareness and improved decision-making process and ultimately making positive impact on the environment. As a result, the National Environmental Information System supports:

- the process of policy creation, planning and decision making;
- identifying effective measures for protecting and promoting the environment;
- on-time and reliable information for the public about the condition of the environment and active participation of the public in the environmental protection; and
- fulfillment of the requirements and the obligations for informing the national and international organizations and institutions.

Therefore, the development and coordination of unique national environmental information system is one of the key Canter's priorities. It is an electronic system for environmental data collection and management. It provides optimized flow of data between relevant institutions and integration of all data into a single operational structure. Hence, environmental indicators, provided by this Centre, are useful tool in the process of environmental reporting. Properly selected indicators, based on properly selected time-series, show the key trends and enable rapid and appropriate intervention of all stakeholders involved in the process of environmental protection. Following the European Environmental Agency concept, the Government of the Republic of Macedonia adopted 40 environmental indicators, which follow the DPISR framework: **Driving forces – Pressures – State – Impact – Response**, where each phase has its own meaning and significance.[9]

Hence, the data base already created by the Macedonian Environmental Information Centre will be a solid foundation for the preparation and implementation of the NAMEA approach in the Republic of Macedonia. However, in order to be able to implement this approach in the Republic of Macedonia, one should apply additional methodological solutions in the separate national accounts, namely:

-   the *production account*: to include data on the activities aimed at pollution reduction (ex.: solid waste management, etc.), i.e. data on expenses related to the activities for the pollution reduction undertaken by the separate production and institutional sectors in the national economy;
-   the *production account*: to include data on the value of environmental taxes paid by the separate production and institutional sectors in the national economy, as well as the value of subsidies paid to the production and institutional sectors for the activities related to decreasing the air polluting gases;
-   the *capital account*: to include data on the investments in so called "clean (environmental friendly) technologies", undertaken by the separate production and institutional sectors in the national economy;
-   in the *Social Accounting Matrix*: to include two additional columns and two additional rows:
    1. row/column for the most significant substances (environment polluters), which should record (in physical terms) the quantities of pollution created in the production and consumption processes in the national economy; and
    2. row/column for the natural resources, which should record data (in physical terms) on the natural resources' depletion, as a result of the production and consumption processes in the national economy.

Consequently, through such designed NAMEA approach for the Republic of Macedonia, one will connect data from the System of national accounts (SNA) and the environmental data in a systematic way (by applying uniform and standard definitions and classifications) and will enable to describe (in physical terms) the quantitative impact of the economic activities on the environment. Moreover, the NAMEA approach for the Republic of Macedonia will be a solid analytical database for preparation and

---

[9] Ministry of Environment and Physical Planning of the Republic of Macedonia, www.moepp.gov.mk

implementation of macroeconomic models for analyzing the environment protection policies and activities.

4. Conclusion

Sustainable development and sustainable development planning are complementary processes which should ultimately lead to increased well-being of the mankind. In order to better understand and implement the whole process, one should take into consideration their methodological basis, their preparation and implementation. Hence, first part of this chapter presented the theoretical foundations of the Social Accounting Matrix and its extension (SESAME and NAMEA), mainly through the work done by other authors. The second part of the paper gave practical recommendations which should, in our opinion, lead to preparation and implementation of the SAM and its extensions (SESAME and NAMEA) in the Republic of Macedonia, for the decision-making process related to creation and implementation of efficient macroeconomic and development policies in the country.

All abovementioned shows that development planning and the market are complementary mechanisms in the developed market economies, and as such should be equally the part of the new socio-economic system of the Republic of Macedonia. The successful combination of the „market's invisible hand" and the „plan's visible hand" will provide a more rational utilisation of the production factors and more dynamic economic development of the national economy. This will lead to a continuous improvement of the economic policy instruments, as well as the other types of planning and programming of the national economy development.

5. Acknowledgment

I dedicate this chapter to my parents, **Divna** and **Aleksandar**, for their continuous and unselfish support, understanding, sacrifice, guidance and unconditional love.

6. References

de Haan, M. (2004). *Accounting for Goods and for Bads: Measuring Environmental Pressure in a National Accounts Framework*, Statistics Netherlands, Voorburg, the Netherlands

de Haan, M. & Keuning, S. (2000). The NAMEA as Validation Instrument for Environmental Macroeconomics, *EVE workshop on "Green National Accounting in Europe: Comparison of Methods and Experiences"*, 4 to 7 March 2000, Milan, Italy

European Commission - EUROSTAT; International Monetary Fund; OECD, UN & World Bank. (1993). System of National Accounts 1993, Brussels/Louxembourg, New York, Paris & Washington D.C

Keuning, S. & Timmerman, J. (1995). An information-system for economic, environmental and social statistics - integrating environmental data into the SESAME, NA-076, Statistics Netherlands, 1995, Voorburg, the Netherlands,

Keuning, S. (1998). Interaction between national accounts and socio-economic policy, *Review of Income and Wealth*, Series 44, Number 3, September 1998

Keuning, S. (2000). Accounting for welfare in SESAME, in: *Household Accounting – Experience in Concepts and Compilation*, Handbook of National Accounting, Studies in Methods, Series F, No. 75/Vol.2, 2000, United Nations, New York

Ministry of Environment and Physical Planning of the Republic of Macedonia (www.moepp.gov.mk)

OECD. (2001). The DAC Guidelines - Strategies for Sustainable Development, Paris, France

5

Broadening Sustainable Development in Praxis Through Accountability and Collaboration

Mago William Maila
University of South Africa, Pretoria
South Africa

1. Introduction

In discussing the four different interpretations of the moral imperative to promote sustainable development, Hattingh (2002, p.14) says in his conclusion about the importance of sustainable development, informed by critical reflexive action, that "any interpretation of sustainable development functions as normative ideas. Such a set of normative ideas can function as guidelines for personal actions and as a baseline in terms of which governments, industry, commerce, consumers and citizens can be held accountable for their actions". As various stakeholders in this debate strive to interpret sustainable development within normative parameters, they need to ground their understanding and implementation strategies of sustainable development guidelines, in critical, reflexive processes of their practice, informed by their theory that guides their practice and action–in-practice that guides their theory. Such a process is pluralistic and collaborative and demands accountability. This chapter therefore, seeks to critically and analytically explore the grounding of sustainable development in praxis as a social, and reflexive informed process directed by collaboration and accountability imperatives.

Praxis is normally equated to practice and theory in social research investigations. A simple clarification of what practice and theory entail, is that practice involves a process-oriented action, and theory involves principles that guide action. May (2005) reaffirms the above notion about practice and theory. He sees the two concepts as simply two sides of the same coin, and he further argues that

"Theory aims at the production of thoughts which accord with reality. Practice aims at the production of realities which accord with thoughts. Therefore common to theory and practice is an aspiration to establish congruity between thought and reality (May, 2005, p. 339)."

In this chapter I theorize about the notion of grounding sustainable development in congruous aspirations of thought and reality within a broader perspective of praxis. However, I intend discussing the role of praxis in this investigation, further down in the chapter. I shall however, turn to accountability and collaboration as obligatory guidelines for both action in development and action in holding governments and civil society structures accountable to sustainability processes within their mandated portfolios.

Sustainable development initiatives need to be built on meaningful partnerships. In such partnerships, principles of accountability and collaboration are but few of the many principles critical in development processes. According to the Catholic Agency for Overseas Development (CAFOD), (2002), the agenda for accountability should be at the foreground of all development processes and the wellbeing of people. To the CAFOD, accountability involves among other things that: the fundamental principle of good governance for a government of a country is to first and foremost be responsible to its people; formal and inclusive monitoring processes be established; and mutual accountability regarding impoverished people be built between governments and donor communities (www.cafod.org.uk/policy_and_analysis/). The crux of this principle is that an inclusive process is imperative for productive partnerships between governments, civil society groups, impoverished communities and the wider community in development initiatives if sustained and meaningful outcomes are envisaged. Needless to mention that, partnerships as collaboration, must be underpinned by commitment, ownership, mutual learning, agency and transparency. I see these underpinnings as tenets of the two principles framing the discussion of this chapter.

Accountability and collaboration call all people to acknowledge both positive and negative causes and effects of their actions, that is a call to, being able to account for one's actions – being responsible for one's doings (Thompson, 1995, p. 10). Reeves (2002) however, cautions that accountability is fraught with danger, but rife with opportunity. Danger in neglecting and marginalizing various communities' other ways of knowing; undermining environmental ethical issues and opportunity for engaging collaboration processes in sustainability matters in order to ensure economic growth that does not compromise the environment and the wellbeing of people. A positive promise by Markandya, Harou, Bellu and Cistulli (2002, p. 15) and, Fien and Tilbury (2002) is the historical document signed at Rio - Agenda 21, by nations of the world, committing themselves to promoting sustainable development through a great variety of transformative educational means, including non-formal, informal and primary and secondary education.

The Millennium Development Goals (2000), World Summit on Sustainable Development (2002), World Declaration on Education for All (1990), the Dakar Framework for Action for Education for All (2000), the Johannesburg World Summit on Sustainable Development (2002), the United Nations Decade of Education for Sustainable Development (2004) are some of the global means put in place to assist nations of the world to better the lives of all people. These global agreements and programmes meant mainly for local implementation processes are also pointers of what the world needs to do in order to address issues of sustainability and poverty eradication.

The understanding that natural resources should be used for the advancement of all people and the eradication of poverty, is founded on the ideal of a democratic world citizenry, primarily loyal to human beings the world over, and whose national, local, and various group loyalties are considered distinctly secondary (Nussbaum, 1998, p. 9). Although this view is seen by some scholars as encouraging 'laziness' and the ability to rely/depend on other people's mercy for survival (Hardin, 1990), the United Nations (UN) and other global initiatives that are worldwide focused to alleviate and eradicate poverty and health-ills through sustainable development programmes, seem to be convinced otherwise. The argument I advance in this chapter therefore, is that sustainable development is integral to economic and human development initiatives (Bell & Morse, 2003) and that it must be embedded in the work human beings do, that is in praxis (Carspecken, 2002). Reiterating the view on the necessity of economic and human growth, Markandya, et al (2002, p. 17) maintains that the quality of economic growth should eradicate poverty over time as the ultimate criterion of sustainable development. Just note that the term sustainability is often used synonymously with sustainable development (Bell & Morse, 2003, p. 3). In this chapter these terms are also used interchangeably.

Note that this chapter starts with a brief introduction illuminating on the discussion of the chapter, followed by a critical exploration and analysis and a suggestion for a better understanding and use of the varied perspectives of sustainable development. Thereafter, I explore an illumination on the links/linkages between theory and practice, praxis and sustainability, with a view to establish and promote interconnectedness and multiplicity of knowledges and knowings, will ensue, followed by a brief discussion on how sustainability research could endeavour to ground sustainability in praxis. Lastly, the chapter discusses partnerships and collaboration processes as underscored by accountability and responsibility in/for sustainability.

2. Varied perspectives

Complex and different views advanced by various scholars about sustainable development as a global trajectory relating to social justice, political justice, trade justice, environmental justice, are contested every day. There is also skepticism when it comes to why humanity needs to develop sustainably (Neefjes, 2000, p. 44). Nations in the North seem to be suspected of using their economic power to demand compliance to issues of sustainability, when in actual fact they are eying the abounding natural resources of the South. The South is also suspected of using its inability to develop sufficiently to uplift itself out of poverty and health-ills, and using the dependency-syndrome as leverage for arguing for more donor funding (Hardin, 1990).

It is for that reason that, defining sustainable development is not simple and easy. As indicated by Neefjes (2000), sustainable development is a complex, broad and vague concept. Neumayer (1999) concurs with Neefjes' observation. Adding to this problem is the fact that sustainable development is a contextually based operationalized activity, with an internationalized mandate (Johannesburg World Summit on Sustainable Development, 2002). Hence, it is described both in terms of historic-cultural dispositions and theoretical dispositions. The complex and diverse environmental problems and risks emerging worldwide as a result of development are seen as the cause of the new approach to development – sustainable development (Hall, 2000a). Reiterating this fact Hall (2000, p. 14) argues that "past patterns of development, especially those based heavily on the use of energy-intensive inputs are destructive of soils, biota, and systems of production, so that new approaches that focus instead on the long-term maintenance of the productive system should be the goal".

Bell and Morse (2003, p. 3) succinctly point out that sustainable development is all about an improvement in the human condition. They further distinguish sustainable development from other approaches that endeavour to improve the quality of human life, but does not emphasize economic growth, nor do they focus completely on people, "but more on the underlying philosophy that what is done now to improve the quality of life of people should not degrade the environment (in its widest bio-physical and socio-economic sense) and resources such that future generations are put at a disadvantage" (2003, p. 3).

Goodland and Daly (1996) and Markandya, et al (2002) maintain that classically sustainable development is broadly portrayed as an interface between environment, economic and social sustainability. As already mentioned that the term sustainability is often used synonymously with sustainable development (Bell & Morse, 2003) and that in this chapter these terms are also used interchangeably, hence, Markandya, et al (2002, p. 17) argue that sustainable development needs a new science of 'sustainomics – this refers to the new science of sustainable development as a trans-disciplinary and multidisciplinary approach. According to Munasinghe (1993) this approach requires a mixture of skills and disciplines.

Such an approach seems to be quite broad in its inclusion of knowledge from a diverse skills and disciplines, it projects a more neutral image which focuses attention explicitly on sustainable development, and avoids the implication of any disciplinary bias or hegemony (Markhandia et al., 2002, p. 18). Neefjes (2000, p. 49) argues that the three dimensions of which sustainable development is founded (economics, social, environment (diverse skills and disciplines) are supposed to be underpinned by environmental sustainability, which is anthropocentric in nature, primarily concerned with human welfare. However, regarding sustainability, all the scholars referred to above concur that politics is the line of the sustainable development discourse. I refer to this process as political sustainability that undergirds development.

Different intellectuals' understand development differently. Hall (2000, p. 14) points out that development means the increased exploitation of resources with the use of fossil energy, and it is by definition not sustainable, because fossil fuels are finite since they are not being made today on any important scale. Other stock resources, including especially soils and many minerals, also fossils, water, many forests, and fishes, are likewise essentially being mined and used once. By this chain of logic one might conclude that sustainable development is an oxymoron, a phrase that is internally inconsistent, such as 'jumbo shrimp'. On the other hand perhaps new technologies can decouple development from resource exploitation and the use of nonrenewable resources.

Eisgruber (1993, p. 4) contest that sustainable development is development that "… is taken to mean a positive rate of change in the quality of life of people, based on a system that permits the positive rate of change to be maintained indefinitely". *Our Common Future* (1987, p. 4) defines sustainable development as development that meets the needs of the present without compromising the ability of future generations to meet their own needs". Again, the emphasis is not just sustaining development for the sake of consumerism and the greedy amassing of wealth tendency by those who have the power resources to do so, but development that provides opportunities and for all people to advance in their wellbeing.

Caring for the Earth (1991, p. 10) concurs with the views of Eisgruber (1993) and *Our Common future* (1987) that perceives sustainable development as that which "improves … the quality of human life while living within the carrying capacity of supporting ecosystems". However, this view largely depends on the kind of actions that people take in the environment regarding the sustenance of their lives, and that these are critical if sustainability is to be maintained. Neefjes (2000, p. 42) argues that sustainable development "requires meeting the basic needs of all people and extending to all the opportunities to fulfill their aspirations for a better life". If the people of the world are to respond positively to halving the poverty of impoverished masses, then it should be 'all hands on deck' as far as achieving this Millennium Development Goal. Nations that have more than enough resources should start and continue to play their role in ensuring that the environment is protected (Neefjes, 2000; Carter, 2001).

Those nations who are impoverished are not off the hook regarding their role in sustainability. They need to ensure that they do not take support for granted, but that support is to be accounted for in a transparent and responsible manner, by ensuring that graft does not undermine the goals of such support in all practices of sustainability. Since the understanding of practice as praxis is a critical strategy in the 'application' of programmes, especially initiatives that are focused to 'support', or foster development programmes, I need to discuss the central thoughts of this strategy within the notion of theory, practice, praxis and sustainability, and point out how the linkages of these can enrich sustainable lifestyles and human wellness.

3. Theory, practice, praxis and sustainability: What are the links?

Does it actually matter what comes first – theory or practice? Yes it does matter, depending whether the research action is based on a deduction or induction approach. May (2005, p. 32) argues that when we consider a general picture of social life and then research a particular aspect of it to test the strength of the theory, that is deduction research and that, theorizing comes before research (2005). On the other hand, when a researcher examines a particular aspect of social life and derives a theory (-ies) from the resultant data, that is known as induction. In this case research comes before theory and seeks to generate theoretical propositions on social life from the data (May, 2005, p. 32).

In order to understand the notion of grounding sustainable development in praxis, I must briefly explain how praxis is different than just practice, as both terms refer to action undertaken regarding a research or just a development endeavour. Some people see praxis as practice or research action. Praxis is more than just an action-in-practice, it involves the continuous reflection on ones work, understanding one's work through critical reflection in/on that work, and being critical involves scrutinizing the theories within one's practice, and the social structures that shape them (Janse van Rensburg, 1998, p. 39). Janse van Rensburg and Le Roux (1998, p. 104), reiterate that, praxis implies a conscious recognition of the relationship that exists between practice and its rationale(s), and that praxis constitutes deliberation on the 'why' question that illuminates meaningful resonance amidst the 'what' and the 'how of one's work (Janse van Rensburg & Le Roux, 1998, p. 104). In the context of a research project it involves asking why we do things the way we do, and this questioning affects what we do next (feed back into the practice). In short, this means that action-in-practice, reflection and praxiological curriculum are the constitutive elements of praxis.

Now, whether theory comes before practice or practice gives rise to theory, the crux of the matter is that the research action or practice must be informed by critical reflexive reflection and praxiological curriculum as constitutive elements of the undertaken practice. Carspecken (2002) reiterates this notion of praxis as embedded within the actions of humanity. Looking at praxis in culture, he argues that action is not determined by structure, but it is rather conditioned by cultural milieu and is always productive of new cultural forms (2002, p. 60). He further observes that, "human beings are strongly motivated to continuously produce themselves" (2002, p. 62).

According to Carspecken, Hegel's philosophy explains how human beings continuously 'produce themselves', through Geist as the agent of praxis, which is the impetus, process, and product of its own self-production. According to Marx (cf Carspecken, 2002, p. 63) self-production is located within the work humans do. This means "humans work to produce useful objects is simultaneously human work to produce themselves". All human beings need to produce themselves through praxis, or the praxis needs. However, realistically, these needs might be denied to them by the capitalist relations of production processes (Carspecken, 2002, p. 64). For praxis needs to be realized certain social conditions are mandatory. These are the control of the conceptualization of production; the control of the tools and resources used for work, and the control of the product in question (2002, p. 64). Let me now unpack this broad view of praxis as practice embedded in the actions of human beings – and how it relates to sustainable lifestyles that are meaningful for improving the quality of human life for present and future generations:

3.1 Need for human beings to produce themselves

Sustainability is a human activity. The unhealthy (overuse, over-exploitation, degradation of natural resources, etc) use of natural resources is a resultant of human beings' actions based either on accelerating profits in the name of development or simple greediness. However, sustainability that is based on engaging and focused on the participation of all people in its processes fosters a better understanding of people themselves and their actions – their practice. It is in the collaboration and engagement with sustainability processes that they see their actions. Those left out become strangers and suspicious of sustainable development.

3.2 Need for human beings to produce worthy goods

Sustainability is about ensuring a better quality of life for all. Goods and services are fundamental to all people's lives. Denying people goods and services is unethical- if people are unjustly marginalized and inhumanly treated the very goods and services produced are probably inequitably provided. So, all social injustices, be it trade injustice, social injustice, political injustice, trade injustice, environmental injustice, economic injustice, push the masses to the peripheral of sustainable development processes. The worthiness of them producing worthy goods and services is marginalized, compromised and denied. Meaning that they end up not been part of sustainable development praxis. Remember that praxis is not just practice, but it is practice that seeks informed actions based on continuous reflexive reflection and reshaped actions.

3.3 Need of human beings to overcome challenges and obstacles that deny one to realize needs of self-production

We many a time say "practice makes perfect", attesting to the fact that being involved in 'action' is better than being 'told' about the action. Those masses participating in sustainable development initiatives or programmes are better positioned in understanding the need to use resources wisely in the present (now) in order to ensure that the future will also be well taken care of.

Not only do people need to produce themselves through praxis needs but, they also need to ensure that they overcome, minimize and eradicate those factors that either deny, or diminishes their chances of realizing their praxis needs. They therefore need to consciously ensure that they understand their actions- in – practice; they need to know and understand how they want to produce themselves; they need to deliberately collaborate their actions and be prepared to be continuously informed by their critical reflections in practice. This process of praxis calls for all people to participate in sustainability actions, and to ensure that factors working against sustainable development are overcome through either minimization or eradication processes. Throughout the processes of praxis resources are critical in ensuring that desired positive outcomes are possible. Hence, resources in the sustainable development action- chain cannot be misused or grafted. Marginalizing human beings in sustainability is equal to the misuse of human resources, and it is just as bad as misusing the employed in their line of duty. Graft is bad, whether practiced by recipients of donor funds or practiced by donors themselves.

A balanced and holistic approach to uprooting these ill practices in praxis should be based on robust anti-corruption strategies, based on international agreements on issues of social justice, political justice, economic justice, environmental justice and human rights and fair practices among donor (countries) and recipients (countries). Normally the term donor refers only to oversees government/NGO/ or company donating some sum of money for a specific development project or activity. However, a broader understanding of this concept should actually refer to both local and abroad donors, be they government, NGOs or industries. Graft is 'graft' whether the money being corrupted is oversees money or local. The result of such deeds is that development is seriously compromised or undermined. The CAFOD (http://www.cafod.org.uk/policy_and_analysis/policy_papers/) policy document seems to offer a better anti-corruption strategy in this regard in order to ensure better sustainability processes:

- *Donors must complement … initiatives to establish peer reviews* by setting up independent assessment mechanisms for monitoring donor performance. The set independent body should judge donor performance against agreed upon principles of donor and recipient country accountability.
- *Such agreed upon principles should foster local control, priority and ownership of aid programmes* and enhance donor responsiveness and accountability to local parliaments, civil societies and taxpayers. Of course, donors must also be accountable to their countries abroad.
- *Donors and governments should conduct a review of the effectiveness of anti-corruption institutions and anti-corruption conditionalities* and build greater local civil society organizations participations, parliament and press in shaping the anti-corruption agenda.
- *International anti-corruption agreements (such as the OECD Bribery Convention) must be strengthened and be supported by all governments.*
- *Binding commitments must be for the pro-active enforcement of regulations* that close loopholes for bribes and open offshore bank accounts for the recovery of corruptly gained assets. Graft should not be allowed to determine the wellbeing of those who need donor support.

- In *countries with a high incident of corruption, development assistance still has a role to play*. It should by-pass corrupt institutions and support, countervailing, anti-corruption influences.
- *There should be enhanced investment in civil service reforms that prioritize local ownership* and finance adequate pay scales for civil servants.
- *Governments should ensure that democratic and participatory guidelines are at the heart of standards of good governance.*
- *An emphasis should be placed on participatory forms of governance in budget and poverty monitoring processes.*
- Institutions such as the World Bank and the International Monetary Fund *should be reformed in order to ensure that developing countries have a greater voice* and influence in the shaping of policies that impact on their lives.

These anti-corruption strategies ideas map out a clear desire to ensure that action in praxis continues to be informed by reflexive, and collaborative decision-making processes, and less by selfish and arrogant inclined decisions. Development actions must be owned by all those meant to be assisted and supported in order to defeat and move out of their poverty status. And those with the capability should continue to ensure that necessary resources are just as mandatory to the attainment of sustainability life styles. Because corruption undermines such endeavours, all people should be called to participate with a renewed energy and determination to fight the scourge of corruption. Such a resolve is not discreet but is pluralistic, ongoing and aggressive collaboration for sustainability.

4. Embedding sustainability in praxis – Is it self explanatory?

I do not believe it is self-explanatory. This kind of action for sustainability cannot be taken for granted and be assumed. Reflexivity in sustainable development processes is supposed to be a deliberate activity, and focused on a rigorous process of improving the lives of all people through reflexive sustainability processes. For that reason, such reflexive action taking processes must be grounded in what people are 'actually-doing' to better the quality of life they want to enjoy. Such actions to be based on deliberate participation processes in matters that impact on their lives.

4.1 Praxis as informed reflexive action in sustainable development processes

Feedback is critical in informing the next step(s) of action in order to ensure maximum utilization of resources and successful achievement of outcomes. Reiterating this observation Lotz-Sisitka (2006, 20), points out that involving people in sustainable processes means that

- participatory approaches and methods are deliberately encourage and further develop in ways that are meaningful to all participants;
- integrated approaches and solutions to poverty, environmental degradation, health risks and other sustainable development challenges, are viewed as legitimate approaches just like educational practice;
- the need to involve people in questioning and critically evaluating the appropriateness of environmental and sustainability education practices is encouraged
- the need to involve people in questioning and critiquing the appropriateness of economic, political, bio-diversity sustainability education practices is mandatory

4.2 Praxis enabling participants to self-production

As a human activity, sustainability ought to engage all people in its processes in order to foster a better understanding of themselves and their actions – their practice. Whether through formal, non-formal or informal education, and whether through government and business enterprises. All people must be involved in order to actualize themselves. Otherwise, if left out, they will become 'strangers to sustainable development' and suspicious to sustainability programmes. According to Lotz-Sisitka (2006), and Moore and Masuku-van Damme (2002) active participatory and people-centred methodologies cannot be overemphasized in sustainability processes. Hence, dealing with complex and uncertain issues, needs people to be involved on personal care levels in order to amicably and appropriately resolve these issues.

4.3 Praxis as a producer of worthy goods and services

Goods and services are basic to all people's lives. For that reason, all people must participate in sustainability programmes that endeavour to enable them to access resources. Unjust and inequitable processes in these programmes that marginalize impoverished communities are to be avoided at all cost. Praxis is not just the action we take in our practice, but it is practice that is based on informed continuous reflections on our actions as we participate using all resources available to us. This observation concurs with the environmental principles adopted at Rio De Janeiro (Rio Declaration 1992) that

- development should be focused on sustainable natural resources use and sound management thereof;
- security of land and resource tenure is a fundamental requirement of sustainable natural resource management;
- long-term food security depends on sustainable natural resource and sound environmental management;
- technologies that are environmentally friendly, socially acceptable should be developed and disseminated for effective use of natural resources;
- facilitating the creation of opportunities for communities and individual resource managers to manage their own natural resources and the environment sustainably should be encouraged

4.4 Praxis as ongoing reflection in sustainable development processes

As participants continue to ask the 'why' and 'how' questions of their practice, they gather information that not only helps them to shape actions, but also to be able to avoid activities and processes that will result in failure and cumbersome challenges. In this regard, I concur with Lotz-Sisitka (2006: 29) that there is great need for research to

- advance the conceptual, theoretical and methodological development of environment and sustainable development;
- strengthen and extend existing environment and sustainability education pedagogies, their relevance in society and their reality congruence;
- strengthen and extend the effectiveness and value of partnerships and networking processes;

- strengthen curriculum development approaches and implementation strategies for mainstreaming environment and sustainable development concerns into education systems;
- inform work-based learning and new approaches to training and professional development to strengthen reflexive practice;
- develop strategies that can address complexities and value based questions in the teaching and learning process in a context characterized by high levels of cultural and linguistic diversity

If praxis is action-oriented processes, then it means that participation and collaboration are inevitable, and such participation and collaboration also have to be grounded in accountability processes if meaningful outcomes are envisaged.

5. Collaboration and accountability in sustainable development

Collaboration is about various groups of NGOs, government, civil society and business coming together and deliberating on various strategies of action in/for sustainable development. Sometimes collaboration is used to refer to actions of 'conniving' and shoddy actions against individuals or operations. However, in sustainable processes, collaboration is about the coming together of various stakeholders with a view to participate at a 'broad-knowledge-base', where all participation endeavours stand to gain. The conviction is that broad-based information 'think-tank' will ensure that the views of all (governments and civil society) are represented. I must say that it is assumed that participation and representation are strategies that are useful when applied effectively. As we know, participation can be 'shallow', tokenistic, and 'endorsement type (Maila, 2006) or can be instrumentalistic and functionalistic (Neefjes, 2000). On the other hand, representation can be (1) biased towards small, well-organized groups with few claims to represent the larger public, (2) undermined by offering opportunities for participation late in the decision-making process when proposals are already developed and accepted by the relevant agency (Tolentino, 1995, p. 142).

In order to either avoid or minimize the effects of the above types of representation through participation, Tolentino (1995) argues for not just participation in sustainable development, but for popular participation. He cites three convincing reasons why popular participation by civil society, business and government should be a requirement for development that is sustainable:

- A number of procedures in the EIA process provide opportunities for interested groups to inform agency deliberations about their concerns and preferences or to contribute formative ideas to the decision-making process.
- Sometimes, dissatisfied concerned citizens use available information to block proposals through protest or litigation.
- Suits are filed against government agencies for failure to regulate activities, which damage the environment and private industries for developing technologies in violation of environmental standards (1995, p. 142-143).

Tolentino (1995, p. 143) further cites other reasons that are constraints to the implementation of popular participation for sustainable development. These are poverty; illiteracy;

inadequacy of political mechanisms; certain agencies of governments view environmental movements as obstacles to development; sometimes the state is the stumbling-block to development itself; absence of laboratories hampers civil society proving their case against environmental degraders and; copious resources of industries can ensure that they do not own up to their environmental ills.

Participation is also integral to democratic values and principles. People cannot be denied this democratic and human right under normal circumstances. Of course, sometimes this value and principle is arbitrarily practiced by those in positions of power, in that the necessity to have people interact with policy issues is ignored. Dryzek (2000, p. 1) argues that democracy is deliberative in its nature and

"As a social process is distinguished from other kinds of communication in that deliberators are amenable to changing their judgements, preferences, and views during the course of their interactions, which involve persuasion rather than coercion, manipulation, or deception."

Of note, is that participation in governance is a democratic right, but also an ethical right. It means that it is morally wrong to deprive people the right to exercise their choices freely in decision matters that impact on their lives. Governance that is ethically correct needs to take the masses on board in decision-making processes.

Good governance relies heavily on the needs of its masses. Civil society normally desires clean governance. No wonder, people throughout the world are rejecting the notion that corruption is inevitable (States News Service, 2006). The abuse of entrusted authority for private or personal enrichment is not acceptable as this undermines all initiatives by governments and nongovernmental organizations geared towards the advancement of society. Various strategies of how to curb corruption are put in place by various governments. For example, raising public awareness on the existence of the problem; helping poor countries develop transparent contracting and audit systems; enforce corruption laws; responding speedily to reports of suspected corruption; training staff to recognize signals of corruption in governments and nongovernmental organization development initiatives (States News Service, 2006). Although all of these strategies seem to be good and focused, their weakness is that they are mainly focused on developing countries and not the developed countries; and focuses on recipients and not the donor. So, any act or doing in corruption would not necessarily be picked up in developed countries' donor agencies, but rather, would squarely be blamed on the recipient and her/his country.

In such a complex situation, the World Conservation Strategy, which proposed the nine principles in Caring for the Earth (IUCN, 1991: 12) seem to provide us with a better lifestyle based on an moral standard for living/lifestyles, and these are

- respect and care for the community of life (an ethical principles defined as duty of care for other people);
- improving the quality of human life;
- conserving the vitality and diversity of the earth;
- minimizing the exhaustion of non-renewable resources;
- keeping within the carrying capacity of the earth;
- changing personal attitudes and practices, in accordance with an ethic for sustainable living;

- enabling communities to care for their own environments;
- forming national frameworks for the integration of development and conservation;
- forming a world alliance to implement sustainability on a global scale

It must be noted however, that the success of the implementation of these principles in meaningful ways, depends on meaningful collaboration and accountability practices in sustainable development.

6. Concluding comment

Sustainable development is a universal peoples-centred policy initiated programme, mainly geared to sustain our environment, economics, and socio-cultural dimensions of life. No one is to be excluded in this programme if we are to change the lives of all people for the better. Hence, the grounding of sustainable development in praxis, continuous informed reflexive action in practice, is enabling participants to meet their praxis needs.

Sustainability initiatives cannot ignore the importance of all people participating in such initiatives. Communities must be part of collaboration initiated for sustainable development. Government, academia and business need to ensure that all participants adhere to both international and local agreed upon principles of rooting and avoiding corruption in sustainability programmes. For that reason, continuous research is imperative regarding the formative, monitoring and evaluation of sustainability processes. Theoretical knowledge and understanding of sustainable development must be a continuous process.

It must be continuous and be applicable to actions taken on the ground. It must direct what needs to be done. It must influence daily activities of governments and civil society and sustainability.   However, conversely, it must be informed by actions that are locally, nationally and internationally situated. As sustainable development shapes actions, it is also shaped by its own actions. Therefore, sustainability programmes that marginalize the masses from participating in development initiatives are both ignorant of what they miss in sustainable development and praxis, and their actions are both irresponsible and dangerous to the environment economics, politics, socio-cultural and biodiversity enhancing lifestyles.

7. References

Bell, S & Morse, S. (2003) *Measuring sustainability learning by doing*. London: Earthscan Publications Limited.

CAFOD (2002) Available at: www.cafod.org.uk/policy_and_analysis/.

Carspecken, P. F.(ed).(2002) The hidden history of praxis theory within critical ethnography and the criticalism/postmodernism, edited by Y.Zou and E.T.Trueba, in *Ethnogrphy and School: Qualitative Approaches to the Study of Education*. Lanham: Rowman & Littlefield Publishers. Pp 55-86,

Carter, N. (2001) The politics of the environment: Ideas, Activism, Policy.

Goodland, R & Daly, H. (1996). Environmental sustainability: Universal and Non-Negotiable, in *Ecological Applications*, 6(4), pp 102-1017.

Cohen, G. A. . Karl Marx's Theory of History: A defense. Oxford: Oxford University Press.

Dobson, A. (1998) *Justice and the Environment: Conceptions of Environmental Sustainability and Dimensions of Social Justice*. Oxford: Oxford University.

Dragun, A. K & Jakobsson, K. M. (1997) *Sustainability and global environmental policy: New perspectives*. Cheltenham, UK: Edward Elgar.

Dryzek, J.S. (2000) *Deliberative Democracy and Beyond: Liberals, Critiques, Contestations*, Oxford: Oxford University Press.

Eisgruber, L.M. (1993) Sustainable development, ethics, and the Endangered Species. *Choices* (Third Quarter), Washington, D.C: U.S. Dept. of Agriculture, 4-8.

Fien, J., & Tilbury, D. (2002) The global challenge of sustainability, in Education and Sustainability, D. Tilbury, R. B. Stevenson, J. Fien and D. Schreider (eds). Gland, Switzerland: IUCN, PP. 1-11.

Ginther, K., Denters, E & DE Waart, P. J. I. M (eds). (1995) *Sustainable development and good governance*. Dordrecht: Martinus Nijhoff Publishers.

Goodland, R., & Daly, H. (1996) Environmental sustainability: universal and non-negotiable. Ecological Applications, Vol. 6, No. 4, pp. 102-117.

Hall, C. A. S. (2000a) The changing tropics, edited by C. A. S Hall, C. L Perez & G Leclerc, in *Quantifying sustainable development: The future of tropical economies*. San Diego: Academic Press.

(2000b) The theories and myths that have guided development, edited by C. A. S Hall, C. L Perez & G Leclerc, in *Quantifying sustainable development: The future of tropical economies*. San Diego: Academic Press.

(2000c). The myth of sustainable development, edited by C. A. S Hall, C. L Perez & G Leclerc, in *Quantifying sustainable development: The future of tropical economies*. San Diego: Academic Press.

Hall, C. A.S., Perez, C. L & Leclerc, G (eds). (2000) Quantifying sustainable development: The future of tropical economies. San Diego: Academic Press.

Hardin, G. 1990 Living on a Lifeboat, edited by A. F Falikowski, in *Moral philosophy: Theories, Skills, and Applications*. Englewood Cliffs, New Jersey: Prentice Hall.

Hardin, G. (1990) 'Living in a lifeboat', in A.F. Falikowski (Ed.) *Moral Philosophy*, Englewood Cliffs, NJ: Prentice Hall.

Hattingh, J. (2002) On the imperative of sustainable development: Aphilosophical and ethical appraisal. In Janse van Rensburg, E., Hattingh, J., Lotz-Sisitka, H. & O'Donoghue, R. (Eds), Environmental education, ethics and action in Southern Africa. EEASA Monograph. Pretoria: HSRC Press. pp. 5-16.

IUCN/UNEP/WWF (1991) Caring for the Earth: A Strategy for sustainable living. The World Conservation Union, United Nations Environment Programme and World Wide Fund for Nature, Switzerland.

Janse van Rensburg, E. 1998. The Educational Response to the Environmental Crisi. Unpublished Paper, Gold Fields Participatory Course in Environmental Education, Course File, Rhodes University Environmental Education Unit. Grahamstown: Rhodes University.

Janse van Rensburg, E. & Le Roux, K. 1998. *Goldfields participatory course in environmental education: an evaluation in process*. Grahamstown: Rhodes University Environmental Education Unit.

Johannesbuirg World Summit on nSustainable Development 2002 United Nations Educational, Scientific, and Cultural Organization (UNESCO) 2004. *United Nations decade of education for sustainable development 2005–14 Draft international implementation scheme*, October, (2004). Paris: UNESCO.

Lotz-Sisitka, H. 2006. Participating in the UN Decade of Education for Sustainability: Voices in Southern African consultation process, in *Southern African Journal of Environmental Education, Vol. 23 (2006):*10-33.

Maila, M. W. (2005) Environmental education as a human capability strategy for poverty reduction, Paper presented at the GRUPHEL Conference in Maseru, Lesotho, in November 2005.

Markandya, A., Harou, P., Bellu, L. G & Cistulli, V. (2002) *Environmental Economics for Sustainable Growth: A Handbook for Practitioners.* Cheltenham: Edward Elgar.

May, T. 2005. *Social research: Issues, methods and process.* New York: Open University Press.

Moore, K., & Masuku-van Damme, L. (2002) The Evolution of People-and-Parks Relationships in South Africa's National Conservation Orgarnisation, in *Environmental Education, Ethics and Action in Southern Africa. EEASA Monograph.* Pretoria: HSRC.

Millenium Development Goals. (2000a) Paris: UNESCO.

Munasinghe, M. (1993) Environmental Economics and Sustainable Development. Washington, D.C: World Bank.

Neefjes, K. (2000) *Environments and Livihoods: Strategies for sustainability. Oxford: Oxfam. Neumayer, E. 1999. Weak versus strong sustainability: Exploring the limits of two opposing paradigms.* Cheltenham: Edward Elgar.

Neumayer, E. 1999 Weak versus strong sustainability: Exploring the limits of two opposing paradigms. Cheltenham: Edward Elgar.

Nussbaum, M. C. (1998) *Cultivating Humanity: A classical defense of reform in liberal education.* Oxford: Oxford University,

Reeves, D. G. (2002) *Holistic Accountability.* Thousand Oaks, California: Corwin Press.

The States News Service (2006) 'Corruption can block economic development', Washington: States News Service.

Thompson, D. (1995) The Concise Oxford Dictionary of Current English, Oxford: Clarendon Press.

Tolentino, A.L. (1995) Guidelines for policiesand programs analysis for small and medium enterprise development. Research and Programmes Development Section, Entrepreneurship and Management Development Branch, Enterprise Development, ILO. Draft 5 May.

UNESCO. (2002) World Summit on Sustainable Development. Paris: UNESCO.United Nations Conference on Environment and Development (UNCED). (1992) *Agenda 21.* Rio de Janeiro: UNCED.

WCED. 1987 Our Common Future. Oxford: Oxford University.

World Declaration on Education for All. 2000a) Paris: UNESCO.

http://www.cafod.org.uk/policy_and_analysis/policy_papers/

6

Raise It, Feed It, Keep It – Building a Sustainable Knowledge Pool Within Your R&D Organization

Wiebke Schone, Cornelia Kellermann and Ulrike Busolt
Hochschule Furtwangen University
Germany

1. Introduction

In light of the increasing globalization and economic competitiveness of the emerging countries, a key to competitive advantages of the western European countries, the United States and Japan lies in leveraging their innovative potential. With regard to the costs of living, social security systems and costs of labor, the industrialized countries can hardly compete with the emerging countries in the producing industry. Since neither Europe, the US nor Japan can beat the low production- and low technical development costs of the emerging countries, it is vital for their economies to focus on their key competences: The stimulation and driving of new innovative products to the market by means of inventions and patents within the R&D (research & development) industry.

Pursuing this target does not only imply high quality education in science and engineering. The actual economic value reveals itself in the researchers' and scientists' industrial careers, when their knowledge is applied to solving technical problems and is transformed into economic return in terms of inventions, patents and products. Hence, we need to realize the importance not only to recruit high quality experts for our research and development, but furthermore to preserve and build on this knowledge throughout the organization in order to gain a sustainable return of the high Western labor costs through innovative new technologies and scientific findings.

This target can be achieved by improving two main aspects within a company's R&D organization: Firstly by creating conditions that secure long term employment of the researchers, and secondly by understanding, improving and nurturing the inventors' communication within the organization.

2. State of research

The Western industry is based on its innovative power in research and development. There is a growing demand for highly skilled personnel in science and technology, such as researchers and engineers (BMBF, 2010). The aim must be to include the whole range of the innovative and inventive potential. Especially women represent a high educated potential labor force, whose potential has not been fully tapped into yet.

In R&D most work structures are team-based as this helps to solve creativity demanding problems and to stimulate innovation. A potential benefit of diversity regarding teamwork in R&D such as differences in education, national background, age or gender is still a research topic that is not fully explored.

The human resources development of female experts and executives in R&D gains rising importance as a critical success factor. In many cases the CEO of a high technology company is the former head of the R&D department (Hartmann 2007, 2009). Looking e. g. at the automotive industry, these individuals were often pioneers with excellent expertise. Within the high technology industry an important indicator for R&D success is expressed in the number of inventions and patents. In Germany and in all European member states women are not as often mentioned as inventors on a European patent as their share of the qualified workforce would indicate. A pronounced gap exists as e. g in Germany about 6% of the inventors of European patents were female whereas about 12% of the qualified personnel (engineers and researchers) were female in 2003 (European Commission, 2006a, Busolt & Kugele, 2009). This gap is a result of a still dominant responsibility of women for family duties, a lower rate of overtime work and more part time work. Moreover women might tend to leave the R&D department earlier than men do as the possibility to combine family duties with a less challenging job in other departments as e. g. quality management or sales might seem more achievable. The latter departments rarely offer any possibilities to generate inventions. Disregarding the detailed reasons why women do not generate as many inventions as their share of the qualified personnel demands, we can state that qualified women represent the largest and most obvious potential to gain sustainable innovative power.

Some studies conclude that gender heterogeneous teams perform better than homogeneous teams as the team members have different ideas and perspectives (Frink, Robinson & Reithel 2003; Hirschfeld, Jordan, Field, Giles & Armenakis, 2005). Other studies indicate a source of friction and a therefore minor performance of gender heterogeneous teams (Jehn, 1995; Pelled, Eisenhardt & Xin, 1999; Randel, 2002). Most studies have forcibly compared homogeneous male teams with gender heterogeneous teams as women are still a minority in R&D (e.g. Burris, 2001). Being aware of this methodological drawback, Pearsall et al. conducted their study by assigning students to foursome teams (Pearsall, Ellis & Evans, 2008). The analysis of the functioning and the performance of homogeneous female teams in R&D is an important aspect on the way to tap the full potential of women in R&D. This is the only constellation in which women might experiment and perform in their own way of teamwork. The results reflect the view of women and lead to valuable improvement recommendations for companies in order to gain higher performing teams (Schone et al, 2011).

Furthermore the working conditions are different for a product development department, where narrower deadlines and high pressure for the achievement of objectives are dominant, and for a fundamental research department where more creativity, lateral thinking and rethinking of old beliefs are tolerated (Schone, 2009). Moreover the measurement of R&D team performance, i. e. creativity, innovation power, efficiency and net output by means of indictors is difficult. The performance of university scientists is measured by their scientific output, i. e. scientific publications (e. g. Martín-Sempere, Rey-Rocha & Garzón-García, 2002). The industry is more interested in the innovative productivity and patents are needed to protect inventions. The number of patent applications is an indicator for the successful inventive achievement of individual researchers in science and technology. Patents, especially if they are not limited to only one country but include more countries as e. g. European patents do, are therefore an output indicator for R&D teams in industry (Busolt & Kugele, 2009). Nevertheless patents have the drawback that they do not differentiate regarding the effort and the creativity of the inventor team and it is therefore barely possible to assess the "real" value of a specific patent. Previous knowledge of the inventors as well as earnings from the products based on the patent is not observable.

3. Study methodology

In our study, we focused, among others, on the knowledge management of male and female inventors in the R&D industry. We observed that a number of highly qualified female researchers drift off their professional field of research during their professional career. How come those highly qualified female researchers seek positions outside their professional expertise and follow jobs in marketing, public relations etc. instead of research? What can be done to preserve the knowledge of these female researchers within the organization? A further research question was: what is the impact on innovation of homogeneous gender teams versus heterogeneous gender teams? The knowledge transfer and its influence on their respective innovative power were investigated as well as knowledge transfer patterns and communication structures. The questions that arise are how to improve the sustainability of knowledge within an organization on the one hand and to find out if inventor teams benefit from diversity within the team on the other hand.

Inventor teams work together for a limited period of time to generate a solution to a specific problem. Therefore they represent ideal conditions for the investigation of team work. In the German industry, there are only a limited number of homogeneous female inventor teams. However few, this allows a comparison between homogeneous female teams, homogeneous male teams and heterogeneous teams. We classify teams as successful when they comprise of inventors who have been granted a European patent, without further assessment of the actual patent value.

We concentrate on the specific determinants of gender impact on R&D inventions in the German industry. Hereby, our main research questions affect the four subgroups:

-   homogeneous female inventor teams
-   homogeneous male inventor teams
-   heterogeneous gender balanced inventor teams
-   heterogeneous male dominated inventor teams

In order to gather data and to contact the inventors, an SQL data base for all European patent applications within Germany was created for the years 2001-2006. Based upon raw data, which were specifically extracted by Eurostat for the EFFINET project, the following steps have been applied to the data base:

-   Gender specific attribution (due to classification of the inventor's first name)
-   Correlation of inventors to institutions or companies
-   Correlation of inventors to specific industry branches

- Determination of inventor team constellation according to the above described team
- Characteristics (by gender specific attribution)

The data base includes the inventor's name, home address (as appearing on the patent application), company or institution and industry branch as well as the differentiation in which team constellation the patent was created.

Our study was conducted in three parts: a statistical analysis of the European patent database, a qualitative interview phase in order to gain insight into the innovation environments and to generate first hypotheses, and a quantitative online survey to verify and deepen our findings.

In the qualitative phase, 21 expert interviews with male and female inventors were conducted. The percentage of interviewees from each of the four target groups described above is distributed evenly. The interview participants have applied for at least one industry patent which was created within a team consisting of at least two inventors.

Our research interest during these interviews focuses on team work, innovative and organizational environment, communication structures and knowledge transfer impacting on the innovation- and patent creation processes.

Based on the hypotheses generated from these expert interviews a quantitative online survey served to verify our findings. A total of 357 inventors participated in this survey. Since the number of homogeneous female inventor teams is rather limited (approx. 300 European patents in Germany within the past 10 years) and a significant amount of those patents can be found in chemistry, pharmacy and medical equipment, our focus for the quantitative research lies on these industry sectors in order to guarantee an equal distribution of participants between the four team constellations (homogeneous female inventor teams, homogeneous male inventor teams, heterogeneous gender balanced inventor teams, heterogeneous male dominated inventor teams).

3.1 Statistical analysis of the European patent database

For over 40 years, the European Patent Organization has been in charge of granting and tracking the patents that have been applied for in Europe. In our study, we analyzed the European patents over the last 10 years according to their technological fields, economic sectors and the gender of the inventors. The aim was to figure out the appearance of female researchers in patenting as base for our research. The analysis of the patent database included the following major steps:

3.1.1 Institutional sector allocation

Patents are applied by industrial companies, universities, research institutions or individual inventors, but can also be applied in cooperation between these actors. Thus, data concerning R&D personnel are usually broken down by institutional sector. The patent database provides the name, country and address, but not the institutional sector of the applicants. To assign patent applicants/inventors to institutional sectors an institutional sector allocation was performed.

3.1.2 Inventor's first name gender assignment

Patent databases of the European Patent Organization do not provide the gender of the inventors. Therefore, an assignment of the inventor's first names to either male or female gender is a necessary precondition to obtain gender disaggregated statistics. First name gender assignment of a large number of names from different countries required a complex, multistep procedure to reach the best results. About 93% of all European names were identified as either male or female with variations between 81% and 100% for the European Member States.

3.1.3 Assignment of patents and inventors to technology fields

An International Patent Classification (IPC), a system of 31 technical units and eight sectors, was developed by the World Intellectual Property Organization. Patent data in the database were treated at the subclass level of the International Patent Classification; thus only the first four digits of the IPC were used for breakdown and aggregation.

3.1.4 Assignment of patents and inventors to economic sectors

Patent and inventor statistics, which usually are presented by technology fields (IPC), do not easily match with data on personnel in R&D, which for the business enterprise sector are usually broken down by economic sectors (NACE). In consequence, one cannot compare data of inventors with data of researchers easily (output-input comparison) without further data transformation from technology fields (IPC) into economic (industrial) sectors (NACE). Hence, methods have been applied to match IPC based technology fields to industrial sectors. As a result, technology fields are shared between industrial sectors.

3.2 Qualitative expert interviews

In the course of the expert interviews with the inventors the problem-centered interview (Wintzel, 1996) has been applied, a theory-generating method ranging between the narrative interview style and structurally guide lined interview. This semi-structured approach allows the experts (interviewees) to share knowledge based on their very own value system within the structural and content boundaries of the research focus.

The interview comprised a warm-up phase including the interviewee's general characteristics, such as professional career and current job position. Furthermore, the description of the specific innovation settings and knowledge transfer of the granted patent is split into an organizational-, team- and individual level with focus on the inventor's perceptions of team work, innovation- and efficiency determinants.

The whole interview has been recorded and subsequently transcribed. Additionally, a postscript of the interview has been generated, in which situation-dependent and non-verbal aspects, interpretation ideas and special characteristics of the interview have been noted.

A summarized and anonymized case description serves as a basis to consolidate the data and investigate central motives in order to generate theory. The interpretation of the data is intended to maintain its explorative, qualitative approach and is not intended to conclude quantitative, generalizing statements. It was, however, the goal to develop first hypotheses to be further tested, verified and developed with the subsequent quantitative online survey among a larger group of participants.

3.3 Online survey

As stated above, the online survey served to test, verify and further develop the hypotheses generated from the qualitative expert interviews. The content of the 35-question survey includes a general section on the innovators' characteristics, such as professional education, career and family situation. This is followed by a generalized section on the participants' own opinion on organizational-, team- and individual influencing factors on the knowledge transfer, innovation culture and efficiency determinants within their R&D environment. Finally, the survey explores the innovation environment of one specific patent in its development and accompanying team processes. The survey questions have been answered in full by 310 inventors, representing team members of the four different inventor team constellations.

4. Do we sufficiently tap our inventors' knowledge?

Analyzing the European patent database, we found that there is a tremendous gap between the headcount of female researchers and scientists in R&D versus the headcount of female inventors. Exemplarily, the figure below shows the European patents of the year 2003 to illustrate the gap.

AT	Austria		IE	Ireland
BE	Belgium		IT	Italy
BG	Bulgaria		LT	Lithuania
CY	Cyprus		LU	Luxembourg
CZ	Czech Republic		LV	Latvia
DE	Germany		MT	Malta
DK	Denmark		NL	Netherlands
EE	Estonia		PL	Poland
ES	Spain		PT	Portugal
EU	European Union		RO	Romania
FI	Finland		SE	Sweden
FR	France		SI	Slovenia
GR	Greece		SK	Slovakia
HU	Hungary		UK	United Kingdom

Table 1. Country abbreviations

The statistical analysis served as a basis for our further research: What are the reasons that female researchers get fewer patents granted than male researchers? Are there conditions that prevent the exploration of their full innovative potential and their complete knowledge? If so, what are the determinants and what should an organization do to tap the full potential and hence expand their sustainable knowledge pool within the organization?

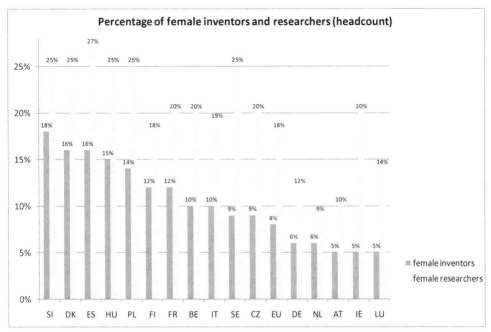

Fig. 1. Percentage of female inventors and female researchers in 2003, based on the European patent database and on EUROSTAT data (European Commission 2006a), 100% corresponds to all female and male researchers or identified inventors

5. The organizational need to nurture

The following section explains how to retain highly educated and experienced researchers within the organization in the long run. It discusses the obstacles and hurdles researchers are confronted with during their professional career, particularly with regard to work-family balance.

Especially female researchers are confronted with major life changes when they decide to start a family. Our study on the patenting- and inventing behavior of researchers shows that if organizational conditions within a R&D environment are managed badly, the inventors often cannot work according to their optimum level and are not able to use their innovative power sufficiently. The consequences of both, good and bad organizational management are revealed in this section. Furthermore, methods to optimize the researchers' conditions within the R&D department are suggested.

5.1 The communication of innovation

Our study revealed that in organizations with more than 1000 employees, there were not only a higher proportion of women in R&D departments, but also innovation systems that have more transparency, networking and support of the individual researcher. In smaller organization of the sample, in contrast, structural measures to increase innovation are hardly existent. However, as stated by the respondents, researchers in smaller organizations have more freedom for research and creativity.

The evaluation of our expert interviews leads to two hypotheses in this specific context that are further explored in the quantitative online survey of the study. In order to establish a base for communication and knowledge transfer, the following findings of the expert interviews are further assessed.

A female researcher (heterogeneous gender balanced team) describes the situation in her company as follows: the employees in research feel well connected. There is a good working atmosphere. Networking and communication takes place as well on a formal (regular project meetings, information from the company) as on an informal level (meeting beyond working hours, spontaneous discussions, e-networking i.e. Xing). Due to the regular presentation of projects within the research department (2 hours every 2 weeks) the employees are well informed about other team projects. These meetings are also attended by part-time employees (from engineer level up).

A male scientist (homogenous male team) states that he summons a weekly meeting but apart from that a lot of spontaneous informal communication takes place in order to exchange knowledge. For this reason a personal network within the team and department is extremely important and stimulated by personal commitment.

Another female researcher (homogenous female team) describes a distinct, formalized communication system in order to encourage knowledge transfer within the company she is working for: Weekly division meetings, weekly group meetings, weekly discussion between employer and employee "jour fixe", telephone conferences and meetings of the project teams (2-3 times per year).

While institutionalized communication structures (such as regularly scheduled team meetings) guarantee a comprehensible flow of information, communication paths among inventors are dominated by informal, spontaneous communication patterns, independent of the composition of the inventor team.

One male researcher (heterogeneous male dominated team) found it especially fruitful when theoretically oriented scientists discuss problems with practically oriented engineers and technicians. When sitting at a table together with different ways of thinking often ideas arise and thus these meetings might become quite productive. Regular team meetings also with experts from outside the project team should therefore be held regularly in order to generate knowledge.

In this context we learned from another interviewee (homogenous female team) that inventor teams are often composed of members from different project teams – "inventions happen". According to the problem at hand an experts is called in from outside the team. The inventor team might be a subgroup of the bigger project team but its team members often originate from different departments. The average size of inventor teams is about 2-3 researchers, depending on the industrial sector. While the constellation of a project team is created for long-term teamwork (typically for at least the duration of one project, but also for several follow up projects) and is frequently determined by the organizational management, the composition of an inventor team underlies spontaneous characteristics, in most cases issued by the inventors themselves: an "ad hoc team" is created to perform one specific task or to solve one specific problem. Once the task is completed and a patent application is filled in, the ad hoc team breaks up back into its assigned different project groups. Obviously the inventor team is not identical with the project team in most cases.

In summary it can be said, therefore, that for one thing networking, social processes and spontaneous discussions are crucial for effective knowledge transfer. For another thing formal communication structures as provided in regular meetings and presentations are equally important.

Results of the online survey regarding communication and knowledge transfer:

Having in mind what information has been given within the expert interviews regarding the importance of regular internal meetings with inventors and presentation of new inventions the answers to the online survey surprise: 72% of the researchers state that regular meetings of inventors within their organization and presentation of new inventions are not offered. Could this be an omission on the part of the organizations to stimulate the sharing of knowledge? On the other hand 77% of the researchers often (partially) make use of the freedom and tolerance for networking and informal communication during their working hours.

Corresponding to this only 47% of the respondents state that institutionalized meetings with participants during the invention process are (very) important to them. A different picture can be seen concerning the spontaneous possibility to discuss with participants (during the

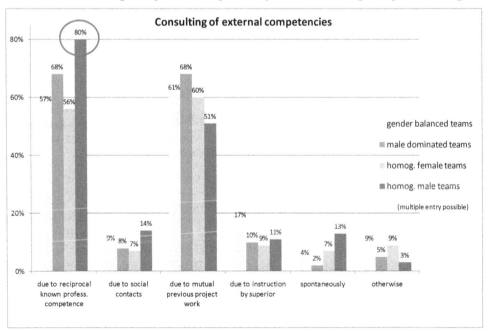

Fig. 2. Consulting of external competences for innovation, 100% corresponds to all interviewees belonging to one of the four groups

invention process): this is (very) important for nearly 100% of the respondents. These results do not show any significant gender difference. Consequently the analyses of our expert interviews have been confirmed by this.

Likewise our result from the expert interviews concerning the consulting of external "knowledge" is confirmed by the participants of the online survey: Approximately 50% of the researchers state that experts from outside the project team have been consulted during the invention process mostly because of the known professional competence and former cooperation. In most cases, this knowledge was gained by informal communication and networking "in the aisles": Researchers knew about competences of colleagues from other projects, asked them for support and later on developed the innovation and patent together. Therefore the inventor team is frequently composed differently from the actual project team (figure 2).

5.2 The participation of female inventors in communication and knowledge transfer

A female researcher (heterogeneous gender balanced team) concludes in the expert interviews that part-time work is difficult to organize: in general she handles the operative workload during the day. Time for invention needs calmness which she has at the end of the day. The company she works for offers part-time work as well as the co-financing of a daycare centre. As these options are quite recent she has only one child in order to arrange family and career.

Inventions and patents are frequently realized by overtime as the priorities during normal work hours focus on project work. Part time employees have less time flexibility (e.g. due to fixed child care hours) and thus have less time for creative brainstorming. For part time employees, it is therefore more difficult to actively participate in the innovation process.

Accordingly a male researcher (heterogeneous male-dominated team) quotes that working part-time in his team is hardly possible as much is discussed informally and spontaneously. During one year of parental leave, of course, much is missed. On the other hand technicians are used to quick changes of the market and thus have the ability to adjust. In his opinion the team only has a short-term knowledge advantage. The compensation of the returner's knowledge deficit is a question of team spirit. He states that the perfect time for returning to work is the start of a new project which is new for all participants.

There was one best practice example, however, a female scientist (homogeneous female team) describing the perfect organization in which it seems possible to work as a part-time executive in R&D and have children at the same time. All members of the inventor team, including the head of the department herself, were part-time employees. Thanks to outstanding support systems within her company this female scientist states that overtime remains the exception. Even when patenting strongly it is possible for her to work part-time. In her opinion this is the result of the following factors: specific corporate culture, support through a patenting department, assistance within and between departments, arrangements for work-life balance, child care and social counseling within the organization. This observation leads to the assumption that the innovative capability of part-time employees is deeply affected by the organization's management competences.

Another female inventor (homogeneous female team) has changed from R&D to part-time consulting for lobby work after having a child. Being head of a team requires commitment and availability within the organization which in her opinion is not compatible with working part-time.

A similar view is held by a female respondent (heterogeneous gender balanced team) who claims that especially in innovative research industries it is most problematic to arrange children and career and not easy to find the perfect time to have children. As a consequence of this conflict she believes that professional paths diverge.

In summarizing it can be stated that male and female inventors show similar innovative potential in the beginning of their professional career. Both typically start their career and create their first patents in R&D departments as development engineers or scientists. They show the same characteristics regarding overtime and devotion to their work. Female researchers though commonly change their professional path when starting a family due to the demands of jobs in R&D and the difficulties of the compatibility of family and work. In conclusion it becomes obvious that parenthood influences the innovative performance of female inventors.

Results of the online survey regarding the participation of female inventors:

The options offered by organization for the compatibility of family and career resulting from the online survey are as follows: 75% of the respondents state that part-time is offered by their organization, 69% indicate flex-time, 45% home office whereas child day care facilities offered by organizations are only mentioned by 29% of the respondents. A wide difference can also be observed when asking by whom the children are being taken care of: 81% of the male researchers state that their spouses take care of the child whereas only 19% of the female researchers state this fact (figure 3).

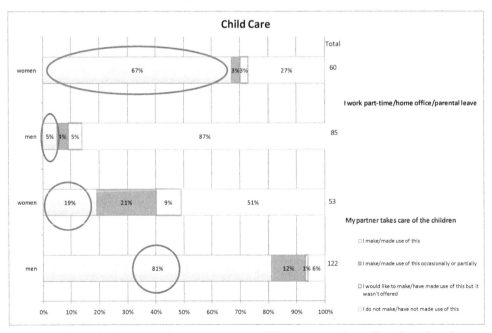

Fig. 3. Child care by female and male inventors, 100% corresponds to all male or female interviewees

Answers to our research question whether there were any gender-specific differences with regard to knowledge transfer during the innovation process show that women as well as men value informal communication and networking within their organization and project teams. Both regard the informal communication and networking as key to their professional success. As opposed to this, the importance of institutionalized meetings with participants during the invention process is very important to 19% of the homogeneous female teams while the other team constellations see this factor as less important (figure 4).

In our study, we found out that if there is a high proportion of part-time workforce the importance of formal communication through regular meetings rises. An increase of formal communication has positive impact on the integration of part-time employers as the randomness of information transfer is reduced.

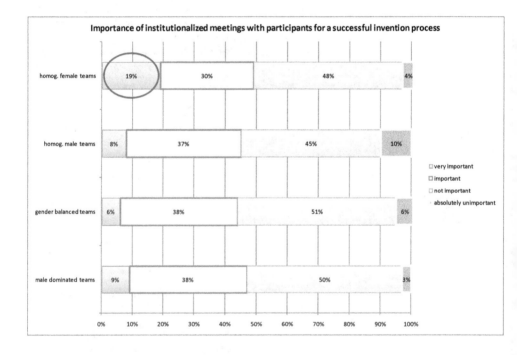

Fig. 4. Importance of institionalized meetings, 100% corresponds to all interviewees of the respective subgroup

In our online survey, the number of men having children differs strongly from the number of children women have: while approximately half the female researchers have no children, 58% of the male researchers have 2-3 children. The question arises: do women drop out of the innovation process as soon as they become mother or do they remain childless for some reason or another. A glance at the child care situation described by the respondents of our online survey shows that 67% of the female scientists take/took parental leave, work part-time/in the Home Office while only 5% of the male scientists do so. Another important result in this context is the fact that 43% of the female researchers state that child care has a great influence on their innovativeness whereas only 13% of the male researchers feel this way.

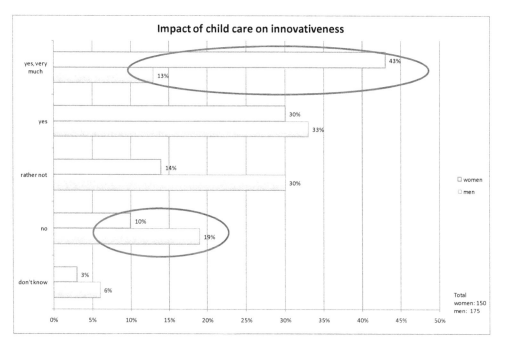

Fig. 5. Impact of child care on the researcher's innovativeness

6. Conclusion and outlook

The results of our study show that in the current innovation environment within the R&D industry, organizations suffer from the lack of sustainability in their knowledge pool. Highly skilled workforces, mainly the female inventors, do not perform to the best of their innovative potential.

The suboptimal support in the area of child care reveals a direct effect on the inventors' innovativeness, especially for female researchers. It creates pressure, whereas this energy could be directed into innovative power provided the organization offered a higher support for childcare. In some cases, it was stated that both, family life and a professional career as a researcher, could not be combined due to inflexibility of the organization or lack of daycare. Several (female) researchers decided to completely drop out of the innovation environment and sought positions in marketing or public relations that were expected to easier allow work-life balance. As a consequence, the knowledge of these highly qualified researchers has left the organization's knowledge pool irretrievably.

Part-time work is offered by several organizations and undoubtedly a successful tool to encourage the integration of highly-qualified female researchers into the organization during parental leave and to prevent them from resigning. However, part-time work reduces the working hours of the female researchers and thereby the period of time that can be used for inventions as well as their participation in networking that are important with regard to knowledge transfer and knowledge generation.

What can an organization do to improve the tapping of their knowledge pool? First of all, management needs to raise an awareness of the importance of a sustainable knowledge pool within an organization. The knowledge that researchers acquired over years during their professional career clearly represents one of the organization's greatest assets. If a researcher leaves the organization or is not able to perform according to his or her optimal level, his or her knowledge leaves the organizational knowledge pool. Oftentimes, leaving the organization is regarded as inevitable by female researchers, when they enter the family phase. Hence, measures such as support for child daycare and flexible work time regulations offered by the organization help the inventors to stay within the organization and focus on their actual research. It is therefore a key to build an organization that understands the needs of its employees and actively sets measures that support the inventors in managing their work-life balance in order to retain the researchers, their knowledge and a sustainable knowledge pool in the organization.

7. Acknowledgment

We would like to express our gratitude to the BMBF (Bundesministerium für Bildung und Forschung) for project funding and thank the following people who have contributed to the project Mr. Felix and Mr. Götzfried from Eurostat for their cooperation, assistance and methodological advice concerning database creation. Special thanks to Mr. Naldi who kindly provided us with the first name database, Mrs. Huber who assisted in technical matters of database preparation, first name analysis and data retrieval and Mrs. Esterle for her valuable support during the expert interviews and the online survey field phase. The project team would also like to thank the members of the advisory board for their valuable advice and suggestions.

Our former colleagues Mrs. Kordula Kugele and Mrs. Pascale Bruno were in charge of the expert interviews and contributed to the questionnaire. We hereby want to thank them very much for their contribution. And last not least, many thanks to all our student team members who assisted in the project.

8. References

BMBF (2010): Bildung in Deutschland 2010. Herausgeber: Autorengruppe Bildungsberichterstattung W. Bertelsmann Verlag GmbH & Co. KG

Burris, Janet, W. (2001): The Impact of Gender Diversity on Technical Team Effectiveness; Dissertation; Ann Arbor: Bell & Howell Information and Learning Company

Busolt, U. & Kugele, K. (2009): The gender innovation and research productivity gap in Europe, Int. J. Innovation and Sustainable Development, Vol. 4, Nos. 2/3, pp. 109 – 122.

Butcher, Jane (2006): Women in Scientific Careers: Unleashing the Potential. ISBN 92-64-02537-5, OECD

Dahlerup, D. (1988): From a Small to a Large Minority: Women in Scandinavian Politics. Scandinavian Political Studies, Vol. 11, Issue 4, 275–298, 1988

Distefano & Maznevski (2000): Distefano, Joseph J/ Maznevski, Martha L. (2000), "Creating value with diverse teams in global management", Organizational Dynamics, Vol. 29 No 1, pp. 45-63, Elsevier Science, Inc.

European Commission (2006a): She Figures 2006 Women and Science. Statistics and Indicators., Luxembourg

Frink, D. D., Robinson, R. K., & Reithel, B. (2003): Gender demography and organization performance: A two-study investigation with convergence. Group & Organization Management, 28, 127–147

Gratton, Lynda (2007): Steps that can help women make it to the top. Financial Times, Published: May 22, 2007, http://www.pleaseadjustyourset.com/pdfs/FT_May07.pdf (download: 2010, Okt. 30)

Hirschfeld, R. R., Jordan, M. H., Feild, H. S., Giles, W. F., & Armenakis, A. A. (2005): Teams' female representation and perceived potency as inputs to team outcomes in a predominantly male field setting. Personnel Psychology, 58, 893–925

Jehn, K. A. (1995): A multimethod examination of the benefits and detriments of intragroup conflict. Administrative Science Quarterly, 40, 256–282

Kugele, K. (2010): Patents invented by women and their participation in research and development. A European comparative approach. In: Anne-Sophie Godfroy Genin (Ed.): Women in Engineering and Technology Research. The PROMETEA Conference Proceedings. No 1, pgs. 373-392

Martín-Sempere, M. J., Rey-Rocha, J., Garzón-García, B. (2002): The effect of team consolidation on research collaboration and performance of scientists. Case study of Spanish University researchers in Geology. Scientometrics, 55, 3 (2002): 377-394

Pearsall, M. J. , Ellis, A. P. J. & Evans, J. M., (2008): Unlocking the Effects of Gender Faultlines on Team Creativity: Is Actvation the Key? Journal of Applied Psychology, Vol. 93, No. 1, 225-234

Pelled, L. H., Eisenhardt, K. M., & Xin, K. R. (1999): Exploring the black box: An analysis of work group diversity, conflict, and performance. Administrative Science Quarterly, 44, 1–28.

Randel, A. E. (2002): Identity salience: A moderator of the relationship between group gender composition and work group conflict. Journal of Organizational Behavior, 23, 749–766.

Schone, W. (2009): How different approaches lead to different results – an explorative case study on the innovation environment of high-tech projects, Masterthesis, Furtwangen

Schone, W., Kellermann C, Busolt, U., Kugele K, Bruno P (2011): Improve – Innovatitvität für Deutschland, In print

Witzel, A. (1996): Auswertung problemzentrierter Interviews: Grundlagen und Erfahrungen In: R. Strobel, A. Böttger (Hg.): Wahre Geschichten? zur Theorie und Praxis qualitativer Interviews. Baden- Baden. Nomos

Part 2

Sustainable Business and Management

7

Sustainable Development as an Aspect of Improving Economic Performance of a Company

Tereza Kadlecová and Lilia Dvořáková
[1]Institute for Sustainability
[2]University of West Bohemia in Pilsen
[1]United Kingdom,
[2]Czech Republic

1. Introduction

The contemporary world, especially its developed part, has been driven by consumerism that is apparent in both the consumption and production. Individual national economies and businesses are under ceaseless pressure for economic growth which goes hand in hand with many negative factors, environmental degradation being one of them.

Over the last twenty years, many of the world's developed economies, public and academic bodies, and industrial corporations, having become more conscious about the unsustainability of current development in terms of the physical capacity of our world, have conceived a range of concepts and models for the sustainable development.

Successful implementation of the sustainable development concept is, however, fundamentally dependent on individual companies. In business practice the concept of sustainable development is applied through **Corporate Social Responsibility** (CSR). The CSR concept is based on three interrelated pillars - economic, environmental, and social.

This chapter deals with the economic and environmental pillars of CSR, or more precisely it looks to examine the nexus between economic and environmental performance of a company. Both aspects of a company's performance intermingle in the **Eco-efficiency** concept applied in the business practice through **Voluntary Environmental Instruments**.

The Eco-efficiency concept and related voluntary environmental instruments (proactive strategy) go beyond the legal framework (reactive strategy) and fall fully within the competence of company management.

Needless to say, companies would only buy into the concept of sustainable development and the related environmental responsibility if they were cognisant of the economic benefits of such approach. It is therefore crucial to convince businesses of the advantages and benefits of proactive environmentally responsible behaviour and motivate them to adopt it.

A proactive approach to environmental protection, applied in practice through the voluntary environmental instruments, features not only the expected positive impact on the environment, but, as practice shows, results in a range of financial (reducing operating costs, increased revenues) and non-financial benefits contributing to business value creation.

2. Eco-efficiency

2.1 The eco-efficiency concept

Eco-efficiency is a management philosophy that challenges businesses to pursue environmental improvements yielding concurrently economic benefits (Lehni, 2009). This concept, entailing a change in production and consumption patterns, promotes innovation and leads therefore to economic growth and enhanced competitiveness. The term eco-efficiency was coined in 1992 by the **World Business Council for Sustainable Development** (WBCSD) in the 'Changing Course' publication. Eco-efficiency is based on the principle of generating larger amounts of products while consuming fewer resources and therefore creating less waste and pollution. (International Institute for Sustainable Development, 2007). In this context, the prefix eco stands for both environment and economics.

Eco-efficiency falls within a broader concept known as **Sustainable production and consumption** introducing change in production and consumption patterns, and leading therefore to sustainable consumption of natural resources. Businesses play an important role in this concept, both as consumers of raw materials and manufacturers of products. Eco-efficiency focuses on three broad sets of objectives (Lehni, 2009):

- Reducing consumption of resources
- Reducing impact on the environment
- Increasing product value

Opportunities for eco-efficiency

Eco-efficiency can be practically implemented into business processes mainly through search for innovation opportunities particularly in the following areas (Lehni, 2009):

- Re-engineering of processes
- Cooperation with other enterprises
- Redesign of products
- Searching for new ways to meet customer needs

2.2 Eco-efficiency indicators

Currently several approaches are known that enable measurement of eco-efficiency applied in production and business operations.

WBSCD

According to WBSCD, eco-efficiency can be formulated as a ratio of a product value (economic performance) to its environmental impact (environmental performance). WBSCD developed a framework for reporting company data relating to eco-efficiency that distinguishes three levels of information - Categories, Aspects and Indicators (Verfaillie & Bidwell, 2000).

UNCTAD / UN-ISAR

Unlike WBSCD, UNCTAD / UN-ISAR regards eco-efficiency as an environmental burden (environmental performance) per unit of economic value (economic performance). Similarly to WBSCD, UNCTAD/UN-ISAR proposed a framework for reporting company data relating to eco-efficiency consisting of three levels of information - Elements, Items and Indicators – as a performance measure of company specific aspects (Müller & Sturm, 2001).

At this point it might be useful to recall the key concepts of economic and environmental performance that will be encountered in the course of this paper.

2.3 Economic performance – BSC

Immense number of methods has been developed to measure economic performance of a company. Some of them focus purely on an assessment of company financial statements while others consider other aspects of a company's life. These multi-criteria methods analyse financial results in the context of wider non-financial achievements.

In our research, Balanced Scorecard (BSC), one of the models for multi-criteria decision making and valuation of economic performance, has been employed to represent the economic pillar of the eco-efficiency. BSC method was developed by Robert S. Kaplan and David P. Norton and published in 1992.

Within BSC, business objectives are classified into four perspectives - financial, customer, internal business processes, and learning and growth (potential) - that are intended for complex measurement and control of a company's performance. Objectives in these individual perspectives are interconnected with 'cause - effect' relations that are depicted by arrows (Kaplan & Norton, 2002).

2.4 Environmental performance

On the broad level, environmental performance (also profile) reflects the general achievement of a system (product, process, company) concerning cut-down on negative impact on the environment. ISO 14001 defines the environmental profile of a company as *"measurable results of the environmental management system, related to an organization's control of its environmental aspects, based upon its environmental policy, objective and targets."* Similarly, ISO 14031 defines environmental performance as *"an organization's success in managing the relationships between its activities, products, or services, and the natural environment."*

To assess the environmental profile of a system, Environmental Performance Indicators (EPI) are used as measures for Environmental Performance Evaluation (EPE). Over the last twenty years, several concepts have been developed to evaluate the environmental performance of a company.

2.4.1 ISO 14031

ISO 14031 defines two general indicator categories for environmental performance evaluation:

- Environmental Performance Indicators (EPI) - which are further divided into:
  - Management Performance Indicators (MPI)
  - Operational Performance Indicators (OPI)
- Environmental Condition Indicators (ECI)

2.4.2 The Global Reporting Initiative

In 2000 the Global Reporting Initiative (GRI) issued its first guidelines on sustainability reporting. Environmental reporting is organized into nine groups called the evaluation criteria which are structured in such a fashion to ensure easy evaluation of inputs (Criteria:

energy, water, materials), outputs (Criteria: emissions, waste water and waste) and impacts on the environment (Criteria: e.g. Products and services, Transport) (GRI, 2006).

2.4.3 Research foundation of Norway (Oestdold Research Foundation)

In his work from 1999, Johan Thoresen, a member of the Norwegian Research Foundation Oestdold, developed three categories of EPI indicators (Thoresen, 1999):

- Category 1 - Performance of a product lifecycle
- Category 2 - Environmental performance of selected process technology
- Category 3 - Environmental performance of operations

3. Analysis of voluntary environmental instruments

3.1 Preventive vs. reactive strategy for environmental protection

A range of preventive and reactive strategies for the protection of the environment are used in the manufacturing. As the name indicates, preventive strategies aim to preclude origin of environmental damage through searching for and minimising sources of pollution and waste. Reactive strategies, on the contrary, do not anything that goes beyond what is necessary to comply with the environmental regulation. The reason why reactive strategies are not so effective and promising is the fact that they do not focus on the causes of environmental damage, they only try to mitigate the negative consequences of production. Among these "end-of-pipe" technologies are, for example, refuse compactors, collection containers and vehicles, waste heat recovery systems, air pollution filters, noise abatement investments and sewage treatment plants. As a result the quantity of toxic agents drops in one environmental domain, but rises in another one.

A fundamental disadvantage of the reactive type of control strategy is that the end-of-pipe technology can never reach 100 % efficiency. Then obviously despite taking corrective measures, that are often very expensive, with an increasing number of pollution sources environmental degradation rises concurrently. Another problem is that the stated emission limits may not be sufficient given the fate of substances in the environment - all forms in which a substance released into the environment can convert and through subsequent reactions, the so-called secondary effects, impact on humans and ecosystems - cannot be confidently determined (Czech Ministry of Environment, 2003).

Freons (CFCs) eroding the ozone layer represent a classic example of ignorance of the fate of substances in the environment. These materials were originally considered to be almost perfectly non-reactive gases, and therefore used as a carrier gas in spray cans and refrigerators.

All these above mentioned flaws of reactive strategies clearly demonstrate that the only solution to sustain a healthy environment is to focus on preventive elimination of damage sources rather than addressing problems already arisen.

As practice has shown, the preventive strategy has a positive impact on the environment and leads to financial savings, economic profit, cost reductions and enhancement of the competitive advantage at the same time. Preventive strategy is therefore considered a double profit strategy: environmental and economic – a 'win-win' strategy. The preventive strategy to the environmental protection can be practically applied through a range of voluntary environmental instruments.

3.2 Eco-efficiency tools

The following text gives overview of selected management tools - voluntary environmental instruments - that help companies maximize their efficiency, product quality and profit through improved corporate environmental profile.

Environmental management systems (EMS) - EMS represent a systematic management and control of particular business areas that are posing risk to the environment. Within EMS companies set environmental goals to ensure continuous improvement of corporate environmental profile in the future. Currently, there are two commonly used standards for implementing an EMS:

- Technical standards of ISO 14001
- EU regulation Eco-Management and Audit Scheme (EMAS)

Environmental management accountancy (EMA) - EMA allows a company to identify and manage its environmental costs and achieve therefore a reduction in total company costs. Within EMA both financial and material information is tracked and analysed.

Cleaner Production - The concept of cleaner production is being connected with the integral preventive strategy which is applied especially to the sector of production (Remtová, 2003a). The aim of this strategy is to eliminate or reduce sources of environmental degradation with the use of technical and nontechnical solutions (e.g. more efficient use of raw materials and energy, elimination or reduction of toxic and hazardous materials, prevention of waste and pollution at source). Within cleaner production all material and energy flows in a company are monitored in order to identify the sources of undesirable waste.

Ecodesign - Ecodesign incorporates requirements for environmental protection into product design and development. Currently there is no unified definition of ecodesign. In general, ecodesign can be defined as a systematic process of product design and development which puts emphasis not only on common product features like economics, safety, ergonomics, technical feasibility, aesthetics, but that also pursues a minimum negative impact on the environment (e.g. reducing quantity and toxicity of materials, product demountability for easier reuse and/or recycling at the end of its useful life) (Remtová, 2003b).

Life Cycle Assessment (LCA) - LCA is a tool for assessing the overall environmental impact of a product in its entire life cycle. Withing LCA all material and energy flows relating to any life phase of a particular product are analysed. This tool is widely used during a product design (see Ecodesign).

Eco-labeling - Eco-labelling is a certification system for products and services that are friendlier to the environment than substitute products. This system is directed by a third party that has to be independent. Currently there are three different Eco-labeling systems that a company can opt for:

- Ecolabeling - Environmental Declaration (Type I)
- Self-declared Environmental Claim (Type II)
- Environmental Product Declaration (Type III)

3.3 Comparison of voluntary environmental instruments

Comparison of voluntary environmental instruments is not a straightforward and simple task to do. On the contrary, even when comparing the benefits and implementation requirements for the same instrument the final conclusions can differ immensely due to the specific situation of investigated companies. Period of return on investment as well as the subsequent financial benefits of the individual voluntary environmental instruments can be significantly different when compared among companies.

Voluntary environmental instruments can be compared from different perspectives (see Table 1). While some instruments are focused on product environmental performance, others help streamline business processes or influence the management and operation of a company as a whole, i.e. the entire corporate system.

Voluntary environmental instruments also vary in terms of so-called external collaboration. Some instruments can only be used if an appropriate background has been created for them

Comparison criterion	Voluntary methods and instruments						
	EMAS	EMS/ ISO 14001	EMA	Ecodesign	LCA	Cleaner Production	Eco-labelling
Purpose	Regula-tive	Regulative Educative	Informative	Regula-tive	Informa-tive	Informa-tive	Regula-tive
Focus	Systems	Systems	Processes	Products	Products	Processes	Products
Normalization	Yes	Yes	No	No	Yes	No	Yes
Necessary external collaboration	Yes	Yes	No	No	No	No	Yes
Preventive strategy	Yes	Yes	No	Yes	Yes	Yes	can be
Financial requirements associated with an implementation	Yes	Yes	No	Yes, conside - rable	Yes, conside - rable	No	Yes
Labour input intensity	No	No	Yes	Yes, conside - rable	Yes, conside - rable	Yes	No
Economic benefits	Yes, partly	Yes, partly	informative benefits more likely	Yes	No	Yes – conside - rable	uncertain
Intended for	All company types	All company types	All company types	Manufac - turing compa - nies	All company types	All company types	Company with products/ services included in existing product categories
Logo/ certificate	Yes	Yes	No	No	No	No	Yes

Table 1. Comparison of a selection of voluntary environmental instruments, Source: authors

to be implemented. Among the key activities that cannot be carried out by a company itself and are compounding this background are inspection, certification and registration. These activities need to be provided by another entity (e.g. an independent third party, a state body). There are of course financial expenses related to the necessary external collaboration (e.g. costs associated with consultations, advisory on EMS or application fee for eco-labeling) that distinguish significantly from one instrument to another. Voluntary instruments differ also hugely regarding benefits for a company and costs required for their application. While some of the voluntary environmental instruments are generally considerably beneficial in terms of operation costs reduction (e.g. Cleaner production) the contribution of other instruments is mainly informative (e.g. LCA).

Table 1 provides an overview and comparison of a selection of voluntary environmental instruments with regard to different perspectives.

3.4 Classification of environmental benefits of voluntary instruments

On the basis of a thorough analysis of the individual voluntary environmental instruments, a generic classification of their benefits was created. These benefits were divided into two major groups, financial and non-financial. Within our research we then elaborated a catalogue with benefits for each one of the voluntary environmental instruments.

a) Financial benefits

The financial benefits stemming from implementation of the voluntary environmental instruments in a company consist mainly in:

- Operating costs savings as a result of energy saving measures , water and materials efficiency programmes, and a decline in fees and taxes related to environmental damage
- Increased revenue as a consequence of access to new markets, increased demand of existing customers or sale of a new product, e.g. waste materials
- Obtaining state aid or subsidies

b) Non-financial benefits

Similarly, application of the voluntary environmental instruments is associated with a number of non-financial benefits that however ultimately affect the financial benefits for a company. We have proposed the following eight groups to which the non-financial benefits can be allocated:

- Business benefits - improved corporate image, growth of "brand equity" and related customers satisfaction
- Employee relations - increasing morale and involvement of employees as a result of increased employee satisfaction and related retention of key employees
- Public relations – enhanced corporate credibility and overall public functioning, improved position of in negotiations with public authorities
- Risk management - emergency preparedness and reducing the likelihood of environmental accidents;
- Compliance with environmental, health and safety legislation and standards - regular monitoring of legislation ensuing in ability to predict possible legislative changes and prepare for them

- Company management - establishment of order in company operations and documentation, improved internal communications, streamlined production processes and increased product quality
- R & D - stimulating innovative thinking
- Environment, Health & Safety (EHS) - Improved quality of working environment and related betterment of safety and health conditions for employees resulting in decreased absenteeism and increased employee productivity

4. Corporate environmental profile index

Corporate Environmental Profile Index (CEPI) was designed for internal business purposes for the multi-criteria analysis and assessment of a company's environmental performance. CEPI is composed of four broad set of categories that are further divided into a number of criteria. The four categories are as follows:

- RC - Resource Consumption
- ERL - Environmental Releases
- ER - Environmental Remediation
- CO - Compliance (with environmental legislation)

The four index categories were deliberately defined broadly to cover all potential environmental aspects of business operations. For each category neither specific criteria nor their number were determined for the general model (see chapter 5).

The importance of the index categories and criteria, in terms of their contribution to the overall environmental profile of a company, is expressed by weights (v_i for each criterion and V for a given category). The weights can take values from 1 to 5 with the following meaning:

1. Very low importance
2. Low importance
3. Intermediate importance
4. High importance
5. Very high importance

As a result, the importance of an individual criterion in terms of the final index value is not only given by their own weight but also by weight of the category to which they belong.

Each criterion within those four categories is further evaluated with a score (s_i) which expresses the qualitative criteria evaluation in the range:

1. Very good
2. Good
3. Average
4. Bad
5. Very bad

The score represents a formal quantification of the performance evaluation, where each criterion is objectively assigned a score value according to the interval that corresponds with its specific criteria performance.

In general, scores can be determined:

a.   With regard to a specific target for individual criteria
b.   With regard to a trend of development of individual criteria

Mathematical expression of CEPI

When employing the input parameters described earlier the CEPI can be calculated as follows: For each category a category score is computed according to the equation (1) shown below, which mathematically represents a weighted average of individual scores of all criteria in the particular category.

$$\bar{w} = \frac{\sum_{i=1}^{n} s_i v_i}{\sum_{i=1}^{n} v_i} \qquad (1)$$

Where

$\bar{w}$ Category score

s Criterion score, for which $s \in S, S = \{1,2,3,4,5\}$

v Criterion weight, for which $v \in V, V = \{1,2,3,4,5\}$

As a next step, each category score needs to be multiplied with a specific category weight, see the equation (2).

$$W = \bar{w}V \qquad (2)$$

The final CEPI value then corresponds to a weighted average of scores for all categories, as calculated in the formula (3).

$$CEPI = \frac{W_{RC} + W_{ERL} + W_{ER} + W_{CO}}{V_{RC} + V_{ERL} + V_{ER} + V_{CO}} \qquad (3)$$

Where

$W_{RC}, W_{ERL}, W_{ER}, W_{CO}$ Weighted score of each category

$V_{RC}, V_{ERL}, V_{ER}, V_{CO}$ Weight of each category

Using the input parameters of this method the formula for determining the CEPI is as follows:

$$CEPI = \frac{\left(\frac{\sum_{i=1}^{n} s_{RCi} v_{RCi}}{\sum_{i=1}^{n} v_{RCi}} V_{RC}\right) + \left(\frac{\sum_{i=1}^{m} s_{ERLi} v_{ERLi}}{\sum_{i=1}^{n} v_{ERLi}} V_{ERL}\right) + \left(\frac{\sum_{i=1}^{o} s_{ERi} v_{ERi}}{\sum_{i=1}^{n} v_{ERi}} V_{ER}\right) + \left(\frac{\sum_{i=1}^{p} s_{COi} v_{COi}}{\sum_{i=1}^{n} v_{COi}} V_{CO}\right)}{V_{RC} + V_{ERL} + V_{ER} + V_{CO}} \qquad (4)$$

Given the specified range of weights and scores the final CEPI can take any value in the range of <1,5>. The resulting CEPI value can be therefore classified into one of the four below intervals assessing the level of company environmental performance:

1.   <1, 2) excellent environmental performance

2.   <2, 3) good environmental performance
3.   <3, 4) poor environmental performance
4.   <4, 5> unsatisfactory environmental performance

5. B2En Performance Development model

5.1 Introduction

In chapter 2, dealing with the Eco-efficiency concept, the interconnection was highlighted between economic and environmental performance or more specifically between environmental and economic results of a company. Selected voluntary environmental instruments have been characterized in chapter 3 as specific examples of application of the eco-efficiency principles in business practice. From chapter 3 it is obvious that implementing eco-efficiency principles in a company not only results in costs savings and increased sales; but that environmentally proactive behaviour, closely associated with innovative thinking, drives a range of non-financial benefits which in the end contribute significantly to the financial bottom line. In our research we went even further and created a conceptual model integrating environmental and economic performance as two interrelated aspects of business activities, enabling effective management and control of both performance areas.

To put it in a different way, the model shows the interdependence between environmental prevention and protection actions on one side, and achievement of business economic objectives on the other side. The name B2En Performance Development indicates that a proactive behaviour in terms of environmental responsibility has a direct impact on improving a company's economic performance. The main aim of the B2En model was to identify interrelations between environmental and economic performance of a company. For the economic component of our model it was necessary to select a suitable method for measuring and managing economic performance.

For the purposes of the model, it was essential to choose such management method that would:

- consider financial as well as non-financial factors (see the nature of the voluntary environmental instruments benefits)
- allow for integration of the environmental perspective
- enable identification of cause-effect relations between environmental and economic performance

Model Balanced Scorecard (BSC), as a comprehensive management method cutting across the entire business, meets all three above conditions and was therefore selected as an economic component of the conceptual B2En model.

As already mentioned, the B2En model is based on the eco-efficiency concept that lies around the principle of generating more products while consuming less resources and therefore producing less waste and pollution. As described in chapter 2, this is achieved especially through innovation in terms of new production and consumption patterns leading to separation (decoupling) of economic growth from resource consumption.

According to the eco-efficiency principle, environment and economics are two interrelated and therefore interacting aspects of a business. For this reason B2En model was designed in such a way that the environmental perspective is integrated into all four original perspectives of BSC (Fig. 2), forming so the interaction between environmental and economic performance of a company.

| Category | Criteria | | | | Weight (V) | Weighted Category Score (W) |
	Name	Score (s_i)	Weight (v_i)	Weighted Score (w_i)		
Resource Consumption RC	Criterion 1				V_{RC}	W_{RC}
	Criterion 2					
	Criterion 3					
	⋮					
	Criterion n					
	Total					
	Category Score $\overline{w}_{RC}$					
Environ-mental Releases ERL	Criterion 1				V_{ERL}	W_{ERL}
	Criterion 2					
	Criterion 3					
	⋮					
	Criterion n					
	Total					
	Category Score $\overline{w}_{ERL}$					
Environ-mental Remediation ER	Criterion 1				V_{ER}	W_{ER}
	Criterion 2					
	Criterion 3					
	⋮					
	Criterion n					
	Total					
	Category Score $\overline{w}_{ER}$					
Compliance CO	Criterion 1				V_{CO}	W_{CO}
	Criterion 2					
	Criterion 3					
	⋮					
	Criterion n					
	Total					
	Category Score $\overline{w}_{CO}$					
	Total					
	CEPI					

Table 2. Corporate Environmental Performance Index

5.2 Model structure

For the B2En model three different diagrams were created that represent different levels of approximation:

a. Level 1 (Fig. 1) - General concept – displays the very essence of the B2En model consisting in integration of the environmental perspective into the original four perspectives of BSC model (economic perspective);
b. Level 2 (Fig. 2) - Structure of the model - provides more detailed information on individual components of the model and mutual relations between them;
c. Level 3 (Fig. 3) - Identification of relations - offers a detailed look at the cause-effect relation between environmental and economic perspectives.

Fig. 1. General Concept of the B2En Performance Development model

When looking closer, B2En model consists of three components: Environmental Perspective (I.), Balanced Scorecard (Economic Perspective) (II.), and Eco-efficiency indicators (III.), which are interconnected by logical links (see Fig. 2). Environmental Perspective of the B2En model comprises the following three components:

1. Environmental performance - presents four broad categories of environmental objectives: resource use, environmental releases, environmental remediation, compliance with environmental legislation;
2. Environmental Performance Indicators (EPI) – these are used as metrics for environmental objectives evaluating a company's environmental performance.
3. Voluntary environmental instruments - enable application of the eco-efficiency concept in business practice. In B2En model voluntary environmental instruments are divided into four groups representing a management level to which they relate: products, processes, systems and strategies.

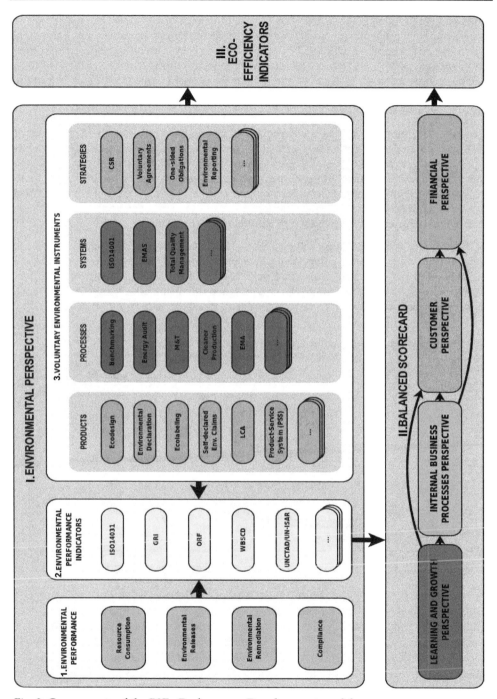

Fig. 2. Components of the B2En Performance Development model

Links and connections between the respective components of the model are expressed by arrows and the principle of the scheme is as follows. A company sets environmental objectives and priorities within the four categories in order to improve its own environmental performance. Once the objectives are set appropriate metrics (EPI) need to be identified and allocated to each of them to measure achievement of these objectives. Voluntary environmental instruments serve as means to influence a specific EPI and achieve those environmental objectives.

Given the nature of the voluntary environmental instruments based on the elimination of waste and pollution prevention at source, achievement of the environmental objectives has a positive effect on one or more perspectives of the BSC. Voluntary environmental instruments hereby ultimately impact on the financial perspective of the BSC model and therefore contribute to the financial performance of a company. Achieving environmental objectives is therefore positively reflected in the economic perspective of business performance which refers to a win-win or a double victory situation.

Fig. 3. provides a detailed look at the cause - effect relationship which interconnects the environmental and economic perspective (representing four original perspectives of BSC) at the level of an individual environmental objective of a company. As the arrows indicate, achievement of an environmental objective has a positive impact on the key economic performance indicators (KPIs).

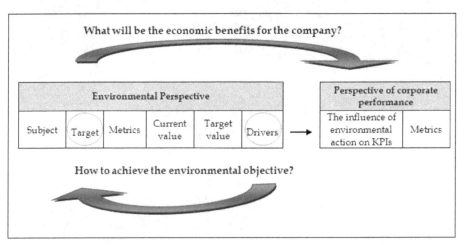

Fig. 3. Cause-effect relationship between environmental and economic perspectives

For each environmental objective, not only the way (drivers – voluntary environmental instruments) to its achievement needs to be identified, but also economic consequences resulting from this accomplishment. Managing eco-efficiency is therefore a two-fold task consisting in reaching the stated environmental objective and achieving the greatest economic benefits, both financial and non-financial. Mutual relationship and linkages between the environmental perspective and individual perspectives of the BSC will be addressed in the following text and the diagram in Fig. 3 will be elaborated for the individual perspectives in greater detail (see table 3).

BSC Perspective		Environmental perspective					Economic perspective	
	Subject	Target	Metrics	Current value	Target value	Drivers	Impact of the environmental action on KPIs	Metrics
Financial Perspective	GHG emissions	Reduction of GHG emissions in the process / product or for the company as a whole	Tons of CO_2 equivalent per 1 million sales Energy consumed (EUR/GJ) per number of employees			Energy efficiency programmes (CP, EMS) Using renewable energy	Cost savings on emissions charges Saving energy costs Revenues from sales of surplus emission allowances	Operating costs (EUR) Turnover (EUR)
Customer perspective	Eco-label	Over the next 12 months an eco-label for two key products to obtained	Eco-label certificate			Environmental specialists to be involved in the technical development teams	Product features and image Acquiring new customers in the European market	% contracts from new customers in targeted EU segments
Internal business processes perspective	Waste	Reduction of solid waste in the process / product or for the company as a whole	Disposed waste (volume, weight, or EUR) per added value			Waste prevention programs (CP, EMS) Programmes to increase material efficiency Reuse of waste	Cost savings associated with the efficient handling of production materials	Operating cost efficiency
Learning and growth perspective	Consumption of toxic substances	Cut down on the volume of toxic substances in the process / product or for the company as a whole	The volume of toxic substances (l) per number of products The volume of toxic substances per added value			Change in Technology Change of major raw materials Change in consumables	Reduced absenteeism due to workplace accidents and illness	Average absenteeism measured as missed working days per 1 employee

Table 3. Links between environmental and BSC perspectives

For whom is the B2En model intended?

The B2En model was designed mainly for internal business purposes to enable efficient management and control of environmental and economic performance with the aim to achieve a win-win situation. In principle, the B2En model can be implemented in all businesses regardless of size and industrial sector. However, for the B2En model to be implemented, the BSC model needs to be applied to certain extent in business practice. Due to this fact and the potential scale, the model is envisaged to be implemented particularly in medium and large companies, whether at the corporate level, individual departments or for a specific product. The win-win principle resulting from a proactive business approach to environmental protection - basic idea of the B2En Performance Development model - is of course essential for the change in general attitude of business community towards environmental responsibility and relating rethinking of the consumption and production patterns. The proposed model is therefore clearly important even for small businesses, although it will not be applied in the proposed form.

5.3 Relationship and interdependencies between the economic and environmental perspective of the B2En model

5.3.1 Financial perspective vs. environmental perspective

In this chapter-part, relationship and interconnection will be scrutinised between the environmental perspective and the individual BSC perspectives. Environmental activities applied in a company through respective voluntary environmental instruments have a wide scope to all levels of business. In the following text, specific areas will be identified in which environmental activities are directly impacting on long-term financial objectives of a company.

a) Increased sales

The increase in business turnover is influenced, among others, by introduction of a new product, new customer acquisition and new market penetration. Many voluntary environmental instruments have the potential to positively affect the growth in turnover of a company through searching for new opportunities, as they offer a different perspective for business decision making.

New Customers - Many enterprises confirm increased sales as a result of implementing EMS - a systematic approach to environmental protection in all aspects of business. An environmentally responsible company with an EMS certification gets new opportunities in commercial sector.

New Product – Waste represents another field of opportunities for further growth in sales. Selling waste as a secondary resource or for recycling, rather than landfilling, constitutes another source of sales. Also, environmental instruments (e.g. Ecodesign, LCA and Eco-labelling) may positively affect the marketability of products.

New markets - Environmentally responsible behaviour improves overall corporate reputation and positively impacts on public perception of a company. This is very important nowadays, especially if a company aims to penetrate or retain foreign markets. An EMS certification or a product eco-label is often an entrance card for companies to export their products to foreign markets.

b) Cost reductions / productivity improvements

Prevention of waste - Environmentally responsible behaviour is primarily associated with savings in operating costs. Preventive voluntary instruments, such as cleaner production, EMS, and Ecodesign seek to minimize waste as a non-product output with a negative or zero market value.

Compliance - A company that comply with environmental legislative requirements is spared fines, fees, denials of permits, closures and other unpleasant consequences of contravening applicable laws. Within EMS, a company is obliged to monitor current environmental laws and regulations on a regular basis to stay ahead of them.

Environmental Risk Management - Through proactive management of environmental risks the likelihood of environmental accidents, leakages, and oil stains can be minimized. Preventive measures, applied through e.g. cleaner production or EMS, can reduce costs associated with repairing damages and penalties for failure to comply with legislation.

5.3.2 Customer perspective vs. environmental perspective

Within the customer perspective, companies in particular strive for growth in the market share and sales in targeted customer segments. These objectives are closely linked to retaining existing and acquiring new customers, and increasing their satisfaction. Among the many aspects influencing customer satisfaction are product features, good relations with customers, and business image.

Product Features - Products manufactured in accordance with the eco-design principles or eco-labelling criteria are easy to disassemble, provide greater security, and can be sometimes purchased at lower prices as a result of recycling some sub-components. Voluntary environmental instruments enable improved product quality in terms of reliability, longevity and ease of maintenance and repair.

Customer Relations – Many customers nowadays aim to improve their own environmental performance and require therefore environmentally friendly products from environmentally responsible suppliers. Voluntary instruments enable inter-company cooperation focused on reducing negative environmental impact across respective supply chains. Industrial symbioses represent another example of such cooperation. Within these clusters, waste of one company becomes raw material input for another one and a closed/ continuous loop is achieved.

Business image and reputation - Environmentally responsible behaviour enhances corporate reputation, positively affects public perception of a company and increases business credibility which is essential when dealing with banks, insurance companies, municipalities and other public institutions.

5.3.3 Internal business processes perspective vs. environmental perspective

An internal value chain consists of three different levels - innovation processes, operational processes and after-sale services (Kaplan & Norton, 2002). Voluntary environmental instruments not only significantly impact on innovative solutions aiming to manage environmental and economic performance at the same time, but they also contribute to streamlining business processes.

Innovation Processes

The Eco-efficiency concept boosts innovativeness through challenging companies to find creative solutions leading to enhanced environmental and economic results. Companies have a plethora of opportunities to innovate, among the most obvious being

- New markets or new customers in existing markets (e.g. Industrial Symbiosis)
- Products innovations (e.g. Eco-design)
- Technological innovations focused on energy and material efficiency

Business Processes

At an operational level, a company should look to 'do more with less', as explained in chapter 2 on eco-efficiency. This does not relate only to economic efficiency, rather the aim is to mitigate harm (emissions, waste, leakages, consumption of material and energy) to the environment when generating more products and services and improving economic benefits. As to resource consumption, eco-efficiency is focused mainly on:

- Streamlining production processes
- Cutting down on resource consumption
- Restriction of hazardous materials
- Reducing unwanted by-products

5.3.4 Learning and growth perspective vs. environmental perspective

Intangible assets of a company represent a prerequisite for flexible and efficient internal processes oriented on achieving objectives of the customer and financial perspectives. A company should also strive to nurture its employees' satisfaction as this is an essential factor in securing customer satisfaction.

Competence of staff

A company can only achieve outstanding results in mitigating negative environmental impact of its business processes, if necessary competence and skills are secured for its employees. However, the excellent performance of employees is not only given by their knowledge and skills, but is significantly influenced by such soft factors as motivation and work attitudes.

Employee Satisfaction

As a result of improved working environment, stemming from an environmentally responsible behaviour of a company, a drop in absenteeism due to illness, fewer workplace accidents, and increased productivity can be expected.

Company image constitutes another factor affecting job satisfaction. A company characterized by environmental responsibility and safe working environment is likely to be more successful in attracting and retaining good employees with a positive environmental attitude.

5.3.5 Colligation of environmental perspective with individual perspectives of BSC

From the above text a clear relationship between environmental management and economic benefits, whether it be financial or non-financial, is obvious. As explained, voluntary preventive environmental approach not only benefits the environment but is connected with a range of positive impacts on all levels of business.

Table 3 was designed as a template for examining the interrelationship between environmental actions and economic performance of a company as a direct consequence of these actions. In the original research, these interrelationships were scrutinised for each BSC perspective in greater detail. Within this chapter only one example is presented for each BSC perspective. These examples are illustrative only; each company will choose its own objectives, metrics and KPIs according to its individual needs.

6. Eco-efficiency Statement

The Eco-efficiency Statement was developed by the authors to assess the impact of a company's environmental activities, more precisely its environmental profile, on its economic performance. The Statement enables evaluation of the relationship between the environmental and economic results achieved within a specific time period.

As displayed in table 4, information included in the proposed Eco-efficiency Statement is classified into three groups:

- Under the **Key economic performance indicators** key items are shown from the profit and loss statement and financial analysis, that get most affected by the environmental profile of a company. The economic indicators include:
  a) Key indicators of Profit and loss statement
     - Indicators relating to business operation: e.g. Sales, Operating income, Operating expenses, Gross profit, Value added
     - Key profit categories: e.g. EBITDA, EBIT, Profit after tax
  b) Key indicators of financial analysis: e.g. ROCE, ROE, ROI; EVA, CFROI

- **Key environmental performance indicators** give an overview of core indicators identified in the GRI guidelines that have the greatest impact on the environmental profile of a company. However, it is necessary to add that the Eco-efficiency Statement is by no means meant to replace the full environmental performance reporting, that is far more comprehensive and gives thorough overview of the total environmental profile of a company. The environmental indicators include:
  a) Input indicators: e.g. Materials or energy used
  b) Output indicators: e.g. Emissions released, water discharge
  c) Impact indicators: e.g. Fines and sanctions

- **Eco-efficiency indicators** represent the third group of indicators included in the Eco-efficiency Statement. These indicators are particularly important as they document the performance of a company and its trend, help identify and prioritise improvement opportunities, and identify cost savings and other eco-efficiency related benefits. Eco-efficiency indicators can also testify that, in a specific business area, there are only limited opportunities for improvement and requirements posed by stakeholders are impossible to achieve. In this Statement, the UNCTAD's approach to eco-efficiency (see chapter 2.2) was applied.

Eco-efficiency Statement								
Key indicators of economic performance	Unit	I.	%	II.	%	III.	%	IV.
Key indicators of profit and loss statement								
Sales	thousand EUR							
Operating income								
Operating expenses								
Gross profit								
Value added								
EBITDA								
EBIT								
Profit after tax								
Key indicators of financial analysis								
ROCE	%							
ROE	%							
ROI	%							
Dividend per share (gross)	EUR/share							
Operational CF	thousand EUR							
EVA	thousand EUR							
CFROI	%							
Key indicators of environmental performance according to GRI								
EN1: Materials used by weight or volume	thousand tons							
EN3: Direct energy consumption by primary energy source	GWh							
EN8: Total water withdrawal by source	thousand m3							
EN16: Total direct and indirect greenhouse gas emissions by weigh	CO2 tons equivalent							
EN19: Emissions of ozone-depleting substances by weight	CFC-11 tons equivalent							
EN20: Nox, SOx, and other significant air emissions by type and weight	tons							
EN21: Total water discharge by quality and destination	thousand m3							
EN22: Total weight of waste by type and disposal method	thousand tons							
EN24: The amount of hazardous waste	thousand tons							
EN28: Monetary value of significant fines and total number of non-monetary sanctions for noncompliance with environmental laws and regulation	EUR							
Key eco-efficiency indicators	Calculation							
Energy efficiency (energy consumption / sales)	GWh/sales							
Consumption of water in relation to added value	m3/value added							
Waste in relation to the added value	tons/value added							
CO2 emissions in relation to sales	tons/sales							
Leakage of toxic substances into the atmosphere in relation to sales	tons/sales							

Table 4. Eco-efficiency Statement

Eco-efficiency Statement enables an assessment of correlation between environmental results (e.g. energy consumption) and economic outcomes (e.g. operating expenses, EBIT). For achieved results to be interpreted, it is especially important to consider time series data and indexes (increase / decrease compared to last year). Again, the content of the Eco-efficiency Statement was not meant to be fixed. Companies will need to adhere to the proposed structure but should include such key indicators that are essential for their specific business situation and accounting standards. An example of the Eco-efficiency Statement is introduced in the table 4.

7. Methodology for B2En performance development model implementation

Phase 1 – Setting objectives for improving environmental profile

Internal analysis

Identification of current and potential business impacts on the environment should be the starting point for any environmental action and effort for improving environmental performance. The multi-criteria index CEPI (see section 4) is a useful tool for identifying problem areas of the environmental profile of a company.

Feedback from external and internal stakeholders

Over and above the results gained during the internal analysis, a company should also consider views of its external and internal stakeholders, as this can be an invaluable source of information. A company should not rely solely on the results obtained from the internal analysis. It is always invaluable to consider opinions of a company's external and internal stakeholders. Through e.g. regular questionnaire surveys, a company should identify and assess business environmental issues important for its stakeholders. Based on the results from internal analysis, opinion polls and analysis of stakeholders' views, an environmental manager sets targets and objectives for improving company's environmental performance.

Phase 2 – Identification of eco-efficiency opportunities

Once a company has a good understanding of its business actions and production processes and their impact on the environment, it can start identifying the eco-efficiency opportunities leading to enhanced business value through an improved environmental profile. It is crucial that each environmental objective is attained with the best economic outcome. The scheme outlined in table 5 can be used as a selection mechanism for business priorities considering both, their environmental and economic importance. Once respective areas for environmental improvements and value creation opportunities are identified, they need to undergo a prioritization process and be allocated into the following groups:

- High priority issues (H)
- Medium priority issues (M)
- Low priority issues (L)

Phase 3 - Application of voluntary environmental instruments

After goals for improving environmental performance have been set and links to the various perspectives of economic performance have been identified, chosen voluntary environmental instruments can be rolled out.

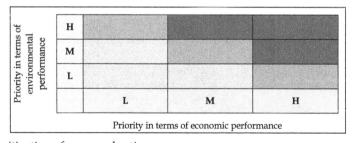

Table 5. Prioritization of proposed actions

Phase 4 - Objectives assessment

During the application of voluntary environmental instruments all results need to be constantly monitored, not only at the end of the project but throughout the entire implementation process. All significant deviations from forecast need to be thoroughly analysed and appropriate corrective actions need to be implemented. The Eco-efficiency Statement (see chapter 6) can be used to evaluate the overall success of the voluntary environmental instruments application.

Phases 5 - Demonstration and communication of results

Demonstration and communication of results achieved through application of selected voluntary environmental instruments follows as the final phase of deployment of the B2En Performance Development model. The economic and environmental benefits ensuing from implementation of individual voluntary instruments shall be communicated to various interest groups in different ways. This process comprises both, internal and external communications.

8. Case study - 3M Corporation

At the end of our research we undertook a case study to verify the functionality of the proposed B2En Performance Development model and its implementation methodology.

The model was applied, in line with the proposed methodology, on real environmental and economic data and targets of 3M Corporation[1] (3M). For the purposes of this paper only a summary of the full case study is introduced.

8.1 Definition of goals for improving environmental performance

3M is characterized by its proactive approach to environmental issues and implementation of preventive solutions. Based on thorough analysis 3M set and prioritised the following goals for the period 2005 and 2010:

- Energy efficiency improved by 20%
- Production of waste materials reduced by 20%
- 800 projects of pollution prevention completed
- Emissions of volatile compounds reduced by 25%

8.2 Application of voluntary environmental instruments

3M apply their responsibility towards the environment mainly through

- Use of Environment, Health and Safety (EHS) management systems (e.g. EMS, Cleaner production)
- Application of product life cycle management systems for permanent protection of the environment, health and safety (e.g. LCA, Ecodesign, Eco-labeling)
- Prevention of environmental pollution through development of new technologies and products.

---

[1] http://solutions.3m.com/wps/portal/3M/en_US/3M-Sustainability/Global/

8.3 Evaluation of achieving stated objectives

8.3.1 Environmental results

Energy - Since 2005, 3M has achieved energy savings to the tune of 37 million USD, especially through implementation of 1400 projects proposed by its staff.

Waste - In 2008, 616 pollution prevention projects were completed, which prevented 55.5 thousand tonnes of waste representing savings of nearly 91 million USD.

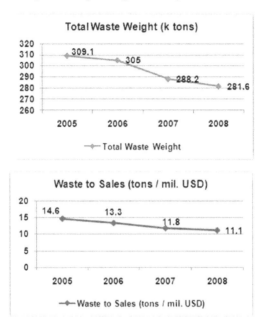

Fig. 4. Total waste production and waste production in relation to sales in years 2005-2008

Emissions – Through a range of energy efficiency programmes, 3M managed to reduce emissions of greenhouse gases by 16% in 2008 compared to 2006 and 69% compared to 1990.

Water - Total water withdrawal, similarly as total water use in relation to sales, showed a negative trend between 2005 to 2008. This was mainly due to introduction of a programme for monitoring and controlling water consumption in 2005 at the corporate level.

8.3.2 Economic results

Operations (See Fig.5.) - Long-term application of programmes for energy efficiency and waste prevention has proved to be particularly important in conditions of surging and fluctuating prices of raw materials. Other positive consequences of these programmes included a decline in recorded work accidents by 7% compared to 2007, with the associated 3.8% decrease in time loss.

Employees – There are many non-financial aspects having direct impact on 3M's employee satisfaction, among the most significant being improved workplace safety, internal communication, motivation, and the overall business image. These factors have been therefore closely linked with the lower turnover of 3M employees.

Customers (See Fig.5.) - 3M's economic success is mainly based on building long-term business relationships with customers who appreciate the world-famous brand of 3M products representing quality, innovation, reliability and sensitivity towards the environment. Since 2003, demand for 3M's products has been growing, with an average growth rate of 6.1% from 2005 to 2008.

Financial results (See Fig.5.) – Due to controlled growth of operating costs and the positive trend in sales, there was a positive trend in 3M's profit in the period from 2005 to 2007. The decline occurred in 2008 due to global economic crisis. This decrease, however, was not so devastating for 3M, again thanks to implemented programmes for energy efficiency and waste prevention. 3M's business results are also translated into interesting gains for investors. In years 2003-2008, the net profit per share rose by 69.8% with an average growth of dividends per share of 51.5%.

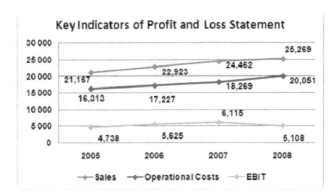

Fig. 5. Trend of key operational business indicators in years 2005-2008

Indicators of eco-efficiency

3M has been successful in improving its eco-efficiency, respectively in reducing the ratio of resource consumption / production of waste and achieved sales.

9. Conclusions

In the current business environment, there is still plethora of companies spurning their environmental responsibility that, as they believe, only represents another financial burden and is in contradiction with economic profit. Therefore, the aim of our research was to testify and demonstrate the positive impact of environmentally responsible behaviour on economic performance of a company. Within our research we undertook a questionnaire survey of some 200 companies in the Czech Republic, carried out an inventory of current environmental performance concepts and tools, and analysed benefits stemming from voluntary environmental instruments applied in business and production processes. Having been cognisant of the contemporary situation, we started developing a conceptual model to illustrate the interrelation between environmental and economic performance of a company. The B2En Performance Development model proposed by the authors, as presented in this book chapter, links the Balanced Scorecard model with the Eco-efficiency concept. The proposed model, comprising multi-criteria index CEPI , Eco-efficiency Statement, and a spread sheet based software tool, enables identification of relations between environmental objectives and KPI of individual BSC perspectives and demonstrates so a positive impact of proactive environmental actions on business performance. The developed model is essential for a complete attitude change of business-community towards environmental responsibility and re-thinking production and consumption patterns, playing therefore an important role in theory and practice of economic performance.

Given the breadth and depth of this topic, however, there are still significant research gaps that need to be addressed, e.g. modification of the B2En model for application in small companies, application of the model on a specific process or product, or detailed look at correlation of links between the environmental and economic perspective at the level of a specific environmental objective.

10. References

Czech Ministry of Environment. (2003). *Eco-labelling*, Czech Ministry of Environment, Retrieved from
<http://www.mzp.cz/osv/edice.nsf/FD52821067CF70CFC1256FF9003D4A7D/$file/Ecolab-final.pdf>
GRI. (2006). *Indicator Protocols Set Environment (EN)*, Global Reporting Initiative, Retrieved from http://www.globalreporting.org/NR/rdonlyres/2FB8358D-293C-4F13-85B6-668292487667/0/G31EnvironmentIndicatorProtocols.pdf International Institute for Sustainable Development. (2007). Eco-Efficiency, In: *Business and Sustainable Development*, 26.4.2010, Available from
<http://www.bsdglobal.com/tools/bt_eco_eff.asp>
Kaplan, R.; Norton, D. (2002). *The Balanced Scorecard: Strategic business performance measurement system* (3rd edition), American Management Association, ISBN 80-7261-063-5, New York
Lehni, M. (2009). *Eco-efficiency - creating more value with less impact*, World Business Council for Sustainable Development, Retrieved from
<http://www.wbcsd.org/web/publications/eco_efficiency_creating_more_value.pdf>
Müller, K.; Sturm, A. (2001). *Standardized Eco-efficiency Indicators: Report I – Concept Paper*, Ellipson AG, Retrieved from
<http://www.ellipson.com/files/studies/EcoEfficiency_Indicators_e.pdf>
Remtová, K. (2003a). *Cleaner Production*, Czech Ministry of Environment, Retrieved from http://www.env.cz/osv/edice.nsf/e26dd68a7c931e61c1256fbe0033a4ee/820af3233682e83ec1256fc0004eaf10?OpenDocument Remtová, K. (2003b). *Ecodesign*. Czech Ministry of Environment, Retrieved from
<http://www.env.cz/osv/edice.nsf/da28f37425da72f7c12569e600723950/7907a38f19e1d57ec1256fc0004fe74d?OpenDocument>

Thoresen, J. (1999). Environmental Performance Evaluation - Tool for Industrial and Improvement. *Journal of Cleaner Production*, Vol.7, No.5, (October 1999), pp. 365-370, ISSN 0959-6526

Verfaillie, H.; Bidwell, R. (2000). *Measuring Eco-efficiency: a guide to reporting company performance*, World Business Council for Sustainable Development, Retrieved from <http://www.wbcsd.org/web/publications/measuring_eco_efficiency.pdf>

8

Embedding Sustainable Development in Organizations Through an Integrated Management Systems Approach

Miguel Rocha[1] and Cory Searcy[2]
[1]ITESM Campus Querétaro
[2]Ryerson University
[1]México
[2]Canada

1. Introduction

The concept of sustainable development (SD) was popularized by the publication of the World Commission on Environment and Development's (WCED) report Our Common Future in 1987 (WCED, 1987). There has been considerable debate regarding the meaning of SD since the publication of Our Common Future. However, the definition provided in that report remains the most widely-cited definition: "development that meets the needs of the present without compromising the ability of future generations to meet their own needs" (WCED, 1987). Building on that definition, there have been many efforts to elucidate the key components of SD. The WCED suggested that SD involved the simultaneous pursuit of economic, environmental, and social goals. These three areas are commonly referred to as the "three pillars" of sustainable development. Gladwin et al. (1995) proposed five principal components of SD: inclusiveness, connectivity, equity, prudence, and security. Additional conceptions on the key principles of SD are widely available in the literature (see, for example, Dresner, 2002).

Although early efforts focused on applying SD to the national and regional levels, it is increasingly being applied at the organizational level (Shrivastava, 1995). Several theoretical frameworks have been used to explore why organizations commit to SD. For example, Bansal (2005) demonstrated that both resource-based (Barney, 1991) and institutional (DiMaggio and Powell, 1983) factors influence SD at the corporate level. Perhaps the most widely-used theoretical framework for explaining organizational SD is stakeholder theory (Freeman, 1984). Stakeholder theory recognizes that organizations have obligations to many individuals and groups, including (but not limited to) shareholders, customers, employees, and the wider community. Building on these theories, several authors have sought to clarify why organizations would operate in environmentally- (Bansal and Roth, 2000) or socially-friendly (Campbell, 2007) ways. These motivations have provided a basis for research on the business case for SD (Salzmann et al. 2005).

Recently, research on organizational SD has begun to shift from why SD should be implemented at the organizational level to how this can be accomplished. In this light, there is a growing stream of research on standardized management systems for SD. The literature highlights that one possibility is developing a stand-alone standard for SD. Singh et al. (2007) provide an example of how this may be accomplished. The literature also highlights the possibility of integrating the principles of SD with existing management system standards (MSS), such as ISO 9001, ISO 14001, and OHSAS 18001, among others. This research builds on wider research on integrated management systems (IMS). Examples are provided by Rocha et al. (2007) and Oskarsson and Malmborg (2005), among others. However, while much research has been conducted on how SD can be implemented at the organizational level, work remains.

This paper contributes to these efforts. The paper argues that an IMS-based approach can be used to embed SD in organizations. The focus on IMS is in recognition of the point that existing MSSs, such as ISO 9001 and ISO 14001, may provide needed leverage points for integrating SD with mainstream organizational issues. It also recognizes that SD should not be seen as a stand-alone initiative, which may be encouraged through the development of a separate MSS focused on SD. An IMS approach provides opportunities to explicitly link SD with existing organizational goals, policies, programs, processes, procedures, and resources. However, research on the application of an IMS-approach to organizational SD is still in its relatively early stages. While several IMS models have been proposed in the literature, they have not been systematically evaluated with respect to their potential to embed SD in organizations.

2. Literature survey

The concept of IMS initially emerged about 15 years ago. Early efforts focused on the integration of ISO 9001 and ISO 14001, though other MSS are increasingly being taken into account in the IMS literature. An increasing body of knowledge is available in the specialized literature containing information about the potential benefits and limitations of IMS; IMS models; and empirical results of implementing IMS in specific organizations (see, for example, Wilkinson and Dale, 2001; Karapetrovic and Willborn, 2002; Scipioni el al., 2001; Rocha et al., 2007; Asif et al., 2009). Further details on the concept of integration, models, methodologies, potential benefits, and lessons learned are provided below.

2.1 The concept of integration

There are many different definitions of IMS in the literature. These differences reflect different approaches and strategies for integration. The differences start with the concept of "integration" and the possible equivalent use of the terms "alignment" and "merge". For instance, integration was defined as the "degree of alignment or harmony in an organization - whether different departments and levels speak the same language and are tuned to the same wavelength" (Garvin, 1991). Alignment has been at the center of the ISO approach in developing updated versions of ISO 9001 and ISO 14001. It has been noted that this has in turn created additional opportunities for an aligned Environmental + Quality MS (Scipioni et al. 2001). Integration through the merging of two standards into one has been explored in the integrated auditing guidelines provided in ISO 19011:2002. However, recent efforts on industry specific standards, such as ISO/TS 16949 for the auto industry or ISO 22000 for food safety, indicate a further proliferation of individual standards rather than a move towards a consolidated set of standards. In recognition of this trend, a comprehensive approach for integration has been developed around the concept of systems theory. In their seminal work on IMS, Karapetrovic and Willborn (1998) defined integration as "linking two systems in a way that results in a loss of independence of one or both means that these systems are integrated" (Karapetrovic and Willborn, 1998). In a similar manner, Bernardo et al. (2008) explained that integration is "a process of linking different standardized MSs into a unique MS with common resources aiming to improve stakeholders' satisfaction". Thus, recent research has generally focused on the integration of management systems, rather than management standards. This is a key distinction.

According to Jonker and Karapetrovic (2004) two elements are required to integrate MSs: (1) a model describing the MS elements and their relevant interactions and (2) a roadmap or methodology showing the process for implementing the model. Although this may seem obvious, relatively few papers actually describe a model for integration of MSs (Karapetrovic and Willborn, 1998; Scipioni et al. 2001; Wilkinson and Dale, 2001; Rocha et al. 2007; Asif et al. 2009; Lopez-Fresno, 2010) and fewer still elaborate proposals for the second requirement (Lopez-Fresno, 2010; Asif et al. 2009; Rocha and Karapetrovic, 2006). These issues are briefly explored in further detail below.

2.2 IMS models

One stream of research on IMS focuses on the development of IMS models. The underlying emphasis on IMS models is generally on achieving integration that goes beyond the development and use of a unique MS manual and supporting documentation system towards the integration of selected functional requirements into the organization structure. To accomplish this, IMS models generally focus on identifying and building on the key management systems elements that are common to all of an organization's initiatives. These elements vary by model. For example, Karapetrovic and Willborn (1998) focused on a systems approach organized around three key elements: goals, processes, and resources. Wilkinson and Dale (2001) proposed a total quality management approach structured around seven key elements: policy, leadership, resources, processes, culture, goals, and stakeholders. Rocha et al. (2007) proposed a model organized around the following elements: stakeholders, resources, leadership, processes, values, objectives, and results. Additional examples are available in the literature. In any case, all models must be able to accommodate the inclusion of current and new MSS, harmonize differing requirements of MSS, and support IMS implementation and improvement.

IMS models are usually designed with a specific scope in mind. The most common starting point for an IMS is using an ISO 9001-compliant QMS that is already in place. This is sensible since more than 1,200,000 organizations worldwide have implemented a QMS based on this standard. According to empirical research done in different countries, EMS and OHSMS have been selected as the preferred MSs to be integrated with QMS (Harjeev et al., 2010; Griffith and Bhutto, 2009; Zutshi and Sohal, 2005; Beckmerhagen et al. 2003; Lopez-Fresno, 2010). There is increasing interest in including social-focused MSSs into the integration mix. ISO 26000, AA1000 and SA8000 have been mentioned as potential candidates for companies willing to tackle the needs of their community (Rocha et al. 2007).

2.3 IMS methodologies

Another stream of research focuses on the development of IMS methodologies. As Karapetrovic (2003) notes, a generic methodology would address (as a minimum) model selection, standard(s) selection, IMS implementation, and IMS audits. Rocha and Karapetrovic (2006) have further noted that having a methodology may increase the attractiveness of integration to companies but more detail is required to address "how to" questions such as: flexibility to cover different starting points (MSs already in place), differences on organizations final scope (QMS/EMS/OHSMS + others), links to overall business strategy, and culture change required for assimilating new roles, among others. As noted above, relatively few papers explicitly address these issues. For example, published methodologies include a PDCA-based implementation process developed by Scipioni et al. (2001); a flexible three-phased IMS implementation process (Rocha and Karapetrovic, 2006); and an implementation process for the PEDIMS model designed by Asif et al. (2009). In any case, all methodologies must be able to illustrate how to put a function-specific MS together while allowing for differing initial organizational conditions and objectives. However, none of the published methodologies have been implemented; thus, they remain unproven. Empirical evidence from experiences of Spanish and Australian companies indicated that IMS implementation requires top management commitment through an appointed integration champion, training to reduce anticipated problems, and the deployment of essential resources (Zutshi and Sohal, 2005). Furthermore, a cellular-like implementation pilot project helps to reduce uncertainty and increase efficiency, while risk assessment enables a reduction of potential problems (Lopez-Fresno, 2010)

2.4 Potential benefits of integrating MSs

Initially, the literature tended to emphasize operational efficiency and effectiveness as the main factors in the promotion of IMS. Over the last several years, additional benefits have been discussed in theoretical and empirical papers. Table 1 shows a summary of the potential benefits of integrating quality, environmental, occupational health and safety management systems, and other management systems. As Table 1 illustrates, the benefits of IMS have been organized in this paper around the three pillars of SD: economic, social, and environmental benefits. A fourth category, operational benefits, was added to include those benefits that serve as enablers for improved performance in those three dimensions.

Economic Benefits	Social Benefits	Environmental Benefits	Operational Benefits
• Reduction in duplication of policies, procedures and work instructions • Time savings • Reduced operational costs	• Increased transparency • Enhanced internal communication • Facilitation of cultural change in the organization • Potential image benefits	• Increased prominence of environmental issues in organizational management • Increased emphasis on compliance with applicable regulatory requirements	• Increased synergy between MSS • Reduced audit fatigue • Explicitly shows how the MSS relates to the rest of the business • Clarification of responsibilities • Improved information flow

Adapted from Zutshi and Sohal (2005); Lopez-Fresno (2010); Harjeev et al. (2010); Griffith and Bhutto (2009)

Table 1. Illustrative Summary of Key Benefits of an IMS Approach

2.5 Lessons learned for integrating MSs

It is important to acknowledge that an IMS approach is not a cure for all problems facing an organization. When integrating MSs organizations may face new obstacles that go beyond their previous experiences with isolated MSs. For example, employees may see an increase in workload and responsibilities. Some of the other key barriers to implementing an IMS include the different nature of individual systems, employee resistance, lack of resources, post implementation difficulties, and organizational culture aspects (Asif et al. 2008, Zutshi and Sohal, 2005, Griffith and Bhutto, 2009).

3. Evaluation of IMS models with respect to sustainable development

The notion of applying an IMS approach to organizational SD has been recognized by a number of authors, including Rocha et al. (2007), Jorgensen (2008), Oskarsson and Malmborg (2005), and Fresner and Engelhardt (2004). This is in recognition of the point that the infrastructure provided by the existing MSS provides opportunities to structure the implementation of SD at the organizational level. However, not all of these papers proposed a specific model for IMS. Nonetheless, the literature survey shows that several IMS models have been developed. The objectives of these models range from the original IMS goal of operational efficiency improvement to current views where SD, corporate responsibility and labor rights are added in an effort to reflect a fast-paced sustainable-oriented market. The IMS models that were found to include a summarized version of key elements and their interactions include: (1) the "Systems approach" developed by Karapetrovic and Willborn (1998, 2001); (2) the "Total quality approach" designed by Wilkinson and Dale (2001); (3) the EQOHSMS model presented by Scipioni et al. (2001); (4) the "Rotor" model developed by Rocha and Karapetrovic (2005, 2006); (5) the "Airline applied" IMS model shown by Lopez-Fresno (2010); and (6) the "Systems approach to integration" model developed by Asif et al. (2010). For a summary of these models see Table 2. To date, no systematic evaluation of the ability of these models to address organizational SD has been conducted.

3.1 The concept of integration

To analyse the ability of these IMS models to embed the principles of SD in organizations, a set of criteria was developed. The criteria were designed assuming a need to be concise and to meet current and future needs of organizations employing an IMS approach to SD. The criteria were divided into two broad categories: management requirements and SD requirements. It should be noted that it is recognized that modifications to the criteria are possible. Additional criteria could be developed and additional questions to guide the analysis could also be created. Nonetheless, the criteria do provide a starting point for structuring the analysis of the existing IMS models with respect to organizational SD. With that in mind, the criteria are introduced below.

1. **Management requirements**: Criteria in this category were focused on ensuring that the IMS model was capable of addressing the diverse requirements of standardized MSSs. The clarity of the model was also an overarching emphasis in this category. With that in mind, several questions were used to guide the assessment of these criteria: Does the IMS model accommodate the requirements of current MSSs? Are the IMS elements clearly explained? Is the scope of the model clear? Does the IMS model provide linkages

to overall business strategy? How do the IMS elements interact to produce the planned objectives? Does the model accommodate different degrees of integration? What lessons have been learned from the application of the model in practice (if applicable)? Does the model address the need for MSs to evolve over time?

To structure both management and integration of MS requirements the criteria were organized around the ISO quality management principles outlined in ISO 9000:2005.

Models (Authors, year)	Key elements	Integration approach
The Systems approach Karapetrovic & Willborn (1998, 2001)	Goals, processes and resources	A generic system connected by a flow of resources transformed by processes to accomplish objectives.
Total quality approach Wilkinson and Dale (1999)	Leadership, stakeholders, integrated processes, resources, goals, infrastructure	A generic system calling for full integration of resources, processes and structure. Special emphasis on cultural issues as foundation for the IMS
The EQOHSMS Scipioni *et al*, (2001)	Structure similar to ISO 9001:2008: Management responsibility; resource management; product realization; measurement analysis and improvement	A system based on ISO 9001:2008 tenets such as processes and systems. Scope limited to Quality, environment and Occupational Health & Safety
The Rotor model Rocha et al (2005, 2006)	Stakeholders providing resources to processes and directed by leadership producing results (rotor movement)	Using a dynamic system where results and stakeholder engagement are included integration is achieved for quality, environment, social and other functions
The Airline applied model Lopez-Fresno (2010)	Global framework and function-specialized modules. The global framework contains: Organisation and policies; planning; resource management; process and activity management; activity evaluation; continuous improvement; Relationship with the authorities	Integration is done at high level processes as described in global framework. However, specific objectives can be managed through specialized programs such as maintenance.
The Systems approach to integration Asif et al (2010)	Stakeholders, requirements, business strategies, management subsystems, documentation, operations, feedback	Integration is achieved at high and low levels on stakeholder requirements and business strategies. Flexibility is provided through management sub-systems.

Table 2. Summary of relevant IMS models

These were chosen due to their wide applicability to MSSs and the inclusion of both the process and the systems approaches, which are the foundation for true integration (Karapetrovic, 2003). Seven of the eight ISO quality principles formed an evaluation category, namely: leadership, systems focus, process approach, human resources focus, building partnerships, factual decisions, and continual improvement. The associated evaluation criteria were further developed based on the literature, particularly that focused on existing MSSs (such as ISO 9001, ISO 14001, OHSAS 18001, SA8000, and AA1000), business excellence models (MBNQA and EFQM), and empirical studies focused on the implementation of MSSs (see, for example, Bernardo et al. 2007). For example, the first principle "leadership" deals with the role of a proactive and dynamic top management leadership. To test how each IMS model deals with this sub-category there are questions focused on issues such as if the model is actually linked to organizational strategic planning; if leadership sets up an integrated policy; if leadership commitment is provided and, if so, how this is done. Integration at this level is vital for the IMS to succeed thus the questions also seek to explore the degree of integration in establishing policies and objectives as well as planning and reviewing the system's performance. The complete set of evaluation criteria for management requirements are provided in Table 3.

2. **Sustainability requirements**: Criteria in this category were focused on ensuring that the IMS model was capable of accommodating the principles of SD at the organizational level. The models' explicit focus on SD was an overarching emphasis in this category. With that in mind, several questions were used to guide the assessment of these criteria: Does the model provide a basis for addressing the key principles of SD? Does the model emphasize the importance of transparency? Does the model explicitly acknowledge the importance of stakeholder participation in the IMS? Does the model accommodate different degrees of stakeholder interaction? Does the model provide a basis for balancing organizational objectives in the decision making process? Is the application of the model to organizational SD discussed?

To provide a structure for addressing these questions, the criteria were organized around three sub-categories: environmental, economic, and social responsibilities. These categories build on the key requirements for SD outlined by the WCED and are closely linked with the "triple bottom line" of organizational SD. They were selected due to their widespread association with SD and their general applicability. The sub-categories and associated evaluation criteria were further developed based on the literature, particularly literature focused on stakeholder theory and corporate sustainability. For example, the social responsibilities sub-category deals with the need to set relevant objectives, to develop indicators to measure progress towards those objectives, to meaningfully consult with stakeholders, and to emphasize the importance of transparency in organizational decision making. The complete set of evaluation criteria for SD requirements are provided in Table 4.

1. Leadership actions
Does the model
1.1. Encourage linking the IMS to the overall business strategic planning?
1.2. Require a balanced and integrated policy?
1.3. Require a leadership system to set up and deploy IMS objectives?
1.4. Ask for a system owner or champion?
1.5. Integrate and balance IMS goals?

1.6. Require top management to provide and deploy needed resources and infrastructure?

2. Systems focus
Does the model
   2.1. Have explicit boundaries?
   2.2. Include elements other than processes and activities?
   2.3. Show interactions among the model elements?
   2.4. Show sub systems and meta-systems?
   2.5. Show type and degree of integration between functional MSs?

3. Process approach
Does the model
   3.1. Require identifying the organizational processes for realizing products and services?
   3.2. Follow the main organizational processes along the supply chain (from supplier to customer passing through stakeholders)?
   3.3. Include supporting processes such as finance, sales, IT and others?
   3.4. Deploy IMS objectives along regular processes without the need for "special programs" when possible?
   3.5. Follow the PDCA cycle to deploy IMS processes?
   3.6. Integrate documentation and activities along the processes?

4. Human resource focus
Does the model
   4.1. Address the need for a skilled human resource through recruitment and training?
   4.2. Require the workforce to become aware of stakeholders needs?
   4.3. Establish the need for roles, responsibilities and authorities for HR?
   4.4. Integrate roles and responsibilities along process roles and responsibilities?

5. Building partnerships
Does the model
   5.1. Include suppliers as part of the IMS?
   5.2. Encourage construction of working relationships with related stakeholders including customers and suppliers?
   5.3. Include two-sided communication paths with relevant stakeholders?

6. Factual decision
Does the model
   6.1. Require a documentation sub system for recording relevant information?
   6.2. Establish a performance measurement sub system aligned to the IMS goals?
   6.3. Include information analysis requirement and possibly a knowledge management sub system?
   6.4. Integrate preventive and corrective actions as regular IMS elements?
   6.5. Provide guidelines for balancing goals in the decision making process?

7. Continual improvement
Does the model
   7.1. Require a feedback loop for continual improvement?
   7.2. Integrate continual improvement in a balanced manner for IMS goals?
   7.3. Require a systematic top management / review of the system performance?

Table 3. Evaluation set of criteria for IMS completeness

1. Environmental responsibilities
 Does the model
     1.1. Explicitly seek to minimize or eliminate negative environmental impacts from organizational activities?
     1.2. Allow flexibility to choose relevant and specific environmental objectives?
     1.3. Deploy environmental objectives into IMS elements?
     1.4. Encourage identification, communication and partnership with environmental-based stakeholders?

2. Social responsibilities
 Does the model
     2.1. Establish specific social requirements both internal and external to organizations?
     2.2. Cover specific social indicators or give flexibility to do so?
     2.3. Integrate social-focused objectives into IMS elements?
     2.4. Require identification, communication and partnership with social-based stakeholders?

3. Financial responsibilities
 Does the model
     3.1. Explicitly establish financial responsibilities for both organizations and community?
     3.2. Allow a flexible range of financial responsibility indicators?
     3.3. Integrate financial-focused objectives into IMS elements?
     3.4. Require identification, communication and partnership with social-based stakeholders?

Table 4. Evaluation set of criteria for SD

3.2 Evaluation and information analysis

Each of the six identified IMS models was evaluated on the basis of the criteria with respect to their ability to act as a potential platform for implementing SD in organizations. A summary of the evaluation is presented in Tables 5 and 6. Table 5 provides an evaluation of the IMS models focused on the total quality approach (Wilkinson and Dale, 2001), the rotor model (Rocha et al. 2007), and the systems approach to integration (Asif et al. 2010). Table 6 provides an evaluation of the IMS models focused on the systems model (Karapetrovic and Willborn, 1998, 2002), the airline model (Lopez-Fresno, 2010), and the EQOHSMS model (Scipioni et al. 2001).

The evaluation of the IMS models provides a general view of the adequacy of the models as a platform for embedding SD into organizations. It is important to emphasize that the evaluation focuses on the features of an IMS model that may enable deployment of SD in organizations. For example, it focuses on the integration of an array of stakeholder requirements into organizational objectives, as well as the characteristics that may hinder that endeavor. The evaluation does not focus on identifying the "best" IMS model for the implementation of SD. With that in mind, some of the key observations from the evaluation are discussed below.

IMS DESIGN	Total Quality approach (Wilkinson & Dale, 2001)	The "rotor" Model (Rocha et al, 2007)	Systems approach to integration (Asif et al, 2010)
1. Leadership	Starts with leadership as a driver for resources, aims and objectives. A single policy is mentioned. No information is given about details on how leadership should be exerted.	Leadership determines organizational values and objectives for processes. An integrated policy is deployed as leadership activity. Also it requires a management representative for the system. No information is given about the definition of a balanced array of goals as a leadership element.	This model draws business strategies from identification of stakeholders' requirements. Goals and business strategy are integrated but at the tactical and operational level the model still shows several MSs, as many as different stakeholders' requirements. No mention is made establishing a policy, management representative and a balanced goal oriented performance measurement.
2. System focus	The entire organization is the system's boundaries. The model includes links with environment and elements such as organizational culture. Processes are integrated around a PDCA based cycle.	The model has a flexible boundary depending on each organization: from two MSs to several and from one location to an entire corporation. The model includes stakeholders as important elements. Full integration is encouraged since no functional sub systems are kept.	The system scope is the entire organization, where stakeholders are the main driver for strategy of "n" management sub systems. Integration of management sub systems happen at the operational level: a single manual and integrated procedures.
3. Process approach	PDCA based processes are at the center of the model. No information is given about the role of supporting processes. Objectives are a single input for the processes with actual outputs as the result. No info about documentation, activities, or programs is included.	PDCA based processes are at the core of the IMS model. The processes follow the supply chain structure too. No special programs are required to isolate stakeholders' needs. Documentation reqs. are deployed as a necessary process supporting subsystem.	Processes are the operational core of the IMS. It seems that only operational processes are included in the system. More detail is required to show how processes are deployed from each management sub system. Procedures and manual are integrated.
4. Human resource focus	No information about HR requirements within the model, however, it does include organizational structure and culture to promote people involvement.	They are included as a subset of the system resources and also as relevant stakeholders. They need to be recruited and trained to fulfill their roles, responsibilities and authorities (shown by the deployed ISO clauses).	No mention is made HR

IMS DESIGN	Total Quality approach (Wilkinson & Dale, 2001)	The "rotor" Model (Rocha et al, 2007)	Systems approach to integration (Asif et al, 2010)
5. Building partnerships	Stakeholders are included as receivers of processes outputs although is left unspecified about the type of stakeholders. No specific communication paths are shown	Stakeholders are explicitly included in two roles: drivers and receivers of the system. Engagement is considered mandatory. Communication is paramount and partnership is sought by inclusion into the system.	Although stakeholders are included in the model no partnership is sought. Rather, stakeholders seem to be solely customers from the system.
6. Factual decision	Measuring, improving and auditing elements are at the center of the processes. No information about documentation or decision making process is found in the model.	Implementation and measurement process steps are dedicated to documentation subsystem. Measurement is a subsystem of product and process performance. No information on decision making process is provided	Composite records are required for the model but no action other than feedback is explicitly included. Information analysis or a performance measurement system are also missing in the model.
7. Continual improvement	The entire system has a continual improvement loop which is also found at the processes core. No details on specific improvement elements are included.	Improvement of processes is included in the Act component of the PDCA cycle. System results are also compared with stakeholder needs and expectations.	To improve the system performance two coordination directions exist: horizontal covering the system scope and vertical taking care of the deployment of stakeholders needs. A feedback loop is also included but no details are provided.
SUSTAINABLE DEVELOPMENT			
8. Environmental responsibilities	Similar to ISO 14001, organizations can define relevant environmental objectives. No information on how the environmental objectives are deployed into IMS elements. Similar case for stakeholders partnership	Similar to ISO 14001 organizations can define relevant environmental objectives. Identified processes are analyzed to identify and implement environmental requirements. Similar case for stakeholders partnership	Similar to ISO 14001 organizations can define relevant environmental objectives. The communication line seems to be one-sided top down as a customer with no partnership required.
9. Social responsibilities	The model's social scope includes solely for worker health and safety. CSR, labor rights and other social responsibilities are out of the model scope.	The model's social scope includes health and safety of workers, CSR and labor rights as defined by MSSs such as OHSAS 18001, AA1000, SA800 respectively.	Ethics, sustainability, and health and safety are included as stakeholders' requirements. However, it is unclear how these requirements are actually deployed and stakeholders are included as part of the system rather than just being system clients.

IMS DESIGN	Total Quality approach (Wilkinson & Dale, 2001)	The "rotor" Model (Rocha et al, 2007)	Systems approach to integration (Asif et al, 2010)
10. Financial responsibilities	Not included in the model	Not included in the model	The model enlists an unlimited number of stakeholders; owners, stockholders, and community. However, no information as to how these requirements may actually be deployed into IMS elements is provided.

Table 5. IMS evaluation results

IMS DESIGN	The "systems model" for IMS (Karapetrovic & Willborn, 1998, 2002)	The "airline applied" IMS model (Lopez-Fresno, 2010)	The IMS model – E/Q/OHS (Scipioni et al, 2001)
1. Leadership	Goal management is the starting point of the model. Linkage with business strategies is missing. Integration happens for policy and targets. Elements missing are: system ownership; balance goal, and leadership tasks.	The system sets organization and policies as set in its global framework, addressing strategic planning from the corporate view. No details about leadership system, system ownership and tasks are included in the model.	Management responsibility is driving the model, including sub elements as described in ISO 9001. Links to business strategy and goal balance are missing from the model.
2. System focus	Boundaries are defined by organizations but the example is limited to QMS/EMS. The system includes goals, processes and resources in a closed loop. No mention about single management subsystems.	The model defines the whole airline corporation as the system. It contains system elements cluster in a global framework + functional sub systems such as maintenance, flight, and security due to the legal relevance.	The system boundaries are quality, environment and health and safety requirements. The structure is highly based on ISO 9001 integrating all requirements around them. The organization is not recognized as the meta system.
3. Process approach	Planning, designing and implementing processes are at the systems core. No discrimination about types of processes is included. In the 2002 version also control and improvement is incorporated, showing deployment of documentation requirements.	Inside the global framework management of processes and activities are included following a PDCA cycle. A corporate single complemented by a number of specific functional manuals exist covering the whole organization.	Processes are found in "product realization" element from design to delivery. As ISO 9001 processes follow a PDCA cycle. No detail is given as to how IMS requirements are deployed into the set of processes.
4. Human resource focus	It is included in resources management. Allocation and deployment are required where training	The model description states that resource management includes HR. The system manual	A single clause for human resource is found similar to the requirement shown in ISO 9001. No detail is

IMS DESIGN	The "systems model" for IMS (Karapetrovic & Willborn, 1998, 2002)	The "airline applied" IMS model (Lopez-Fresno, 2010)	The IMS model – E/Q/OHS (Scipioni et al, 2001)
	and roles + responsibilities are defined.	establishes corporate and functional responsibilities.	provided for the way HR is ready for an array of requirements beyond quality.
5. Building partnerships	Stakeholders are included in the 2002 model. However, no partnership is explicitly sought. Communication lines are setup only for goal management.	No detail is provided about stakeholders beyond compliance with regulatory requirements. Suppliers and other stakeholders are not explicitly included in the model.	Stakeholders are considered only as systems customer but no real partnership is required.
6. Factual decision	Control and improvement is done according to ISO 9001 requirements. No guidelines for balancing goals in the decision making process is included.	Documentation is spread in global framework and specific modules. Given the nature of the air transportation sector preventive and corrective actions are included in the model. No goal balance strategy is provided.	Documentation is maintained as required in ISO 9001 helping to take decisions for corrective, preventive and improvement actions.
7. Continual improvement	The system has a closed feedback loop that assumes continual improvement approach. Management review is performed following ISO 9001 requirements. No mechanism for goal balance is provided.	Continual improvement is engraved into the global framework. No detail on how this is done and whether or not applies to functional modules.	Continual improvement requirement is included in the model. However, no detail is provided on the mechanism to balance this assorted array of performances.
SUSTAINABLE DEVELOPMENT			
8. Environmental responsibilities	Environmental objectives as mentioned by ISO 14001. Processes are used as guidelines to identify and implement environmental requirements. Similar case for stakeholders partnership.	Environmental objectives as mentioned by ISO 14001 which are regulatory for this industry sector. Encourage close relationship with authorities.	Environmental objectives as mentioned by ISO 14001. No information is provided for communication and partnership with environmental stakeholders.
9. Social responsibilities	Although indicated as possible no requirements are shown for social accountability and health and safety of workers.	Air safety is considered as a system objective. No other social requirement is mentioned in the model.	Social responsibilities are reduced to health and safety at the workplace. No detailed information is provided on how safety requirements are deployed into the IMS
10. Financial responsibilities	Although indicated as possible no requirements are shown for financial responsibilities whatsoever.	Not included into the model.	Financial responsibilities are not included. Lateral impact from quality efforts for customer satisfaction and cost reduction.

Table 6. IMS evaluation results (cont)

a. Strong features of the models for SD:

From the evaluation of the management requirements, it was found that most of the IMS models call for strong leadership to drive the system. In all six of the IMS models evaluated, leadership is exerted by setting up integrated policies and objectives, which led in turn to the allocation and deployment of required resources into appropriate structures (Quinn & Dalton, 2009). The models also require top management to define the system scope according to the organizations´ needs and evolution. To ensure management commitment, IMS models typically require the assignment of a management representative capable of working across organizational, national and international boundaries to achieve stated objectives.

To varying degrees, all six IMS models employed a systems approach to develop the holistic vision that sustainability requires to be successful (Goel, 2006). Most of the IMS models have a flexible scope that depends on the current organizational needs and possibilities. However, IMS models also call for scope expansion towards the whole organization and increasing stakeholders´ requirements (Karapetrovic & Willborn, 1998; Rocha et al., 2007; Lopez-Fresno, 2010). However, as mentioned by Senge et al (2007), "systems thinking can be messy and uncomfortable". Inclusion, relationship building and true engagement of stakeholders, as done at different depth levels in all six IMS models, is a direct result of systems thinking and an enabler for sustainability (Roome and Bergin, 2006; Senge et al, 2007; Pepper and Wildy, 2008; Quinn and Dalton, 2009)

Within their system requirements, all IMS models have processes as the building blocks for fully deploying stakeholder requirements in both operational and supporting activities. As mentioned by Lueneburger & Goleman (2010) sustainable development needs a "specific set of business processes geared to manage previously unquantified risks and capture new opportunities". All processes may be organized according to the PDCA cycle, which is common in IMS frameworks. Such an approach emphasizes planning as an important activity before taking any substantive action. As pointed out by Quinn and Dalton (2009) sustainability requires timing and readiness in their activities; an organization that does plan according to the opportunities and positive outcomes has a better chance to succeed.

The IMS models generally emphasize the importance of skilled human resources that are aware of an assorted array of functions (such as quality, environment and so on). Employees are a key stakeholder that must be engaged in collaborative action along the processes mentioned above (Senge et al, 2007; Pepper & Wildy, 2008). Partnerships with stakeholders are included in the IMS framework however at different levels; only the "Rotor" Model explicitly includes stakeholders as part of the system by providing resources (Rocha et al. 2007). Other models include stakeholders, but only as a receiver of the system outputs. Engagement with stakeholders has been identified by several authors as an essential element for sustainability (Roome and Bergin, 2006; Senge et al, 2007; Pepper & Wildy, 2008; Quinn & Dalton 2009).

Driving sustainability throughout an organization requires a deep knowledge of sustainability (Pepper & Wildy, 2008). IMS models can help facilitate a process of embedding sustainability in organizations through enhanced training in SD issues, process documentation, and measurement and analysis of processes outputs. All together these elements help enable factual-based decision making and an increased body of knowledge for SD within the organizations.

From the evaluation of the SD requirements, it is evident that environmental responsibilities are largely accounted for through the explicit incorporation of ISO 14001 into the IMS. Social requirements are also included in several of the IMS models, though this is generally to a lesser degree than environmental issues. Social accountability, occupational health and safety, labor rights and decent work are the most common social requirements included. All of these issues have international standards that facilitate their inclusion into the IMS models.

b. Weak aspects of the models for SD:

From the management side the analysis shows that, although an IMS does have *leadership* requirements to drive sustainability, it still falls short of the level required to succeed. Most of the IMS models lack of guidelines for objectives that balance priorities between financial, social and environmental dimensions. To make sustainable development sustainable Quinn & Dalton (2009) indicate that there are two options: organizations should look for "solutions both sustainable and economically profitable" or change the objective measurement from economic based to sustainability based. Either way SD requires a strong integrated performance measurement which is non-existent in all six models. Furthermore, sustainable development requires being part of the strategic planning of the organization (Pepper & Wildy, 2008); nonetheless only two models, the "airline applied" and the "systems approach", explicitly include strategy planning as part of the IMS elements.

Due to their integrative nature, IMS models encourage synergy and holistic vision. However, sustainability goes beyond companies' walls and even suppliers and customers to include more active relationships with stakeholders into the system (Roome & Bergin, 2006; Quinn & Dalton, 2009). In all six IMS models more detail as to how IMS elements (e.g. processes, documentation, and measurement) are integrated into a unique system is required. Two models, the "airline applied" and the "system approach", describe modules for particular functions, thus allowing certain flexibility, yet more detail on how they are integrated is required.

An element that needs to be integrated into the system is the set of supporting processes (finance, marketing, IT) which helps in engaging internal and external stakeholders. All analysed IMS models include "processes" at the general level, leaving open to interpretation which processes are included in the search for sustainability. IMS models also show a fragile structure for sustainability in their integration of human resources (as partners) and other stakeholders into the system. Most of the IMS models consider stakeholders just as system clients yet their role as enablers, resource providers and doers is not included or at least diminished. Several authors emphasize the importance of stakeholders' integration as partners as an essential element to accomplish sustainable development (Roome and Bergin, 2006; Senge et al, 2007; Pepper & Wildy, 2008; Quinn & Dalton 2009).

Broadly speaking, the IMS models still need to address similar issues as those highlighted in empirical studies: obtaining real top management commitment, aligning with business strategies, focusing on training, integrating around processes rather than divisions, and the creation of new functions. It is interesting to note similarities between the suggested improvements for the IMS models and the results from empirical studies on SD implementation. For example, Luenerburger & Goleman (2010) mention identification of risk and opportunities as a first step of a proposed methodology for SD implementation.

Quinn & Dalton (2009) emphasize the need for implementation processes based on positive outcomes and focusing on areas where early success would facilitate more stakeholders participating in the SD process.

Overall, the evaluation of the SD requirements highlighted the lack of emphasis on economic issues in the context of an IMS. Two models indicated the need to develop financial indicators, however, no detail was given as to how they are going to be used in the overall IMS. It is possible that the lack of MSSs on financial management has contributed to this gap. This is a possible concern given the widely held view that financial objectives typically overshadow social and environmental requirements (Roome and Bergin, 2006; Senge et al, 2007; Pepper & Wildy, 2008; Quinn & Dalton 2009). From the two remaining SD dimensions, namely environmental and social issues, the IMS models go no further than briefly stating requirements that the system must address. There is no indication of a management element that helps to prioritize this array of requirements without leaving any of them unattended. A performance measurement system set with a balanced emphasis on the triple bottom line may be part of the answer to help address these issues.

4. Conclusions

A growing number of organizations around the world have made commitments to apply the principles of SD to their operations. Becoming an organization focused on sustainable principles necessitates addressing specified social, environmental and economic objectives. These objectives must be pursued in an integrated manner while drawing on a common pool of resources. For more than 20 years, organizations have employed MSS, such as ISO 9001 and ISO 14001, to meet a portion of these objectives. The infrastructure provided by the existing MSS may be leveraged to help implement SD at the organizational level. Insight into how this may be accomplished is provided by the growing literature on IMS. The concept of an IMS was created to build synergy among MSS, optimizing resources and focusing on meeting an array of different objectives. There are several models available in the literature and an increasing body of knowledge related to their implementation and operation. However, more research is needed on how an IMS approach may be applied in the context of organizational SD.

The purpose of this paper was to explore how existing IMS models can be used to leverage the implementation of organizational SD. Six prominent IMS models were analyzed with respect to their potential to help embed SD in organizations. An original two-prong set of criteria were developed to help guide the analysis. The analysis showed that the existing IMS models do provide a useful starting point in implementing SD in organizations. However, there are numerous opportunities to strengthen the existing models, particularly regarding their application in practice.

The defined set of criteria explored two dimensions of the IMS: first, it analysed the IMS models for management system strength and coherence; second, it evaluated the feasibility to cover SD principles. For the first category (management requirements), seven sub-categories modeled on the quality management principles in ISO 9000 were employed. Each principle was divided into four to six questions that focused on the depth of the management system elements, their interaction, and their level of integration. For the second dimension (SD requirements), the criteria were divided into three categories closely associated with SD, namely environmental, economic, and social issues. Each responsibility was deployed into questions focusing on the ability of the IMS to address triple bottom line issues and stakeholder requirements.

In general, IMS models were found to be a useful platform to develop SD within an organization. The "process" and "system" approaches that organizations are already familiar with create a mindset for integration and synergy that is necessary for the diverse set of requirements SD demands. In the models, "leadership" is exerted by top management by following the PDCA cycle for the entire system and by allocating and deploying resources needed for processes to operate. Decisions are increasingly taken based on facts and analysis; methods such as lean thinking, six sigma, performance measurement systems and others are all based on "factual decision" principles and thus complement the IMS model. The decision making process is solidly focused on "continuous improvement" which should be deployed to the entire organization. However, while these elements provide a basis for integrating SD into an organization's core infrastructure, areas of improvement were also found in this analysis. One key issue was the lack of MS elements to build partnerships with employees, the community, customers, suppliers, and other stakeholders. Until organizations realize partnering is not a choice but a necessity, SD may prove to be an elusive goal. The IMS models recognized human resources as an important resource, but partnership needs to be built into them more explicitly. Another area of improvement is the identification, maintenance, control and improvement of processes, not only those with direct impact on product realization, but also on those supporting the operation such as sales, finance, and marketing. Lack of integration of these processes into the IMS would leave an isolated IMS with small resources and impact on the company's strategy.

Current IMS frameworks have quality, environmental, health and safety and social responsibility within their scope; meaning that two thirds of the TBL range may already be largely (if not comprehensively) covered. However, little emphasis on economic issues was found in the models. This is a significant oversight, which may contribute to the general lack of application of the models in practice. From the environmental side, the widespread requirement for including ISO 14001-compliant MS elements provides a strong base in the existing IMS models for moving towards more explicit recognition of SD. Social responsibilities in the existing IMS models were addressed to varying degrees. The most common approach was to include solely health and safety at the workplace. Social issues such as CSR, labor rights, and social accountability are other established options, but only one third of the models reviewed considered these as possible requirements. Finally, more detail on how requirements are deployed and controlled in ongoing organizational processes is needed in all models. These will help clarify issues such as the importance of requirements, the risks of not meeting objectives, and the evaluation of employees, among other issues.

5. Recommendations for future research

This paper provided the first systematic review of IMS models and their potential to embed SD in organizations. It is anticipated that the results will be of interest to both academics and practitioners in organizational SD and IMS. However, it is recognized that additional research is necessary. As Bernardo et al. (2009) state, more evidence-based research is necessary to better understand the application of IMS models in practice and how to manage the various degrees of integration. Only five empirical studies were found in the literature and all of them were solely based on surveys, thus limiting the objectivity of results. As seen in quality audits, answers from management may not correspond to the real situation or the perception of people working directly on organizational processes. Based on empirical results, better models for IMS can be developed, thus making them more appealing to organizations. The development of standards or frameworks for financial management and their possible integration in IMS models is another area for future research. Due to the fast-paced, economic-focused market that organizations are facing nowadays, the lack of economic-oriented MS limits the practice of IMS in the real world. Finally, deployment of SD requirements into an IMS will require enhanced performance measurement systems capable of dealing with an increasing array of diverse objectives. This system should facilitate employees working directly in the process to deploy social, environmental, and economic issues into operational objectives.

6. References

Asif. M., de Bruijn, E., Fisscher, O., Searcy, C., Steenhuis, H. (2009), "Process embedded design of integrated management systems ", *International Journal of Quality & Reliability Management*, Vol. 26, No. 3, pp. 261-282.

Bansal, P. (2005), "Evolving sustainably: a longitudinal study of corporate sustainable development", *Strategic Management Journal*, Vol. 26, No. 3, pp. 197-218.

Bansal, P. and Roth, K. (2000), "Why companies go green: a model of ecological responsiveness", *Academy of Management Journal*, Vol. 43, No. 4, pp. 717-736.

Barney, J. (1991), "Firm resources and sustainable competitive advantage", *Journal of Management*, Vol. 17, No. 1, pp. 99-120.

Beckmerhagen, I., Berg, H., Karapetrovic, S., Willborn, W. (2003), "Auditing in support of the integration of management systems: a case from the nuclear industry", *Managerial Auditing Journal*, Vol. 18, No. 6, pp. 560-568.

Bernardo, M., Casadesus, M., Karapetrovic, S., Heras, I. (2009), "How integrated are environmental, quality and other standardized management systems? An empirical study", *Journal of Cleaner Production*, Vol. 17, No. 8, pp. 742-750.

Campbell, J. L. (2007), "Why would corporations behave in socially responsible ways? An institutional theory of corporate social responsibility", *Academy of Management Review*, Vol. 32, No. 3, pp. 946-967.

Daub, C.H. (2007), "Assessing the quality of sustainability reporting: an alternative methodological approach", *Journal of Cleaner Production*, Vol. 15, pp. 75-85.

DiMaggio, P.J. and Powell, W.W. (1983), "The iron cage revisited: institutional isomorphism and collective rationality in organizational fields", *American Sociological Review*, Vol. 48, pp. 147-160.

Dresner, S. (2002). *The principles of sustainability*, Earthscan, London, UK.

Elkington, J. (1998), "*Cannibals with Forks: the Triple Bottom Line of 21st Century Business*", Capstone Publishing, Oxford, MA

Freeman, R.E. (1984), *Strategic management: a stakeholder approach*, Pitman, Boston, MA.

Fresner, J. and Engelhardt, G. (2004), "Experiences with integrated management systems for two small companies in Austria", *Journal of Cleaner Production*, Vol. 12 No. 06, pp. 623-631.

Gladwin, T.N., Kennelly, J.J., and Krause, T-S. (1995), "Shifting paradigms for sustainable development: implications for management theory and research", *Academy of Management Review*, Vol. 20, No. 4, pp. 874-907.

Gloet, M (2006). "Knowledge management and the links to HRM - Developing leadership and management capabilities to support sustainability", *Management Research News*, Vol. 29 No. 7, pp. 402-413

Griffith, A and Bhutto, K. (2009), "Better environmental performance. A framework for integrated management systems (IMS)", *Management of Environmental Quality*, Vol. 20, No. 5, pp. 566-80.

Harjeev, K., Laroiya, S., Sharma, D. (2010), "Integrated management systems in Indian manufacturing organizations", *The TQM Journal*, Vol. 22, No. 6, pp. 670-686.

Hart, S.L. (1995), "A natural-resource-based view of the firm", *Academy of Management Review*, Vol. 20, No. 4, pp. 986-1014

Jennings, P. and Zandbergen, P. (1995), "Ecologically sustainable organizations: an institutional approach", *Academy of Management Review*, Vol. 20, No. 4, pp. 1015-1052

Jønker, J., and Karapetrovic, S., (2004), "Systems thinking for integration of management systems", *Business Process Management Journal*, Vol. 10 No. 6, pp. 608-615.

Jørgensen, T.H. (2008), "Towards more sustainable management systems: through life-cycle management and integration", Journal of Cleaner Production, Vol. 16 No. 10, pp. 1071-1080.

Karapetrovic, S. (2003), "Musings on integrated management systems", *Measuring Business Excellence*, Vol. 7 No. 1, pp. 4-13.

Karapetrovic, S. and Willborn W. (1998), "Integration of quality and environmental management systems", *The TQM Magazine*, Vol. 10 No. 3, pp. 204-213.

Lueneburger, Ch. and Goleman, D. (2010), "The Change Leadership Sustainability Demands", *MIT Sloan Management Review*, Vol. 51, No. 4, 49-55

Lopez-Fresno, P. (2010), "Implementation of an integrated management system in an airline: a case study", *The TQM Journal*, Vol. 22, No. 6, pp. 629-647.

Muhammad Asif, Erik Joost de Bruijn, Olaf A.M. Fisscher, Cory Searcy, (2010) "Meta-management of integration of management systems", *The TQM Journal*, Vol. 22 Iss: 6, pp.570 - 582

Oskarsson, K. and Malmborg, F. V. (2005), "Integrated management systems as a corporate response to sustainable development", *Corporate Social Responsibility and Environmental Management*, Vol. 12, No. 3, pp 121-128.

Pepper, C. and Wildy, H (2008). Leading for sustainability: is surface understanding enough?, *Journal of Educational Administration*, Vol. 46, No. 5 pp. 613-629

Quinn, L. & Dalton, M. (2009), "Leading for sustainability: implementing the tasks of leadership". *Corporate Governance*, Vol. 9 No. 1 pp. 21-38

Rocha, M. and Karapetrovic, S. (2006), "A Modular Three-Phased Alternative for E-IMS Implementation", *Proceedings from 8th International Conference on ISO 9000 & TQM*, National Quality Institute, Montreal, Canada

Rocha, M., Searcy, C., and Karapetrovic, S. (2007), "Integrating sustainable development into existing management systems", *Total Quality Management and Business Excellence*, Vol. 18, No. 1/2, pp. 83-92.

Salzmann, O., Ionescu-Somers, A. and Steger, U. (2005), "The business case for sustainability: literature review and research options", *European Management Journal*, Vol. 23, No. 1, pp. 27-36.

Scipioni, A., Arena, F., Villa, M., Saccarola, G. (2001), "Integration of management systems", *Management of Environmental Quality*, Vol. 12, No. 1/2, pp. 134-135.

Searcy, C. (2009), "Setting a course in corporate sustainability performance measurement", *Measuring Business Excellence*, Vol. 13, No. 3, pp. 49-57.

Senge, P., Lichtenstein, B., Kaeufer, K., Bradbury, K. and Carroll, J. (2007). Collaborating For Systemic Change, *MITSloan Management Review*, Vol. 48, No.2, pp. 44-53

Seuring, S. and Muller, M. 2008, "From a literature review to a conceptual framework for sustainable supply chain management", *Journal of Cleaner Production*, Vol. 16, No. 15, pp. 1699-1710.

Shrivastava, P. (1995), "The role of corporations in achieving ecological sustainability", *Academy of Management Review*, Vol. 20 No. 04, pp. 936-960.

Singh, R. K., Murty, H. R., and Gupta, S. K. (2007), "An approach to develop Sustainability Management Systems in the steel industry", *World Review of Entrepreneurship, Management and Sustainable Development*, Vol. 03, No. 1, 90-108.

Wilkinson, G, and Dale, B.G. (2001), "Integrated management systems: a model based on a total quality approach", *Managing service quality*, Vol. 11, No. 5, 318-330.

World Commission on Environment and Development (WCED), 1987, Our Common Future. Oxford University Press, Oxford, U.K.

Zutshi, A. and Sohal, A. (2005), "Integrated management system: the experiences of three Australian organisations", *Journal of Manufacturing Technology Management*, Vol. 16, No. 2, pp. 211-32.

9

Innovative Sustainable Companies Management: The *Wide Symbiosis* Strategy

Francesco Fusco Girard
Seconda Università Degli Studi di Napoli
Italy

1. Introduction

The main objective of this paper is proposing a theoretical and innovative approach for companies' sustainable strategy development from the *private* point of view and according to a best-practices approach focused on very new market trends.

Nowadays the scientific community widely accept that one of the main pillars for sustainable development effective implementation is the "closed loop economy" objective: energy and material processes shifting from linear (open loop) systems - in which resources move through the economic system to become waste - to a closed loop system where wastes are inputs for new processes.

Starting from the above statement, the key thesis of this paper is that "closed loop economy" approach and concept can be successfully extended from material and energy flows to non-material flows, shifting all the relationships which involve the companies (B2B-Business to Business; B2C-Business to Consumers; B2I-Business to Institution as well as Companies internal relationship) from a type I (linear / hierarchical / one-way relationship) towards a type III - symbioses (cyclic relationships with internal loops and feedbacks). When this symbiosis approach involves simultaneously all the relationships (B2B, B2C, ..) we will name it "wide symbiosis". We will see as this systemic approach can support in sustainable development implementation that is achieving economic, environmental and human/social goals (Fusco Girard, 2009).

In relation with the above aim, in the **second paragraph** we will briefly describe the most important **business trends expected for next years** focusing on the actual global economic crisis. **Then (par. 3)** we will point out **what sustainable company management is** and **how** it can support companies for next years challenges. After that (par. 4) the **sustainability vision is declined** in some **more concrete sustainability strategic objectives**. The next **paragraphs are focused on describing B2B, B2C, B2I and companies internal relationships** aiming at **analyzing how they can contribute to the loop economy promotion achieving the sustainability strategic objectives** as they have been previously described.

The final part of the paper is focused on analyzing **decision making processes** and **evaluation tools** to effectively implement loop economy choices. Indeed, conclusions highlight the **primary role of cultural values** to promote the *Wide Symbioses* relationships.

2. The context

Below are summarized some of the main global business trends as they are expected for next years. Of course the aim is not an in-depth drill down on possible future trends (see also Simon and Zatta, 2011), but just providing some general context evolution tendencies.

- **Global economic crisis is expected to continue**, with some relevant exception (eg. developing countries like Brazil, Far East ...): by this time it seems clear that the wished economic recovery at *ante-2008 level* will not take place soon and presumably the global GDP will evolve according to a "W trend"(see also Hope, 2011)
- **Globalization is expected to increase**: competition is getting more and more global, involving every value-chain phase: world based recruiting; worldwide level sourcing; manufacturing de-localization... This is fostering to increase Asian leadership Vs Western countries (see also Yeung et at, 2001 and Sunley, 2011)
- **Interest in environmental issues is expected to continue to grow** over the next years driven by customers and supported also by Governments. Climate change / greenhouse gas reduction and fresh water scarcity are expected to be the main issues (see also Laszlo, 2009 and Makower, 2009)
- **Raw materials and energy costs are expected to continue their increasing trends**, mainly driven by their scarcity and the widening demand to support developing countries growth. Also the recent nuclear giving up at global level (following the Fukushima nuclear disaster) will foster to increase the energy prices
- **Politics and Government influence on business is expected to remain high:** following the first part of current economic crisis (2008-early 2009) many Governments strongly entered into market dynamics for saving companies and the overall global interests (see also Simon and Zatta, 2011)
- **Customer's behaviours definitely changed**: customer's expectations Vs new products are by far higher than in the past, furthermore web-based communities fostered customers sharing of products experiences, their opinions about performances, ranking, values etc. (see also Boaretto et al, 2011)
- **Growing internet based-connections**: starting from the "home internet" we pass through the internet everywhere (due to PDA / *smartphones* diffusion) to reach the "internet anything", where a number the equipments (eg. TV, cars, washing machines, home anti-theft systems ...) are web-connected (see also Simon and Zatta, 2011 and Boaretto et al, 2011).

3. The vision

In the above-described context, sustainable company management can be the key approach to really win the competition (Werbach, 2009). So, below we will describe first of all what sustainable company management is and then how it can be implemented.

Following Seralgerdin approach (Seralgerdin, 1999), which recognizes four different kinds of capital (economic capital, natural/environmental capital, social capital and human capital), sustainable development can be seen as aimed to maintain or increase all capital stocks at the same time.

According to this approach, it is also possible to express sustainability under the point of view of comprehensive efficiency or "complex efficiency" (Fusco Girard, 2008), which is the extension of the efficiency concept to all the above forms of capital, involving (see also Goodwin, 2003):

- Economic efficiency, which is the efficiency in economic capital usage. It refers to minimizing the economic resources use while maximizing goods and services production aiming to reach the state where nothing more can be achieved given the available resources (O'Sullivan and Sheffrin, 2003)
- Natural efficiency (eco-efficiency) that is linked to the use of natural capital. It refers to minimizing the environment impacts (also in terms of waste and pollution) while maximizing goods and services production (Schmidheiny, 1992) aiming to *de-coupling* economy that is economic growth without corresponding increases in environmental pressure (Bleischwitz and Hennicke, 2004)
- Social efficiency, which involves the efficiency in social capital usage. It refers to the aiming of achieve economic results maximizing the value of social relations and the role of cooperation and confidence (Putnam, 2000)
- Human efficiency, which entails the efficiency in human capital usage aiming to improve stock of competences, knowledge and personality attributes (see also Human Development Report, UNDP, 2011)

with the overall aim of maximizing / balancing them.

Furthermore, complex efficiency enlarges time and space perspectives in managing companies:

- Enlarging spatial perspective means following a **holistic approach** and focusing not just on the single company
- Enlarging time perspective means considering not just short term, but also **medium and long term**

Given the above descriptions, it is clear that complex efficiency requires a multidimensional and systemic approach (Fusco Girard, 2009).

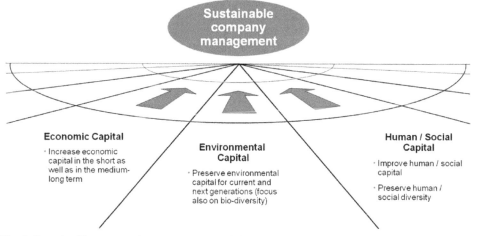

Fig. 1. Sustainable companies management vision

4. Vision and strategic objectives

The above sustainable company management vision can be realized through a **systemic approach** and by the achievement of a number of strategic objectives. Below we have grouped them according to their phase in product lifecycle and summarized, being aware of some significant overlapping.

- Overall:
  - **Achieve good financial/ economic performances also in medium-long term:** long-term approach is based on customer loyalty. Indeed, companies built on a foundation of customer loyalty can grow over the long term through all types of economic climates, while companies with weak customer loyalty face an unstable future: "Bad profits" undermine customer relationships, and include ill-gotten earnings (Reichheld, 2006). The main condition for companies sustainable management is – for sure- that the company itself must exist and long-term approach is *conditio sine qua non* in order to achieve it. Clearly, companies' risks must be managed coherently (see also Laszlo, 2008)
  - **Promote companies integration with eco-systemic dynamics:** companies as well as their products and their by-products must be always considered as part of larger eco-system, avoiding perturbing it (Ayres and Ayres, 2002)
  - **Promote companies integration with social environment (with major focus on** *local* **social environment):** encouraging strong relations with local communities can be a leverage also for increasing employees (that usually are local citizens) well-being and productivity. The integration with local social environment can be an important way to improve companies internal and external social capital
- Production phase:
  - **Follow a lifecycle approach in product design:** in the design phase all the stages of a product's life from-cradle-to-grave must be considered (i.e., from raw material extraction through materials processing, manufacture, distribution, use, repair, maintenance, and disposal or recycling) in order to achieve whished impacts on economic, environmental and social / human capital (see also Costanza, 1991)
  - **Reduce production costs:** production cost cutting (usually achieved leveraging on people/processes/IT improvements) can positively impact economics and also environment, when linked to energy / raw materials savings (see below)
  - **Reduce the quantity of required energy and materials, focusing on renewable:** consider dematerialization opportunities, that is " doing more with less", de-coupling economy wellness from energy and material consumption, maximizing renewable energy and material consumption Vs non-renewable. (Bleischwitz and Hennicke, 2004)
    In many field this trend is gaining room also leveraging on technologies opportunities (eg. in the work environments the use of paper is greatly reduced / eliminated thanks to digital communications) and swapping from product to services. For examples, customers need to illuminate a room (that is the service) while light bulb is just a way (product)
  - **Sell every manufactured product:** production waste is usually referred to product not-fulfilling specific "qualitative" requirements, but production waste can be considered also every unsold manufactured product: as matter of facts, unsold products can be seen as waste of money, energy and material impacting both economics and environment

- **Consider that employers have no capability limits:** the main key success factors to achieve best employers performances are an adequate management, the trust that enterprise successes are their own success, and an effective training
- Lifecycle / using phase
  - **Reduce lifecycle costs for maintenance and operation:** nowadays consumers are more aware of lifecycle costs and their increasing can be seen almost as a robbery when not justified. For example paying more than €400 for a rear mirror on a €8.000 car (about 5%) negatively impacts the customer-carmaker relationship
  - **Follow integrated approach to protect the environment:** considering that real environmental impacts are very-very difficult to be calculated, their global minimization can be very important to protect the environment *leveraging on both new technology opportunities and correct customers behaviors promotion*. For example, new cars CO2 emission are usually reduced VS previous models, but this effort can be invalidated by a more frequent car usage as well as by inappropriate drive style
  - **Satisfy customers:** considering that customers can be sorted in promoters, passives and detractors (Reichheld, 2006), every post-sales effort could be aimed on increasing the number of promoters and strongly avoiding detractors, definitively overcoming a Customers Relationship Management (CRM) approach focused just on IT systems
- Lifecycle end
  - **Extend product lifecycle:** as matter of fact, wasting a "still working product" has two key important issues: a) increasing the waste disposal needs (that can be very hard for electronic waste / hazardous materials ....) b) increasing energy and materials needs for replacing the wasted product. For example, replacement rate of *still working* mobile phones is very high, impacting the needs of waste disposals (that can be hard for some components such as batteries and silicon parts) as well as the need of new material and energy embedded in the new mobile phone
  - **Facilitate reuse:** as stated for the production phase, also the design efforts could be focused on both conventional reuse - where the product (or its parts) is used again for the same function – as well as new-life reuse where the product (or its parts) is used for a new function. As matter of facts, reuse help save time, money, energy, and resources, opening new opportunities to offer quality products to people and organizations with limited means
  - **Facilitate recycling** in order to prevent waste of potentially useful materials, cut the consumption of fresh raw materials, cut energy needs, reduce air and waste pollution by reducing the need for "conventional" waste disposal, and lower greenhouse gas emissions as compared to virgin production.

In the Table 1 every objective is mapped according to its impact on complex efficiency

5. How objectives can be implemented: Symbioses

Implementing sustainable development means shifting a linear economic model to promote a loop economy.

In this context, the paper aims at investigating conditions and consequences of the above well accepted and known statement:

- Extending its focus from material also to non-material flows, partially widen industrial ecology (Ayres and Ayres, 2002) / ecological economics (Costanza, 1991) approaches

- Approaching the issues from the private/enterprise point of view, filling an important gap, since most of studies are mainly focused on a public point of view.

Group	Objective	Impacts on capital		
		Economic	Environmental	Social / Human
Overall	Follow a long-term business approach	●		◐
	Promote companies integration with eco-systemic dynamics		●	◐
	Promote companies integration with social environment			●
Production phase	Follow a lifecycle approach in product design	◐	●	
	Reduce production costs	●	◐	
	Reduce the quantity of required energy and materials, focusing on renewable	◐	●	
	Sell manufactured products	◐	◐	
	Consider that employers have no capability limits	◐		●
Lifecycle / using phase	Reduce lifecycle costs for maintenance and operation	●		
	Follow integrated approach to protect the environment		●	
	Satisfy customers	●	◐	
Lifecycle end	Extend product lifecycle	◐	●	
	Facilitate reuse / recycling	●	●	◐

Table 1. Objectives impacts on complex efficiency

For the above aim, we will follow a 3 steps path:

a. Recognizing of the existing elements and relationships in every node involving the companies (B2B-Business to Business; B2C-Business to Consumers; B2I-Business to Institution as well as Companies internal relationships)
b. Investigate the *wide symbioses* approach opportunities, that is demonstrating how making every recognized relationships "symbioses" (that are cyclic relationships with internal loops and feedbacks) can help to promote sustainable company management matching the above strategic objectives
c. Recognizing the role of specific tools and of cultural factors in order to effectively implement symbioses.

6. B2B symbioses

Hereby we investigate the development of the relationships among different companies (Business to Business), pointing out the system evolution from one-way relationship toward symbioses.

6.1 The relationship evolution and symbioses description

Type I: Before the "revolution of quality", the **relationship** among different companies **was one-way and hierarchical:** main assembler companies provisioned from lot of little suppliers. Assemblers lead auction among suppliers to buy components previously designed by the assemblers and to be manufactured by suppliers. The competition was fully price-based: best is cheaper with very limited care of product quality. Since only assemblers are in charge of the design process, no product innovation could be implemented by suppliers, which were **frequently replaced in order to save money**.

Type II: One of the main "revolution of quality" innovation is the recognizing that **knowledge, know-how, technology and a real understanding of process and production can be allocated also in suppliers**. This recognising is the first step in establishing a new relationship with **suppliers, which became partners**: supply agreements have been lasting longer and different companies have been working in partnership sharing knowledge, technology, and – occasionally - also management strategies.

Type III: New and useful relationships (symbioses) among different companies are established when collaboration involves not only business goals, but also environmental and social goals, such as the implementation of a common Environmental Management System or the extension of LCA (Life Cycle Analysis) also to partner process.

Top symbiosis level is **industrial symbiosis,** which is based on resource exchanges: although there is not a general accepted industrial symbiosis definition, in general three primary opportunities for resource exchange are considered (Chertow, 2007):

1. By-product reuse—the exchange of firm-specific materials between two or more parties for use as substitutes for commercial products or raw materials
2. Utility/ infrastructure sharing—the pooled use and management of commonly used resources such as energy, water, and wastewater
3. Joint provision of services—meeting common needs across firms for ancillary activities such as fire suppression, transportation, and food provision.

In such a way partnership relationship do not involve only products, but also by-products, waste, emissions and whatever is no more functional for a company but could be useful for another one (Rutten and Boekema, 2004).

Below the industrial symbiosis is described by one of the main example: Kalundborg (Danmark).

6.2 Kalundborg: An industrial Symbiosis example

The most well-known example of industrial symbiosis is Kalundborg (Denmark), where co-operation has developed spontaneously over a number of decades and currently involves about 20 different projects. By products exchanges are schematized in figure 2.

Main participants in the Kalundborg Industrial Symbiosis are:

- DONG Energy Asnæs Power Station,
- Gyproc plasterboard factory,
- Novo Nordisk pharmaceutical plant
- Novozymes enzyme producer A/S,

- Statoil oil refinery
- Kara/Noveren waste company
- Kalundborg Municipality

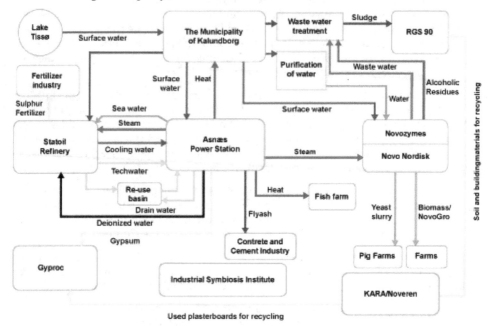

Fig. 2. Kalundborg by-products exchanges

Examples of Symbiosis Projects: Energy savings

The Asnæs Power Plant is a coal power plant (one of the main electricity plant in Denmark, producing about 10% of the electricity consumption in the Country); its excess heat is utilised as process steam and central heating. More in detail, Statoil Refinery, Novo Nordisk and Novozymes A/S receives annually about 1,5 mio. GJ (equivalent to more than 75.000 family houses yearly electricity consumption and to around 240.000 tons CO2).

The symbiosis will optimise the process steam cooperation and return up to 150.000 m3 steam condensate to the Asnæs Power Plant.

Currently a DONG Energy technology company is setting up a bio-ethanol plant next to the power plant operating on straw - a by-product in the agricultural sector. The use of steam and delivery of biomass as fuel in the power plant improves the overall CO2 account with more than 25.000 tons CO2 – not including the CO2 reduction from the replacement of bio-ethanol in gasoline and diesel. (http://www.symbiosis.dk/industrial-symbiosis.aspx)

6.3 Symbiosis sustainability evaluation

Below we will analyse the most important symbiosis impacts on complex efficiency in achieving vision's objectives; furthermore there are many other impacts which are more in depth analysed in industrial symbiosis and industrial ecology literature. (Ayres, 2002)

a. Achieve good financial/ economic performances also in medium-long term: symbioses are usually shaped by market hand, driven by economical reason and usually require stable business partnerships, which is very consistent with the long-term business approach

b. **Promote companies integration with eco-systemic dynamics:** as matter if fact, industrial symbioses follow the nature approach (see also Hawken, Lovins et al, 1999), where one company's products waste is an input for an other company

c. **Promote companies integration with social environment (with major focus on *local* social environment):** despite the appearances, industrial symbiosis is strictly linked with (mainly local) social capital. In literature many scholars point out that some industrial symbioses fail to grow despite economic profitability of material exchanges. The reason is that cultural background is a key factor to implement symbiosis. It seems – from practical experiences - that realizing material exchanges require, on its turn, the existence of non material exchanges (Fusco Girard, 2009)

d. **Reduce production costs:** as Kalundborg experience points out, many symbiosis projects have been stimulated just by economic saving opportunities. Indeed in some cases the industrial symbiosis was not "discovered" by outsiders / scholars just because the exchanges have been shaped by the invisible hand of the market rather than from conscious direct involvement

e. **Reduce the quantity of required energy and materials, focusing on renewable opportunities:** a positive impact on the environment due to symbiosis is based on natural resources improved management, since energy, by-product exchange and common waste management increase the efficiency in resources using (Fusco Girard, 2008)

f. **Facilitate reuse / recycling:** these are, by definition, the industrial symbioses key characteristics

7. B2C symbioses

Over the past decade, the **relationship between customer and companies** has been substantially developing. The following analysis shows this **evolution from mass production toward co-creation**, focusing on material and immaterial loop closing.

7.1 The relationship evolution

Type I: During the so-called *mass-production* period, companies produce as much as they can in order to profit from scale economy according to a push model. Of course models are always **standard** with no (or very limited) customization opportunities, quoting Henry Ford "We produce any colour cars - so long as it's black".

This "**push**" approach is completely **linear** and implicitly implies that companies know how to fully satisfy customers.

Type II: The spread of the so-called **Toyota spirit** radically changed the relationship between customers and companies. The push approach is gradually replaced by the pull one: *just in time* and *lean production* make possible the achievement of scale economy, although products are customized, so customers began to contribute to product features. Due to this customization opportunities the **B2C relationship** is no **more only one-way: customer**

requiring a tailored products (Customer to Business input) **to be provided by companies (Business to Customer feedback).** The new customer-focused approach implies also that, according to Kano quality model (Watson, 2001), companies put into practice a new proactive attitude, trying to anticipate, and not just to follow, customers requirements. (Ohno, 2004)

7.2 Symbiosis description

Type III: Implementing symbioses with customers means **recognising customer relationship as the most important factor in company development, promoting continuous flows and feedbacks with customers, potential customers and non-customers.**

B2C relationship as key success factor

The importance of this relationship is pointed out by many scholars, highlighting as in nowadays markets the key **success factor is not a requirement but is the relationship with customers.** Reichheld (Reichheld, 2006) stated that the relationship allows tracking promoters and detractors distinguishing good from bad profits. . The discriminator between good and bad profits is "the ultimate question": "treat the customers as if you were a customer yourself". As matter of facts, bad profits boost short-term earnings but alienate customers. They undermine growth by creating legions of detractors-customers who complain loudly about the company and switch to competitors at the earliest opportunity. On the contrary, company makes good profits through enthusiastic customer participation which can be achieved joining material requirements (as they are Kano's model) to not-material requirements, such as environmental compatibility, health care, respect of human rights etc (which can be also important drivers for product and process innovations).

B2C relationship for the Co-Creation symbiosis

For a long time, the market was the place of *value exchange*, while companies' premises were the place of *value creation*.

According to Prahalad and Ramaswamy (Prahalad and Ramaswany, 2004) some new trends (currently costumers are informed, aware, and networked and the product value is not linked just with provided services or products) are important drivers to radically change the relationship with customers **enabling co-creation, a new form of value creation.**

Times companies simply sell their products and services are over. Successful companies allow their customers to take part in the designing process, to keep track of production tasks, to participate in customers community: **customers take part in value creation, which that is not *out*, but *in* the market.**

Under current trends, consumers "seek to exercise their influence in every part of the business system," and companies have to accommodate, by designing "experience environments" for creating new value creation spaces.

Wikinomics can be seen as the peak of the above new B2C relationship. Indeed, according to Tapscott and Williams (2006), the word *Wikinomics* identifies how some companies are using **mass collaboration** (also called *peer production*) and open-source technology to be successful.

The use of mass collaboration in a business environment, in recent history, can be seen as an extension of the trend to outsource: i.e. externalizing formerly internal business functions to other business entities. The difference however is that instead of an organized business body brought into being specifically for only one function, mass collaboration **relies on free individual agents to come together and cooperate to improve a given operation or solve a problem** (A.T. Kearney, 2009).

The new B2C Symbioses examples can be seen along all the value chain

- **Research and development:** for example, P&G (Procter and Gamble, Fortune 500 American multinational corporation manufacturing a wide range of consumer goods) leverages on external networks (NineSigma and InnoCentive), where consumers develop and put forward suggestions for technical / scientific issues submitted by P&G. This allows an overall R&D investments reduction, while innovation success rate significantly increased. So P&G is planning to source 50% of their new product and service ideas from outside the company, where 90,000 scientists around the world can help solve tough R&D problems for a cash reward (Tapscott and Williams, 2006)
- **Purchasing:** Goldcorp (a Canadian gold producer), published geological data about an area of 200 km2 on the Internet and offered awards for the best potential sources of gold. Scientists and researchers from all over the world used that data for identifying 110 locations out of which half were new for Goldcorp. At over 80% of the listed sources, Goldcorp discovered a total of 227 tons of gold (Tapscott and Williams, 2006)
- **Design:** Companies can allow customers to design their own individualized product, such Sumerset, a US houseboats manufacturer that aims to leverage on "emotional bonding with... the company" and "a greater degree of self-esteem". (Prahalad and Ramaswamy, 2004). Other interesting practice is NikeID where, according to the slogan "you design it. We build it", customers can customize shoes in terms of both look (materials, colours and personal id adding ...) and performance (wide and narrow sizing, independent left and right sizes, outsoles picking ...)
- **Marketing/Communications:** here many companies are focusing their efforts frequently leveraging also on new trends like Facebook. One of the main example can be Danone (a French food-products multinational corporation): Consumers vote for the targeted flavour of a new pudding by SMS or on a website or by facebook: more than 1 million consumers voted and were already familiar with the product before it went to market.
- **Sales/Distribution:** here frequently online companies are the most involved. For example, eBay set up an online community which is free of charge, optional offer with discussion forums, news, tips for all eBay members. Users of the eBay community bid twice as much in auctions, pay up to 24% higher prices, and spend 54% more than eBay members who are not part of the community.
- **Post-sales / CRM** (Customer Relationship Management): this broad term covers all the concepts which are used by companies to manage their relationships with customers, including collecting, storing and analyzing customer information.
  Since getting new customer costs about 5 times more than keeping current ones (Farinet and Ploncher, 2004), CRM is a key tool to really understand customers' requirements in order to design new products and to adjust current products in this perspective. Once again, this points out that the best company key factor is the best possible relationship with customers, rather than the best possible products. One of the main example of post sales B2C symbiosis can be H3g (UMTS-based mobile phone company) which provides on-line free customer assistance fully leveraging on other customers knowledge (see www.lesaitutte.it)

7.3 Symbiosis sustainability evaluation

The B2C symbioses impacts on complex efficiency can be summarized as below:

a. **Achieve good financial/ economic performances also in medium-long term:** establishing and maintaining a strong customer relationship through B2C symbioses is a good way for generating long-term "good profits"; According to Reichheld, good profit implies high level of success: for example, every Dell not satisfied customer costs about 57 USD, whereas satisfied customer breeds about 328 USD (Reichheld, 2006)

b. **Promote companies integration with eco-systemic dynamics:** Recognising that customers personal values are appreciated also by the companies could be an important key factor in the B2C symbiosis perspective. For example, ISPO survey points out that 77% customers state to avoid to buy products made by company which are not involved in social or environment campaign

c. **Promote companies integration with social environment:** Active and proactive customer participation foster to improve human and social capital. Looking for and reaching information is useful to improve skills, but also to establish new networks, through internet or customers associations.
Recognising that important values are shared also with a company could be a main factor also in purchasing process: in a 2004 survey over an half of customers declare to know fair trade and 6,6% declare to buy only from fair trade

d. **Reduce production costs:** B2C symbiosis allows many interesting savings in production costs as above pointed out in many examples along the value chain

e. **Sell every manufactured product:** the B2C symbiosis approach implies that only sold products have been manufactured

f. **Reduce lifecycle costs for maintenance and operation:** co-design experience by definition foster lifecycle costs reduction according to customers needs

g. **Satisfy customers:** co-design and value co-creation can strongly help companies in making customers satisfied. Furthermore, likewise CRM surveys point out that (Farinet and Ploncher, 2004) 76% of satisfied customers buy again the same product, 33% buy again a not fully satisfying product, but 89% of customers buy again a not fully satisfying product if customer service is excellent. Once again, this points out that the best company key factor is the best possible relationship with customers, rather than the best possible products (Ronchi, 2003)

h. **Extend product lifecycle:** through co-creation, it is possible to establish an emotional relationship with the product that can foster to prolong product life cycle, encouraging also a correct maintenance Vs thruway approach.

8. B2I symbioses

Below we investigate the evolution of the relationship among companies and Institutions, starting from command and control approach/ policies, which is a one-way and hierarchical relationship, towards voluntary agreement that needs feedbacks and reactions (symbiosis).

The relationship evolution

Type I Command and Control was the first step towards modern sustainability policies. At the beginning these laws were focused on safety in the working environment, and then have been widened also to environmental emission and waste management. These kinds of laws reflected a one-way relationship between Public Institutions and companies so that, in many cases, companies do not appreciate them seeing mostly their additional management costs (Fusco Girard, 2009).

8.1 Symbiosis description

Type II and III: below we will analyse the type II and III relationship between Companies and Public / Private Institutions. We will analyse both types in the same time because both are based on considering sustainability as a development opportunity, through active involvement instead of passive laws acknowledge (Fusco Girard, 2009) and hereby the difference is not in the instruments, but in the level of involvement.

In this perspective, command and control instruments have been integrated by:

- **Economic instruments, in order to internalize external costs**. This goal is usually achieved through **emission trade system set up**. Here, a central authority sets a limit or cap on the amount of a pollutant that can be emitted. Companies are required to hold an equivalent number of credits or allowances which represent the right to emit a specific amount. The total amount of credits cannot exceed the cap, limiting total emissions to that level. Companies that need to increase their emissions must buy credits from those who pollute less. The transfer of allowances is referred to as a trade. Therefore, at least in theory, those who can most cheaply reduce emissions most cheaply will do so, achieving the pollution reduction at the overall lowest possible cost.
- **Green / Sustainable Procurement**: integrating also environmental and social criteria into procurement decisions in addition to the conventional criteria of price and quality.
- **Education**: education is a fundamental instrument to spread the sustainability concept, in order to increase the demand for sustainable product/services as well as a coherent use.
- **Voluntary agreement** is a contract between the public administration and a company in which the firm agrees to achieve a certain environmental or social objective and receives a subsidy to change its technology through R&D and innovation. The agreement is bilateral, between firm and administration, and requires a voluntary element on both sides.

It is very important to consider that best results can be achieved only by well combining the above instruments, in particular: (Carminio et al., 2002)

- Command and control instruments are necessary to guarantee minimal requirements
- Economic instruments are very useful to change implant behaviour
- Education is finalised to promote sustainable behaviour also through best practices.
- Voluntary agreement can be implemented only where there is a social background that promotes sustainability concepts.

8.2 Symbiosis sustainability evaluation

Symbioses promote sustainable organization of companies with positive impacts in achieving the sustainable objectives

a. Achieve good financial/ economic performances also in medium-long term: Sustainability is the key strategy that European Union Institutions identified in order to improve European companies competitiveness. Furthermore, it is important to notice that also "do nothing" option has its own costs: Munich Re (2009) pointed out that economic loses due to climate change related extreme events glowed up from less than 5 Millions USD in 1950 till to current over 70 Millions USD

b. **Promote companies integration with eco-systemic dynamics; Follow a lifecycle approach in product design; Reduce the quantity of required energy and materials, focusing on renewable; facilitate reuse / recycling:** The reduction of the impacts on natural capital is frequently the aim of such instruments, so we can assume that this goal is always - at least partially - achieved. Here it is also important to consider that these kinds of impacts are typically extra-national, so involving as many countries as possible is the key strategy for increasing effectiveness

c. **Promote companies integration with social environment; Consider that employers have no capability limits:** Beyond the impacts on safety in the working environment, the impacts on human and social capital involve both offering and demand (Fusco Girard, 2009)

   • Offering-side policy could be achieved to spread new skills and knowledge in strategic fields. Similarly this kind of intervention can foster the cooperation among different firms to promote also knowledge spread through the creation of new network.

   • Demand-side policies are fundamental to educate customers to choose sustainable products. In this kind of intervention it is very important also the communication: for this reason these policies must be completed by a labelling system laws

9. Company internal symbioses

Below we will analyse the evolution of the internal organization of the firm, focusing on the employees relationships and on the relationship among company structures to analyse how the setting out of symbioses relationships could be useful in fostering towards sustainable development.

9.1 The relationship evolution and symbioses description

Type Ia: before **Scientific Management** internal relationships were **one-way and hierarchical.** Procedures and methods were not explicit, so companies management/white collars did not have to coordinate workers activities, but mainly to convince them to make available to the company their own knowledge (which usually remained tacit).

Type Ib: When the Scientific Management (also called *Taylorism*) spread, companies internal relationships have **usually been still one-way,** with no feedbacks between management and workers. Tacit knowledge was partially coded in procedure and methods, so that management duties were mainly focused on requiring and checking that workers effectively implemented them.

Type II: quality revolution led to enhanced productivity through improving workers role; the best way to assure that **a job gets done right is not to increase worker supervision, but to educate, empower, and trust the person assigned to the job.** Concurrent opinions and feedbacks finished the passive-workers era to foster a new one in which **every worker can contribute to the company** by defining his own duties and organization improvement.

Internal relationships developed similarly to customer-supplier relationship: if the quality is in the process, end process inspection is useless.

Type III: since the goals of companies are complex and do not involve only the economic sphere in the short term, **top management know-how, skills and knowledge cannot be enough.** *Top management* **leadership must be integrated with** *workers* **leadership.**

So, top management have to foster vertical collaboration - among different company hierarchical levels - and horizontal collaboration - among different company departments -. The aim should be to improve knowledge sharing through formal and informal networks and feedbacks (Galgano, 2004). This approach is becoming more and more common across both small and larger companies (Foray, 2006).

For example, Toyota – the car world leading manufacturer -   implemented a program to stimulate employees to submit new ideas to management. The results is that in 2008 management approved nearly 100% of the 400 new ideas coming from the employee, achieving benefits in terms of quality of the final products, total costs and in employee satisfaction.

For implementing so high a collaboration level it is necessary to deeply understand that human capital has no limits capability if it is well run, if it trusts that enterprise success is its own success, and if it is conveniently trained.

Another instrument to implement symbioses in **internal relationship is the workers participation in company capital.** In such a way it is possible to extend the internal symbiosis to strategic aspects as company owners. In the USA, for example, institutional investors are increasing their stock participation in raising number of firms. (Caselli, 2006)

9.2 Symbiosis sustainability evaluation

Companies internal symbioses contributions to promote sustainable objectives achievements can be summarized in the following way. (Fusco Girard, 2009)

a. **Achieve good financial/ economic performances also in medium-long term:** Revenues are the main driver in fostering company towards internal symbioses. First of all, the sharing of goals and strategies all around the firm is the only way to really implement them. Besides, it is universally recognised that the essential requirement for continuous improvement is the full involvement of every employee. Furthermore, workers participation in company capital allows to achieve:
    • workers stability within the firm (minimizing turn-over)
    • keeping tacit knowledge within the firm
    • trust and collaborative company atmosphere
b. **Promote companies integration with social environment:** internal symbioses promote human development through the implementation of social networks, strengthening internal links. Indeed workers trust that, apart from their formal links with the company, their future is strictly linked with company future, which contributes to foster collaborative and constructive company internal climate. Trust is stimulated by cooperation and, in its turn, it promotes cooperation

10. Multi criteria evaluation approach for symbioses design

Multi criteria evaluation approaches and tools (Shi and Zeleny, 2000; Saaty and Vargas, 2006; Bogetoft and Pruzan, 1999) must have a primary supporting role for symbioses design and strategy implementation:

- *Ex ante* multi criteria evaluation should have a central role since symbioses require the involvement of a number of different stakeholders (eg. private companies, public institutions...) with different (and in some cases even conflicting) interests to be adequately evaluated and prioritized through multi criteria approach. The same approach can successfully support also coordination of actions of all partners/stakeholders to create synergies and positive interdependences in innovative management toward symbioses (see also Freedman, 2007)
- *On-going* multi criteria evaluation to monitor different steps effective achievements of goals and strategy redefinition, when need (Saaty and Vargas, 2006)
- *Ex post* multi criteria evaluation of pilots / experimental projects have a significant role to highlight costs and benefits and identify best practices as well as lesson learned providing benchmarks and strategies (Nijkamp and Rietveld, 1990), in order to support the definition of a path from "one shot" experiences towards ordinary practice

An important issue about the above evaluation is related to the indicators: as matter of fact, a consistent set of both quantitative and qualitative indicators is required to compare multidimensional aspects, defining targets and monitor project outcomes.

11. The role of culture to promote the wide symbioses

As we have seen above, in some case symbioses relationships are increasing and growing up, fostered mainly by economic drivers (Estes, 2009; Makower, 2009). According to that, can we assume that global economy is running toward sustainability? Can only market rules promote sustainability?

Unfortunately the most likely answer is negative.

What is needed is a **strong intentionality and a systemic approach** (Meadows, 2008): it is necessary to recognise that lot of aspects which seems to be like chalk and cheese, are just different aspects of the same issue.

So, what could be done in order to promote the **wide symbioses** and the **systemic approach**?

Symbioses are – by definition- constituted by two main elements: nodes, which are deputy of exchanges, and connections which are the mean to let the exchange be.

Currently we are living the "communication age": internet, mobile phones, PDAs, as well as "old media" such as radio and TV make available billions of information to the public, so it can be assumed that what is needed are not new connections.

The need is to improve the nodes of the current network, making them really able to exchange information. Symbioses, in facts, can happen only among active *subject nodes* and never among *passive objects nodes*: **closing both material and immaterial loops through feedbacks and immaterial exchanges have sense only among subjects.**

Thus, there is not an issue related to infrastructures, but related to **culture**. So, in other words, **the first step in implementing symbioses is to "reintroduce the subject"** (Scola, 2006).

The subject reintroduction does not involve only the firms, but all the system of values they are part of. **Reintroducing the subject** means focusing on "why do?" and on "what to do?" instead of "how to do something" (Zeleny, 2005). In other words, it means to **promote critical thinking and wisdom**. The real challenge is recognizing links that others are not able to recognise, understanding the connections among different form of capital and the following impossibility to long term maximize economic capital giving up other forms of capital.

But which are, practically, the values that promote the subject re-introduction, that means to promote the symbioses implementation?

Even though in a different context, F. Capra points out that there is a link among ethics, values and thought (see table below), so we need to **promote integrative thought instead of self assertive one,** (Capra, 2005) **that implies to promote integrative values for indirectly supporting the symbioses implementation** (see also Fusco Girard, 2009).

Thought		Values	
Self-assertive	Integrative	Self-assertive	Integrative
Rational	Intuitive	Expansion	Conservation
Analytical	Synthetic	Competition	Co operation
Fragmented	Holistic	Quantity	Quality
Linear	Not linear	Domination	Association

Table 2. Values and Thought

12. Conclusions

In the new global context - characterized by a growing complexity and uncertainty, increasing scarcity of natural resources and energy, climate destabilization, in which public institutions, consumers and NGO are demanding a better environment quality - business sector is charged by new responsibilities, **companies play a central role in the strategies for sustainable development effective implementation.** (World Business Council for Sustainable Development, 2009)

In the above context, we pointed out, in order to achieve sustainable company management, the importance of closing both material and not material flows, that is the wide symbioses generating/promoting.

All the nodes of the network which the company is part of could promote the wide symbiosis model: innovative relationships with other companies, with customers, with public institutions as well as company internal relationships. This should be the master way to real implement sustainability.

To the above aim, cultural aspects play a great role: we need to promote integrative thought instead of self assertive one that implies to promote integrative values.

13. References

A.T. Kearney (2009) *"A.T. Kearney Customer Energy:* The empowered consumer is revolutionizing customer relationships". Obtained through the Internet http://www.atkearney.com/index.php/Publications/harnessing-customer-energy.html [accessed 13/01/2011].

Ayres, R. U. and Ayres, L.W. (2002), *A Handbook of Industrial Ecology*, Edward Elgar Publishing Inc., Northampton, MA

Bleischwitz R. and Hennicke P. (2004), *Eco-efficiency, regulation and sustainable business*, Edward Elgar Publishing Inc., Northampton, MA

Begg K., F.Van der Woerd, D.Levy (2005),*The Business of Climate Change*, Greenleaf Publishing Limited, Sheffield

Boaretto A., Noci G., Pini F. M. (2011) Marketing reloaded, Il Sole 24 Ore Libri

Bogetoft P., Pruzan, P. (1991), *Planning with Multiple Criteria*, North Holland, New York

Capra F. (2005), *La rete della* vita, BUR, Milano

Carminieo G., Frey M. and Araldo C. (2002), *Gestione del prodotto e sostenibilità*, Franco Angeli, Milano

Caselli, L. (2006), *Democrazia economica ed azionariato dei lavoratori* in Rusconi, G., Dorigatti, M., *Impresa e Responsabilità Sociale*, Franco Angeli, Milano

Chertow, M.R. (2007), *Uncovering Industrial Symbiosis* in Journal of Industrial Ecology, Volume 11, Number 1

Costanza, R. (1991), *Ecological economics: the science and management of sustainability*. Columbia University Press, New York.

Estes J., (2009) *Smart Green: How to Implement Sustainable Business Practices in Any Industry - and Make Money*, Wiley

Farinet A. and Ploncher E. (2004), *Customer Relationship Management*, ETAS ,Milano

Foray D. (2006), *L'economia della conoscenza*, il Mulino, Bologna

Fusco Girard, F. (2009) 'Symbioses strategies for sustainable company management', *International Journal of Sustainable Development, Volume 12, Number 2-4 / 2009*

Fusco Girard, F. (2009) Extending industrial ecology principles to nonmaterial flows: the "Widen Symbiosis"' *Book of Abstracts 2009 International Society for Industrial Ecology Conference*

Fusco Girard, F. (2009), 'L'approccio della ecologia industriale per l'efficienza energetica nei processi produttivi', *Rivista AEIT num.1, 2009*

Fusco Girard, F. (2008) 'Impresa e sviluppo sostenibile: strategie innovative per la sostenibilità dei prodotti', *Rassegna ANIAI ,10, 2008*

Freedman R.E., Rusconi G. and Dorigatti M. (2007),*Teoria degli stakeholder*, Franco Angeli, Milano

Galgano, A. (2004), *I Sette Strumenti della qualità totale*, Il Sole 24 ore Edizioni, Milano

Goodwin, N.R. (2003) *Five Kinds of Capital: Useful Concepts for Sustainable Developmen*, Global Development and Environment Institute Tufts University

Green, K. and Randles, K. (2006), *Industrial Ecology and Spaces of Innovations*, Edward Elgar Publishing Inc., Northampton, MA

Hawken, P., Lovins, A., and Lovins, L.H. (1999), *Natural Capitalism*, Rocky Mountain Institute, Snowmass, CO

UNDP, *Human Development Report 2011Sustainability and Equity: A Better Future for All*, (2011) UNDP Press

Hope, W. (2011) Crisis of temporalities: Global capitalism after the 2007–08 financial collapse. *Time & Society* March 2011 20: 94-118

Kronenberg, J. (2007), *Ecological economics and Industrial* Ecology, Routledge, Abington

Laszlo, C. (2009) *WorldShift 2012: making green business, new politics, and higher consciousness working together*, Inner Traditions, USA

Laszlo, C. (2008) *Sustainable Value: How the World's Leading Companies Are Doing Well by Doing Good*, Stanford Business Books

Makower, J. (2009), *Strategies for the Green Economy*, McGraw-Hill, New York

Marchesini G.C. (2003), *L'impresa etica e le sue sfide*, Egea, Roma

Meadows D.H. (2008), *Thinking in Systems*, Chelsea Green Publishing

Michelini L. (2005), 'L'approccio consumer-ethic driver nell'innovazione di prodotto' *Paper presented at "Le tendenze del marketing", Ecole Supérieure de Commerce de Paris – EAP, January 2005*

Munchener Ruck *(2009) Topics Geo Annual review: Natural catastrophes 2008*, Munich ReGroup, Munich

Nijkamp P., Rietveld P. (1999) Multicriteria Evaluation in Physical Planning, North-Holland

Ohno T. (2004) *Lo spirito Toyota*, Einaudi, Torino

O'Sullivan, A., Sheffrin, S.M. (2003). *Economics: Principles in action*, Upper Saddle River New Jersey

Pine J.B. and Gilmore J.H. (1999) *The experience Economy*, Harvard Business School Press, Boston

Prahalad, C.K. and V.Ramaswany (2004), *The future of competition*, Harvard Business School Press, Boston

Pruzan, P. and Pruzan Mikkelsen K. (2008), *Leading with Wisdom: Spiritual-Based Leaders in Business*, Greenleaf Publishing Limted, Sheffield

Putnam, R. (2000) *Bowling Alone: The Collapse and Revival of American Community*, Simon and Schuster

Reichheld F. (2006), *The ultimate question. Driving good profits and true growth*, Harvard Business School Press, Boston

Ronchi, M. (2003) *CRM per tutti*, Franco Angeli, Milano

Rutten R.,F.Boekema(2004), 'A knowledge Based View on Innovation in Regional Networks', in De Groot H.L.F.,P. Nijkamp and R. Stough (editors), *Enterpreneurship and Regional Economic Development*, Edward Elgar Publishing, Cheltenham

Saaty T.L.,Vargas L.G.(2006), *Decision Making with Analytic Network Process*, Springer Science, New York

Schmidheiny S. (1992), *Changing Course: A Global Business Perspective on Development and the Environment*, World Business Council for Sustainable Development (WBCSD), Geneve (CH)

Schumper J.(1954), *Capitalism, Socialism and Democracy*, Allen and Unwin, London

Scola, A. (2006) '*Antropologia, etica ed affari*' in: Rusconi, G., Dorigatti, M., *Impresa e Responsabilità Sociale*, Franco Angeli, Milano

Serageldin I. (1999) *Social Capital. A Multifaceted perspective*, The World Bank, Washington

Shi y., M.Zeleny(2000), *New Frontiers of Decision Making for the Information Technology Era*, World Scientific Publishing, Singapore

Simon H.,Zatta D. (2011) *I trend economici del futuro*, Il sole 24 ore

Sunley, P. (2011) The consequences of economic globalization, In: *The SAGE handbook of economic geography* , Leyshon A., Lee R., McDowell L.

Tapscott, D. and Williams A.D. (2006) *Wikinomics: How Mass Collaboration Changes Everything*, Portfolio Hardcover, USA

Watson, G.H. (2001), *La visione strategica dei clienti: il modello di Kano"* in Conti, T., De Risi P., *Manuale della qualità*, ed. Il Sole 24 Ore

Werbach A. (2009), *Strategy for Sustainability: A Business Manifesto*, Harvard Business Press

World Business Council for Sustainable Development (2009), *Transforming the Market*, Ata Roto Press, Geneva

Yeung A., Xin K., Pfoertsch W., Liu S.(2011) The Globalization of Chinese Companies: Strategies for Conquering international Markets, John Wiley & Sons, Singapore

Zeleny M. (2005) *Human System Management*, World Scientific Publishing, Singapore

Part 3

Sustainable Environment

10

Innovation Ecosystem for Sustainable Development

Kayano Fukuda and Chihiro Watanabe
National University of Singapore
Singapore

1. Introduction

Innovation has significantly contributed to growth and development of society in the 20th century. Advances in science and technology have boosted the productivity and competitiveness of industry, and also have vastly improved living standards and the quality of life. At the same time, however, rapid growth of industrial production has increased consumption of energy and natural resources enormously. Subsequent impacts on the environment have caused various threats to endanger the survival of life such as global warming and excessive use of energy, land and water resources.

Sustainable development is of the most concern to the global society today (Millennium Ecosystem Assessment, 2005; Intergovenmental Panel on Climate Change [IPCC], 2007; World Business Council for Sustainable Development [WBCSD], 2008). While it has often focused on environmental concerns, sustainable development has three dimensions: economic, environmental and social (World Commission on Environment and Development, 1987; Senge & Carstedt, 2001; Sheth et al., 2011). The challenge of sustainable development necessitates fundamental changes in the way of growth and development in the world. Industries should develop their business model not only based on economic performance, but also taking account of the environment and social impacts as well. Governments should adopt legislation and regulation on economic activities and environmental issues to enhance sustainability consideration and social awareness.

Achieving these goals requires innovation for improving the triple bottom line: economic, environmental and social well-being. Innovation is an essential driver for sustainable development. It enables industries to increase productivity while decreasing resource uses and environmental impact. It also delivers new value to satisfy high standards of living.

With the growing awareness of the significance of innovation for sustainable development, a concept of an 'innovation ecosystem' has been postulated (President's Council of Advisors on Science and Technology [PCAST], 2004; Council on Competitiveness, 2004; Industrial Structure Council, 2005). In this concept, innovation is considered a comprehensive system interacting closely with surrounding environment rather than a linear and mechanical progression. This consideration means not only optimizing internal innovation processes but also optimizing externally. An innovation ecosystem needs to adjust all public and private sector stakeholders as well as adapt itself to changes in the external environment.

This chapter attempts to analyze dynamics of innovation systems from perspective of sustainable development. An empirical analysis focusing on stability properties rather than process complexity of ecosystem was conducted. Section 2 reviews the dynamics and features of innovation ecosystem to sustain stability. Section 3 describes the mutual inspiration cycle between Japan and the US in the last three decades. Section 4 explains the new global trends in innovation for sustainable development. Finally, Section 5 briefly summarizes and concludes this chapter.

2. Properties of innovation ecosystem

The concept of innovation ecosystem often stresses that innovation occurs through interactive networks at various levels (Council on Competitiveness, 2004; Iansiti & Levien, 2004; Industrial Structure Council, 2005; Organisation for Economic Co-operation and Development [OECD], 2008). These networks are a broad and complex array of stakeholders in both of public and private sectors. An important function of innovation ecosystem consists of governmental organizations that fund R&D activities, many areas of policy which impact the effectiveness of innovation, large and small firms who transform research and new knowledge into the market place, universities, research institutes, and different kinds of infrastructure such as transportation and telecommunications. All stakeholders are related to one another in a complex manner in innovation processes as a part of an innovation ecosystem. Their behaviors improve the performance of an ecosystem and, in doing so, improve individual performances.

Strong performance of an innovation ecosystem requires reduction of uncertainty in innovation processes (Klein & Rosenberg, 1986; Iansiti & Levien, 2004). The changes in a highly turbulent environment increase the uncertainty not only on technological performance but also on the market response and ability of stakeholders to absorb and utilize the requisite changes effectively. This correlation between change and uncertainty in an innovation ecosystem necessitates autonomic reaction of each stakeholder and coordination of the network of stakeholders. This combination of autonomy and coordination enables an innovation ecosystem to improve its performance enough to survive uncertain global circumstances, achieving sustainable development.

A natural ecosystem provides a suggestive analogy for sustaining an innovation ecosystem. Stability is an important characteristic of a natural ecosystem reflecting complex homeostatic processes, especially in the face of environmental variability (Millennium Ecosystem Assessment, 2005). As with a natural ecosystem, stability in the face of external shocks is an important goal of an innovation ecosystem sufficient to meet demands for sustainable development. Ecosystem stability requires three factors: resistance, resilience and functional redundancy (Allison & Martiny, 2008). Resistance is the capability of a system to remain in the same state in the face of disturbance. Resilience is the rate at which a system returns to its initial state after being disturbed. Functional redundancy is the ability of a system to carry out a functional process in a similar rate regardless of disturbance. Resistance can be described as inertia, and resilience has some aspects including elasticity (rapidity of restoration following disturbance) and amplitude (zone from which a system will return to a stable state) (Westman, 1978). Allison & Martiny (2008) explains the potential impacts of disturbance on an ecosystem with these three factors. When a disturbance is applied to an ecosystem, it might be resistant to the disturbance and not change. Alternatively, if an ecosystem is sensitive to the disturbance and does change, it could be resilient and promptly recover to its initial state. Finally, an ecosystem which is not resilient might perform like the original state if the constituent members of an ecosystem are functionally redundant. This process to absorb external disturbance is deeply correlated with internal interaction process between its components. Marten (2001) pointed out three emergent properties of interaction in an ecosystem: co-existence, co-evolution and co-adaptation. Co-existence is built into an evolutionary game between species. Co-adaption (fitting together) is a consequence of co-evolution (changing together). Species in an ecosystem has an ability to change as circumstance demands. They change the way in which they interact with other species, and as a consequence, organize themselves through co-adaption. Co-evolution is essential to coordinate an ecosystem internally in a stable manner. These processes to maintain stability of ecosystem both internally and externally are combined in such a way that the ecosystem as a whole continues to function on a sustainable basis.

The natural ecosystem analogy suggests that a competence to address rapidly changing circumstances is necessary for sustainable development. Like a natural ecosystem, an innovation ecosystem is characterized by various participants interacting with each other. They co-exist, co-evolve and co-adapt with each other, and through this interaction, improve performance of the innovation ecosystem as a whole. Meanwhile, the innovation ecosystem resist, resilient and functionally redundant to external disturbances. This function to maintain both of internal and external stabilities has been recognized by the current global society facing an increasingly uncertain future. Serious demands for sustainable development in economic, environment and social dimensions are a new challenge for an innovation ecosystem.

3. Co-evolutionary cycle between the innovation ecosystems in Japan and the U.S.

While the U.S. was the primary leader in innovation over the course of the 20th century, it confronted a challenge from Japan in the 1980s (Council on Competitiveness, 2004). After the two energy crisis of the 1970s, Japan achieved notable energy efficient improvement in the 1980s that contributed to high economic growth driven by manufacturing technologies (Watanabe, 1995). The emergence of Japan initiated mutual inspiration between both countries leading to contrasting success in innovation during the last three decades (Fukuda & Watanabe, 2008).

The challenge from Japan triggered efforts to restore the U.S. competitive position. Both public and private sectors released proposals on competitiveness, including the 1985 report of the President's Commission on Industrial Competitiveness, the Report by Council on Competitiveness analyzing competitiveness problems in 1987 and *Made in America* published in 1989, and the federal government enforced new innovation legislation such as the Bayh-Dole Act of 1980, the National Cooperative Research Act of 1984, the National Competitiveness Technology Transfer Act of 1989. Through these efforts, the U.S. established a foundation for a new economy driven by information and communication technology (ICT) in the 1990s.

While the U.S. enjoyed economic success in the 1990s, Japan experienced economic stagnation known as the Lost Decade. During the period, both of public and private sectors in Japan rushed to construct the ICT infrastructure and accelerated R&D and dissemination of telecommunication equipments. Besides, the government enacted the Science and Technology Basic Law in 1995 to support national development through science, technology and innovation, and in response to the Basic Law, adopted the First Science and Technology Basic Plan in 1996 for a period extending to the end of Japanese Fiscal Year 2000. As a result of these efforts, Japan began to show signs of recovery in the early 2000s. However, it confronted the reality that the revitalization of its manufacturing industry is not whole industry-wide, which resulted in bi-polarization in profitability among high-technology firms.

This cyclical reversal of competitive dominance between Japan and the U.S. suggests that the natural ecosystem analogy explains the dynamism of mutual interaction between both countries. Both countries have co-evolved each other in the rise and fall of competitive advantages over the last three decades, and have survived to maintain stability of their national innovation ecosystems respectively. Their reactions to external disturbances varied over time.

3.1 Japan's success in the 1980s

Japan's conspicuous economic achievement in the 1980s can be attributed to its success in technology substitution for constrained production factors. This is similar to a function of biological ecosystems, where some species slows down and other species speeds up to compensate, in order to maintain homeostasis.

Both of public and private sectors in Japan accumulated efforts to reduce energy dependency after the energy crisis in 1973 (Watanabe, 1992, 1995a). The industry efforts to increase energy technology significantly contributed to improvement of energy efficiency. The government appropriated its R&D budget for energy R&D on a priority basis to induce vigorous energy R&D efforts by industry in the late 1970s and the early 1980s, and national R&D program projects such as the Sunshine Project initiated in 1974 and Moonlight Project initiated in 1978 encouraged networking among industries as well as between the government and industry.

	Production	Energy efficiency	Fuel switching	CO_2 emissions
Japan	3.97	-3.44	-0.59	-0.06
U.S.	2.78	-2.62	-0.11	0.55

Source: Watanabe, 1999.

Table 1. Comparison of development path in Japan and the U.S. (1978-1988) – average change rate; % per annum

The combination of autonomic efforts by industry and coordination by the government achieved a dramatic energy efficiency improvement (Watanabe, 1995b, 1999). As tabulated in Table 1, Japan recorded the highest economic growth (3.97% per year) with a 0.06% decline in CO_2 emissions in the 10 years following the second energy crisis in 1979. This was possible due to conspicuous energy efficiency improvements (3.44% per year). During the same period, the U.S. attained 2.78% economic growth, and CO_2 emissions increased by 0.05%, although its energy efficiency improvement remained at 2.62%. This notable success of Japan enabled the national innovation ecosystem to be resilient to energy shocks and to improve its performance by means of the fusion of efforts by public and private sectors.

3.2 Reversal of competitive advantage between Japan and the U.S. in the 1990s

Contrary to its economic success in the 1980s, Japan suffered a serious economic downturn and declined its competitiveness severely in the 1990s (IMD, 2002). This dramatic decline in competitiveness can be attributed to resistance of the national innovation ecosystem to emergence of ICT.

Japan continued to cling to a growth-oriented development trajectory in which economic growth leverages further growth, largely because its successes in the decades of high economic growth owed to the traditional development trajectory (Watanabe, 1995a; Fukuda & Watanabe, 2008). As a consequence of this strong resistance, the contribution of technological progress to economic growth decreased dramatically during the decade (European Commission, 2001). Table 2 demonstrates trends in TFP (total factor productivity) growth rate, R&D intensity (the ratio of R&D investment to GDP) and marginal productivity of technology (MPT) in Japan and the U.S. during the period from 1975 to 2001. TFP growth rate in Japan was 2.8 % per year in the late 1980s, significantly higher than that of the U.S., 0.9 % per year. However, the reverse occurred in the 1990s. The growth rate in Japan fell to negative 0.3% per year in the first half of the decade and slightly recovered to 0.2 % per year in the second half. During the decade, Japan maintained a high level of R&D intensity around 3.0 % (Ministry of Ministry of Education, Culture, Sports, Science and Technology of Japan, 2002, 2003). Since TFP growth rate is measured by the product of R&D intensity and MPT, the dramatic decline in TFP growth rate notwithstanding such high R&D intensity can be attributed to the remarkable decrease in MPT.

		1975-1985	1985-1990	1990-1995	1995-2001
Japan	TFP growth rate	1.4	2.8	-0.3	0.2
	R&D intensity	2.2	2.8	2.9	3.1
	MPT	0.6	1.0	-0.1	0.1
U.S.	TFP growth rate	1.0	0.9	0.9	1.5
	R&D intensity	2.3	2.7	2.6	2.6
	MPT	0.4	0.3	0.3	0.6

Table 2. Trends in growth rate of TFP, R&D intensity and Marginal Productivity of Technology (MPT) in Japan and the U.S. (1975-2001) – average change rate; % per annum

In contrast, the U.S. increased its TFP growth rate from 0.9 % per year to 1.5% per in the 1990s due to doubling of MPT as tabulated in Table 2. While Japan suffered from a misleading option, the U.S. successfully shifted its growth trajectory to a new trajectory which maintains sustainable growth based on developing new functionality (Watanabe, 1995; Fukuda & Watanabe 2008). ICT has a feature that closely interacts with individuals, organizations, and society during the course of its diffusion (Watanabe et al., 2004). This feature enhanced the U.S. industry efforts to expand outsourcing, disseminate products and services quickly, and improve relationships with customers. Furthermore, intensive R&D efforts in both the government and industry achieved dramatic progress in ICT. Thus, the U.S. innovation ecosystem successfully substituted ICT for manufacturing technologies and promptly increased adaptability to a new paradigm of ICT.

3.3 New reality for Japan and the U.S. in the 2000s

Facing the new century, Japan and the U.S. confronted a new reality of global competition in innovation.

While the U.S. enjoyed the benefits of ICT in the 1990s, threats to its competitiveness emerged internally and externally. During the decade, the U.S. capacity of innovation stagnated (Porter et al., 1999). One of the causes underlying the stagnation was emerging shortage in the R&D talent pool. The number of R&D workers as a percentage of the total workforce declined. Many international R&D talents were trained in the U.S. and returned their home country on completion of their studies. Another cause was declining investment in R&D by cutbacks at the Federal level. Total spending on basic research has declined steeply as a percentage of GDP. In addition, private research showed clear signs of becoming much shorter term. Meanwhile, catch-up competitors in emerging countries achieved rapid growth to threaten the U.S. global competitiveness. China and India accelerated their GDP growth faster than major advanced countries (OECD, 2010). These disorders suggest inertia of the U.S. innovation ecosystem which caused it to lose adaptability to rapid changes in the global economy.

On the other hand, Japan made every effort to learn from the accomplishment of the U.S. in the 1990s. Japanese high-technology firms learned from and assimilate the experiences of the U.S. to achieve greater gain from ICT thorough competition in the global markets. Even as high-technology industry overall strengthened learning efforts, some firms improved its learning ability more effectively than others. Consequently, a discrepancy firms endeavoring to learning from the global markets and clinging to traditional business behaviors increased (Fukuda & Watanabe, 2008; Watanabe, 2009). Fig. 1 illustrates trends in learning coefficients in four leading electric machinery firms. The coefficients in Canon and Sharp reversed to steadily increase in 1992 and 1997, respectively. While Hitachi and Panasonic also started to increase their coefficients in the end of the century, those increases were not significant. The difference in their learning results led to a difference in profitability. The operating income to sales (OIS) of these four has changed in two different trajectories as demonstrated in Fig. 2. Canon and Sharp increased their average operating income to sales (OIS) steadily over all three periods examined. Comparing the average OIS in the 2000s with that of the 1980s, Canon doubled its OIS and Sharp increased it by 50%. On the other hand, Hitachi and Panasonic decreased their OIS sharply. Hitachi reduced its average OIS by 48% in the 1990s and 66% in the 2000s, and Panasonic decreased its average OIS by 41% in the 1990s and 86% in the 2000s.

These trends imply that the Japanese innovation ecosystem was resilient to the economic downturn in the 1990s. Although the national innovation ecosystem revitalized its performance through learning from competitors, its composition changed to being more heterogeneous after the downturn.

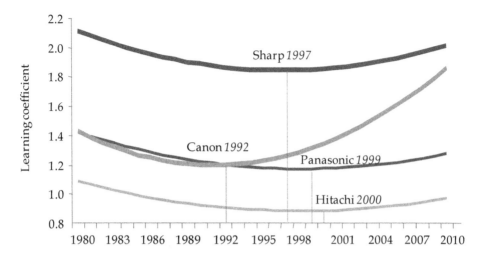

Source: Fukuda & Watanabe, 2008.

Fig. 1. Learning coefficients of Japanese four leading electric machinery firms (1980-2003: estimate; 2004-2010: prediction based on trends in 1980-2003).

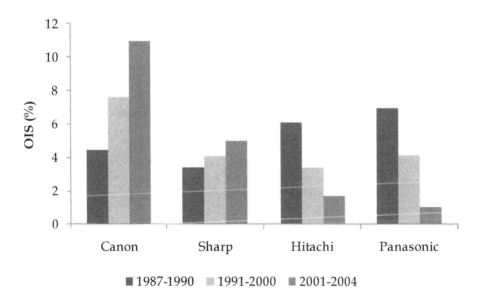

Source: Nikkei Financial Data, 2005.

Fig. 2. Operating income to sales (OIS) of Japanese four leading electric machinery firms (1987-2004).

4. Co-evolutionary dynamism in innovation ecosystem for global sustainability

In the 2000s, the global economy entered an unprecedented new era of global competition due to the continuous rapid growth of emerging countries. Their growth became a new engine of new global growth. New demand from emerging countries created broader markets to offer greater business opportunities. However, at the same time, it became threats to global sustainability. Global production growth increased consumption of energy and natural resources as well as widened a gap between rich and poor.

Confronting such sever circumstances, major developed countries promoted innovation activities to secure their growth and welfare. The U.S. took various measures to activate its innovation performance including fostering R&D talents, stimulating high-risk research and catalyzing alliances between stakeholders. The European Union was launched the Lisbon Strategy in 2000 to build more creative innovation systems during the decade. Individual countries, such as the United Kingdom, Germany and France, also established national innovation strategies. Japan, Taiwan, Singapore and South Korea systematically focused on the direction of innovation to drive their growth.

While developed countries made intensive efforts to maintain their competitiveness, emerging countries expanded their markets and productivity. This was particularly the case in China and India as tabulated in Table 3. Their annual GDP growth rates exceeded or approached two-digit levels until the global financial crisis after the collapse of Lehman Brothers in 2008. In contrast, the growth rate in Japan maintains low level and in the U.S. remains lower than that in the 1990s.

	Average 1990-1999	Average 2000-2008	2007	2008	2009
China	12.5	13.1	17.6	12.0	10.2
India	8.1	9.6	13.1	8.5	7.7
Japan	3.8	4.0	5.4	1.0	-5.4
U.S.	5.5	4.9	4.9	2.2	-1.7

Source: International Monetary Fund, World Economic Outlook Database, April 2011.

Table 3. Growth of GDP (PPP) in four countries – average change rate; % per annum

These contrasting growth trajectories depend on development and utilization of ICT. Fukuda et al. (2011) conducted an empirical analysis to examine ICT contribution to the economic development in 40 countries consisting of both of emerging and developed countries including OECD members, ASEAN original members, Taiwan and BRIC. The results reveal that emerging and developed countries take contrasting approaches to functionality development by ICT as demonstrated Fig. 3. In developed countries, increase of ICT driven functionality development results in decrease of marginal productivity of ICT (*MPI*), whereas it improves *MPI* in emerging countries. This distinct difference suggests that developed countries have fallen into the paradox of ICT development which resulted from a vicious cycle between ICT driven functionality development and its marginal productivity improvement while emerging countries have maintained a virtuous cycle during the global financial crisis, resulting in significant contribution of ICT to their economic growth.

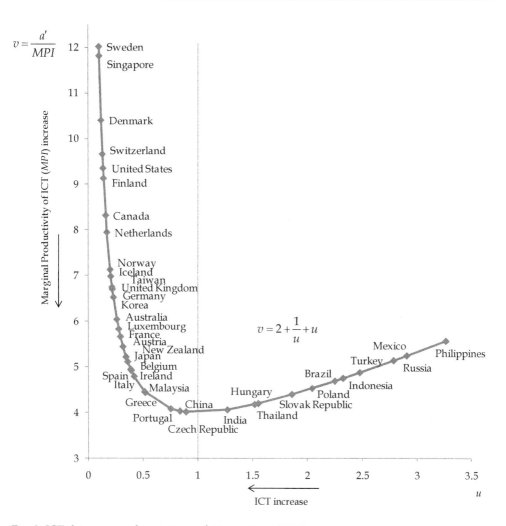

Fig. 3. ICT driven growth trajectory of 40 countries (2009).

While emerging countries are expected to continue to increase consumption steadily, their propensity to consume is not necessarily below that of developed countries as demonstrated in Table 4. Contrary to the gap in GDP per capita, household consumption as a share of GDP in emerging countries is almost at the same level as in developed countries. This contrast suggests that there should be certain structural impediments to consumption-led growth in emerging countries.

Inspired by this observation, Fukuda & Watanabe (2011) conducted an empirical analysis to identify the optimal trigger for the consumption effects on economic growth in 37 out of 40 countries examined above (except 3 countries whose data was not available). Since investment in both public and private sectors is the large component of GDP by expenditure as well as consumption, the analysis emphasized the role of investment. The results indicate contrasting approaches to development trajectory of marginal productivity of investment per capita (*MPi*) induced by GPD per capita increase between emerging and developed countries as illustrated in Fig. 4. While economic growth in emerging countries largely depends on consumption growth rather than investment increase, they confront a vicious cycle where per capita GDP growth leads to *MPi* decrease. Developed countries, on the other hand, leverage investment for their growth.

	HFCE per GDP (PPP) (%)	GDP per capita (PPP) (current $)
Emerging countries [a]	56.9	3945.1
Developed countries [b]	58.2	39458.7

[a] 8 countries: Brazil, China, India, Indonesia, Malaysia, Philippine, Russia and Thailand
[b] 29 countries: Austria, Belgium, Canada, Czech Republic, Denmark, Finland, France, Germany, Greece, Iceland, Ireland, Italy, Japan, Korea, Luxembourg, Mexico, Netherlands, New Zealand, Norway, Poland, Portugal, Slovak Republic, Spain, Sweden, Switzerland, Taiwan, Turkey, United Kingdom and United States.
Source: International Monetary Fund, World Economic Outlook Database, April 2011.

Table 4. Household final consumption expenditure (HFCE) per GDP and GDP per capita in 37 countries (2009) – median in each economic development level

The analysis also reveals that emerging and developed countries take contrasting approaches to investment driven development. Emerging countries have encountered an autarky cycle of consumption driven development. Although they have increased GDP by consumption growth strongly, they simultaneously suffered from the drop of *MPi* along with GDP growth. As a result, they cling to an autarky cycle where consumption contributes to life improvement and then brings GDP growth. On the contrary, developed countries enjoy a virtuous cycle between GDP, consumption and investment growth. Here, GDP growth induces consumption increase. Increased consumption, in turn, increases GDP to induce investment. Investment stimulates further GDP growth, which increases consumption demand for more attractive goods and services. The new demand contributes to a better quality of life and then leads to GDP growth.

A possible trigger for inducement of investment by growth in emerging countries can be 'frugality'. Frugality does not just mean second-rate or low cost (The Economist, 2010), but satisfies new demand on the ground in emerging countries. Their new demand come from their own unique economic, environmental and social situations which are completely different from those in developed countries, and implies the necessity of new functionality to improve their life. Emerging countries will necessitate more in-market, low-cost innovations that make new products are services satisfying frugality for their sustainable development. This necessity urges firms to change their business strategy in emerging countries. Historically, they have relied on local adaptation strategy to deliver their products and services and make a few adaptations for local markets. However, it is shifting to in-market development starting with local innovation to create new global products and services. Developed countries need to be a greater focus not on activities oriented toward their own perspectives but on demand of emerging countries from their own perspective

(Jose, 2008; Landrum, 2007). Frugality is the requirement to satisfy new demand of emerging countries from their own perspective for more attractive products and services, which would trigger a shift from a closed cycle to an investment driven cycle of growth.

$y = \alpha' \cdot MPi$

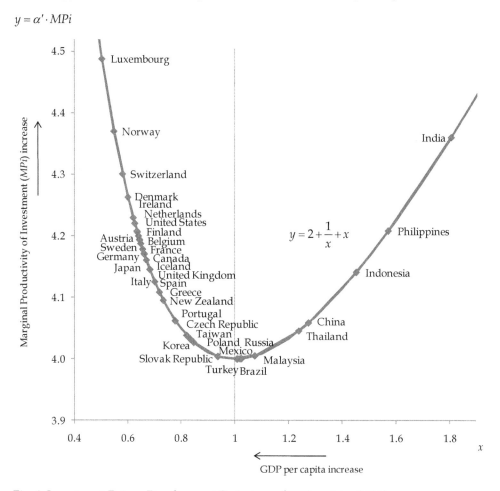

Fig. 4. Investment Driven Development Trajectory of 37 Countries (2009).

All the above results imply the necessity of co-evolution between emerging and developed countries in the new era of global competition. The co-evolution will resolve the paradox of ICT confronted by developed countries, and will enable emerging countries to shift from an autarky cycle to an investment driven cycle of growth. ICT is an essential tool for new functionality development, and effective growth driven by investment is necessary for life improvement in sustainable way. Frugality is a key for fusing the ability of emerging countries to leveraging ICT and the approach of developed countries to effective growth driven by investment. It could lead a way to exploring the new functionality which induces investment for their economic and social well beings, and further contribution to global sustainability.

The co-evolution between emerging and developed countries toward frugality will improve adaptability of innovation ecosystems of both countries to the era of global competition. Emerging countries are facing limitations in their growth cycle heavily depending on consumption growth whereas their vigorous growth is becoming a key driver of the global economy. On the other hand, developed countries are suffering economic stagnation and seeking an opportunity to boost their growth in emerging countries. Both emerging as well as developed countries need new partners to co-adapt to rapidly changing circumstances through co-evolving each other. The co-evolution with new partners will improve the performance of a national innovation ecosystem to survive the global competition as well as overcome threats for global sustainability economically, environmentally and socially.

5. Conclusion

Sustainable development is a serious global concern today and requires new innovation urgently. Innovation creates new value through new products and services, and contributes to long-term wealth creation and higher living standards. Continuous innovation is necessary for sustainability of each country as well as the global society.

Given innovation is a complex and multidimensional, its dynamics is best seen as an ecosystem where various stakeholders are interacting and networking each other. A natural ecosystem provides a suggestive analogy for sustaining an innovation ecosystem. A biological ecosystem maintains stability by resistance, resilience and functional redundancy to external disturbance, and species co-exist, co-evolve and co-adapt to address changing circumstance demands. This combination of autonomy of species and coordination of an ecosystem achieves sustainable development of ecosystems.

A techno-economic analysis can provide insight into mechanisms of an innovation ecosystem. It explains what kind of and how production factors including labor, capital, energy and materials, and technology stock contribute to productivity increase for sustainability corresponding to the environmental condition. It also demonstrates substitution possibilities between production factors for maintaining sustainable growth in the face of the constraints of certain production factors. Thus, a mechanism enabling Japan's success in overcoming energy crises can be analyzed as technology (which is constraints free production factor) substitution for energy (which is critical constraint for Japan). These mechanisms suggest dynamism between inputs, outputs and the external environment of both of them. A nation's policy environment, common infrastructure, culture and tradition influence dynamics between input and output where they both affects and are affected each other. A techno economic analysis incorporating these institutional factors helps to elucidate a complex and dynamic innovation ecosystem.

Japan and the U.S. have developed their national innovation ecosystem through the co-evolutionary cycle with the U.S. during the last three decades. The reversal of competitive dominance between both countries corresponds to a paradigm shift in each decade as follows:

- In the 1980s, Japan has achieved conspicuous economic development. This achievement can be attributed to its success of technology substitution for constrained production factors, primarily energy, after the energy crises in the 1970s.

- In the 1990s, however, the U.S. recovered its dominant competitive position over Japan. The U.S. successfully substituted ICT for manufacturing technology. This success resulted in timely switch from a traditional trajectory in which economic growth leverages further growth to a new trajectory which maintains sustainable growth based on developing new functionality.
- Facing the 2000s, the U.S. is again confronting a new reality due to the emergence of catch-up competitors such as India and China, as well as a move in Japan toward new innovation. This confrontation can be attributed to resistance of the national innovation ecosystem to rapid changes in the global economy.
- Japan began to show signs of recovery in the early 2000s due to learning from competitors in the 1990s. However, at the same time, the recovery sheds light on a profitability gap in Japanese high-technology. High-profit firms endeavor to learning from the global markets whereas low-profit firms cling to traditional business behaviors.

During the above mutual co-evolutionary competition with the U.S., Japan has maintained the system stability by means of transferring external disturbance to a springboard for innovation. This maintenance of ecosystem stability can be attributed to a sophisticated combination of autonomic efforts by industry and coordination by the government, which contributes to fusing indigenous structural strength in the ecosystem and learning from competitors. However, the national innovation ecosystem exposes limitations in the function due to resistance to the new era of global competition, coinciding with the U.S. confrontation with a new reality.

These limitations of both national innovation ecosystems necessitate co-evolution with new partners of innovation to address rapidly changing circumstances. The global financial crisis after the Lehman Brothers collapse in 2008 explicitly indicated that emerging countries is driving the global economic growth. Amid a shift of the center of innovation gravity from developed countries to emerging countries, co-evolution with emerging countries is becoming a big challenge for not only Japan and the U.S. but also other developed countries. Meanwhile, threats to global sustainability are growing seriously. Rapid growth of emerging countries is increasing industrial production as well as consumption in the world enormously, and in turn, endangers the economic, environmental and social survival of life.

Frugality is a key to new functionality satisfying local demand of people in emerging countries. Most of people in emerging countries are rising as the new middle class. They contribute to sustainable development of emerging countries, which affects global sustainability amidst the economic structural shift to emerging countries. The new middle class's contribution to global sustainability comes through its consumption growth for life improvement and its investment inducement to further economic growth leading consumption demand for more attractive goods and services. Frugality would trigger the shift from an autarky cycle between the consumption and GDP increases to investment driven development

The co-evolution between emerging and developed countries would generate frugality oriented new functionality development trajectory. Developed countries have accumulated their efforts to serve demands in emerging countries. On the other hand, emerging countries have leveraged ICT for economic growth and promoted innovation in their own unique economic, environmental and social situations. Fusing these efforts would realize the co-evolution between them leading to sustainable development in emerging countries and global sustainability as well.

Japan has overcome the oil crisis twice and a severe energy shortage and has subsequently improved energy efficiency in the 1970s by joint efforts of government and industry. These efforts accelerated learning and assimilation of spillover technology which contributed to dramatic decrease of energy consumption in each segment of manufacturing industry. The success in declining unit energy consumption has enabled Japan to establish an energy efficient and eco-friendly society. Its primary energy consumption and carbon-dioxide emission in 2003 are 106 and 68.9 tons of oil equivalent/GPD respectively, both of which are lower than half of those of the world's average. This conspicuous energy efficiency will remain an indigenous strength of Japan's innovation ecosystem and a driving force of innovation toward sustainable development.

A competence to address rapidly changing circumstances in the world is crucial to sustainable development. Perspectives on innovation ecosystems provide precious suggestion for continuous improvement of the competence. Mutual inspiration between stakeholders as well as stakeholders and an innovation ecosystem achieve this improvement. Furthermore, mutual co-evolution between innovation ecosystems enhances each performance enough to survive severe competition and threats for global sustainability. Sustainable development is a big challenge for all innovation ecosystems.

6. References

Allison, S.D. & Martiy, B.H. (2008). Resistance, Resilience, and Redundancy in Microbial Communities. *Proceedings of the National Academy of Sciences of the United States of America*, Vol. 105, Supplement 1, (August, 2008), pp. 11512-11519, ISSN 0027-8424

Council on Competitiveness. (2004). *Innovate America: National Innovation Initiative Summit and Report*, Council on Competitiveness, ISBN 1-889866-20-2, Washington, DC, USA

European Commission. (2001). *European Competitiveness Report 2001*, Office for Official Publications of the European Communities, ISBN 92-894-1813-3, Luxembourg.

Fukuda, K. & Watanabe, C. (2008). Japanese and US Perspectives on the National Innovation Ecosystem. *Technology in Society*, Vol. 30, No. 1, (January, 2008), pp. 49-63, ISSN 0160-791X

Fukuda, K., Zhao, W. & Watanabe, C. (2011). *Dual Hybrid Management of Technology: Co-evolution with Growing Economies*, Journal of Technology Management for Growing Economies, Vol. 2, No. 1, (April, 2011), pp. 9-26, ISSN 0976-545X

Fukuda, K. & Watanabe, C. (2011). *A Perspective on Frugality in Growing Economies: Triggering a Virtuous Cycle between Consumption Propensity and Growth*, Journal of Technology Management for Growing Economies, Vol. 2, No. 2, (October, 2011), in print, ISSN 0976-545X

Iansiti, M. & Levien, R. (2004). *The Keystone Advantage: What the New Dynamics of Business Ecosystems Mean for Strategy, Innovation, and Sustainability*, Harvard Business Publishing, ISBN 1-59139-307-8, Boston, USA

IMD. (2002). *The World Competitiveness Yearbook 2002*, IMD, ISBN 2-970-01216-2, Zurich, Swizerland.

Industrial Structure Council. (February 2005). Science & Technology Policy Inducing Technological Innovation, In: *Ministry of Economy, Trade and Industry*, 17.06.2011, Available from: http://www.meti.go.jp/report/downloadfiles/g50223a01j.pdf

Intergovenmental Panel on Climate Change. (2007). *Climate Change 2007 : Synthesis Report*, IPCC, ISBN 92-9169-122-4, Geneva, Switzerland

Jose, P.D. (2008). Rethinking the BOP: New Models for the New Millennium. *IIMB Management Review*, Vol. 20, No. 2, (June 2008), pp. 198-202, ISSN 0970-3896

Klein, S.J. & Rosenberg N. (1986). An Overview of Innovation, In: *The Positive Sum Strategy: Harnessing Technology for Economic Growth*, Landau, R & Rosenberg N. (Eds.) National Academy Press, pp. 275-306, ISBN 0-309-03630-5, Washington, DC, USA

Landrum, N.E. (2007). Advancing the "Base of the Pyramid" Debate. *Strategic Management Review*, Vol. 1, No. 1, (May 2007), pp. 1-12, ISSN 1930-4560

Marten, G.G. (2001). *Human Ecology: Basic Concepts for Sustainable Development*, Earthscan, ISBN 1-85383-714-8, London, UK

Millennium Ecosystem Assessment. (2005). *Ecosystems and Human Well-being: Current State and Trends*, Island Press, ISBN 1-55963-227-5, Washington, DC, USA

Ministry of Education, Culture, Sports, Science and Technology of Japan. (2002). *Annual Report on the Promotion of Science and Technology 2002*, MEXT, ISBN 4171520770, Tokyo, Japan.

Ministry of Education, Culture, Sports, Science and Technology of Japan. (2003). *White Paper on Science and Technology 2003*, MEXT, ISBN 4171520789 , Tokyo, Japan.

Organisation for Economic Co-operation and Development. (2008). *Open Innovation in Global Networks*, OECD Publishing, ISBN 978-92-64-04767-9, Paris, France

Organisation for Economic Co-operation and Development. (2009). *Tackling Inequalities in Brazil, China, India and South Africa: The Role of Labour Market and Social Policies*, OECD Publishing, ISBN 978-92-64-08835-1, Paris, France

Porter, M.E., Stern, S. & Council on Competitiveness. (1999). *The New Challenge to America's Prosperity: Findings from the Innovation Index*, Council on Competitiveness, ISBN 1-889866-21-0, Washington, DC, USA

President's Council of Advisors on Science and Technology. (January 2004), Sustaining the Nation's Innovation Ecosystem: Information Technology Manufacturing and Competitiveness, In: *PCAST*, 17.06.2011, Available from: http://www.whitehouse.gov/sites/default/files/microsites/ostp/pcast-04-itreport.pdf

Ravallion, M. (2010). The Developing World's Bulging (but Vulnerable) Middle Class. *World Development*, Vol. 38, No. 4, (April 2010), pp. 445-454, ISSN 0305-750X

Senge, P.M. & Carstedt G. (2001). Innovating Our Way to the Next Industrial Revolution. *MIT Sloan Management Review*, Vol.42, No.2, (January 2002), pp. 24-38, ISSN 1532-9194

Sheth J.N., Sethia N.K. & Srinivas S. (2011). Mindful Consumption: a Customer-centric Approach to Sustainability. *Journal of the Academy of Marketing Science*, Vol. 39, No. 1, (February 2011), pp. 21-39, ISSN 0092-0703

The Economist. (2010). First Break All the Rules. *The Economist*, Vol. 394, No. 8678, (April 2010), pp. 3-5, ISSN 0013-0613

Watanabe, C. (1992). Trends in the Substitution of Production Factors to Technology. *Research Policy*, Vol. 21, No. 6, (December 1992), pp. 481-505, ISSN 0048-7333

Watanabe, C. (1995a). The Feedback Loop between Technology and Economic Development: An Examination of Japanese Industry. *Technological Forecasting and Social Change*, Vol. 49, No. 2, (June 1995), pp. 127-145, ISSN 0040-1625

Watanabe, C. (1995b). Mitigating Global Warming by Substituting Technology for Energy. *Energy Policy*, Vol. 23, No. 4/5, (June 1995), pp. 447-461, ISSN 0040-1625

Watanabe, C. (1999). Systems Option for Sustainable Development: Effect and Limit of the MITI's Efforts to Substitute Technology for Energy. *Research Policy*, Vol. 28, No. 7, (September 1999), pp. 719-749, ISSN 0048-7333

Watanabe, C., Kondo, R., Ouchi, N., Wei, H. & Griffy-Brown C. (2004). Institutional Elasticity as a Significant Driver of IT Functionality Development. *Technological Forecasting and Social Change*, Vol. 71, No. 7, (September 2004), pp. 723-750, ISSN 0040-1625

Watanabe, C. (2009). *Managing Innovation in Japan: The Role Institutions Play in Helping or Hindering how Companies Develop Technology*, Springer, ISBN 978-3-540-89271-7, Berlin, Germany

Westman, W.E. (1978). Measuring the Inertia and Resilience of Ecosystems. *BioScience*, Vol. 28, No. 11, (November, 1978), pp. 705-710, ISSN 0006-3568

World Business Council for Sustainable Development. (2008). *Sustainable Consumption Facts and Trends From a Business Perspective*, WBCSD, ISBN 978-3-940388-30-8, Geneva, Switzerland

World Commission on Environment and Development. (1987). *Our Common Future*, Oxford University Press, ISBN 0-19-282080-X, New York, USA

11

The Environment as a Factor of Spatial Injustice: A New Challenge for the Sustainable Development of European Regions?

Guillaume Faburel
Urban Planning Department (Institut d'Urbanisme de Paris)
University of Paris Est – Coordinator of Research Unit Aménités
France

1. Introduction

The poor are much more subject and vulnerable to environmental degradation, or risks of its occurring; they are also more strongly affected by the negative impact of certain international, national or local policies. This has long been the case, in France and abroad, in the North as well as, of course, in the South (see for instance Schroeder and al., 2008). Similarly, so-called pro-environmental attitudes and practices (relative to food, energy or mobility, for example), which have recently made their appearance, particularly in western European countries, prove to be no less non-egalitarian or inequitable. This issue nowadays represents a major stake for social and spatial justice, at various levels: from the continental and intercontinental (e.g.: ecological debt, environmental refugees…), to the local level (e.g.: socially precarious energy resources), and encompassing the urban scale (e.g.: gentrification and environmental segregation). Yet, the equitable rights of individuals to a healthy and quality environment have been set down in a number of texts, some of them constitutional, both international (Aalborg and Leipzig Charters, in 1998 and 2007, Declaration of Istanbul in 1996) and national (e.g. Environmental Charter in the French Constitution in 2005).

The values (moral, social and/or esthetic references embraced by a given group at a given time) and the principles that found its action (social norms and rules of implementation), embodied in public planning, environmental and even social policies, are direct queries, both as concerns their contribution to these contradictions, and with a view to bring about change for sustainable development. Starting out from sustainable development, we note that for a long time the official discourse relative to these values and principles limited itself to the "prophetic horizons" by formulating the famous ecological, economic and social pillars: livable, viable… and equitable. These are what sustainable development was supposed to guarantee. In fact, it is only as part of eco-neighbourhood (or so called sustainable neighbourhood) projects, and more generally within the framework of so-called sustainable urbanism and planning projects, that these considerations on values and principles are today more clearly highlighted. It is also true that they pose increasingly concrete questions concerning the various forms of socio-environmental segregation, i.e. social and environmental inequalities in certain places, to which these new neighbourhoods may have given rise abroad (BedZed in London, BO01 in Malmoe, Vauban in Freiburg, Germany).

Thus, the theme of environmental inequalities, and the forms of injustice they generate, appear to be anything but neutral for the practical implementation or practice of sustainability. In France for instance, this is illustrated by the updating in 2006 of the National Scheme of Sustainable Development, which places environmental inequalities squarely in the center of the approach. For Europe, we find similar initiatives in Scotland, with the Strategy for Sustainable Development (Section 8, 2005), as well as the earlier official report of the UK Environmental Agency on *Poverty and the Environment* (2003), which subsequently introduced a poverty indicator into environmental accounting (UK Environmental Agency, 2007). In fact, it is in the United States that official recognition of this issue goes back the furthest. Born of the civic rights movement and the fight against discrimination, *Environmental Justice* is based on early proof (General Accounting Office in 1979 and 1983 ; the United Church of Christ, in 1987 ; Bullard in 1983, 1990 and 1994 ; Wenz in 1988) of a non-egalitarian distribution, first ethnic (especially Blacks, Amerindians and Hispanics), then economic, of populations relative to the major forms of infrastructure and equipment that have a major impact on the environment (health risks, mortality rates). On 11 February 1994, the Federal Administration institutionalized *Environmental Justice* pursuant to Executive Order 12898 : *Federal Actions to Adress Environmental Justice in Minority Populations and Low-Income Populations*. This Executive Order decreed that all federal agencies including the EPA or Environmental Protection Agency should: *"identify and remedy the effects of measures that disproportionately affect the health and living conditions of the poor or those who belong to ethnic minority groups"*.

More recently, as we will discuss below, developments focus particularly on regions or cities, which increasingly concentrate these environmental inequalities, and challenge social and spatial justice. For instance, as mentioned by the Interministerial Delegation for Cities in France 2006, it is becoming difficult to call for social mixity in neighbourhoods with a strongly degraded environment. It is true that, as confirmed by experiences with eco-neighbourhoods, such inequalities are particularly damaging when considering the city, i.e. lifestyles that are strongly affected by socio-spatial divisions and forms of socio-spatial segregation that are historically constituted but also subject to powerful market mechanisms (e.g.: scarcity of property and building costs / acquiring housing). If one adds a few recent challenges and the ecological considerations they feed into (e.g. 'shrinking cities', even 'urban decline'), it is easy to admit that environmental inequalities theoretically represent major social and spatial stakes for territorial governance and urban regulation.

Hence, if the subject of environmental inequalities or injustices is today an increasingly vital question addressing the sustainable development that underlies a growing number of actions, it continues to be globally ignored or overlooked in the public policies. Admittedly, the subject closely interlinks environmental, social and economic aspects, a combination that is theoretically at the basis of all sustainable thought and action, but often finds it difficult to fully realize them. In fact it requires that we overcome sectorial approaches that have developed historically and that are often implemented rationally from the top down. The first reason being that they address major questions relative to the technical approaches and normative answers developed to date throughout the world to solve these problems, and to reposition them within the universe of socio-environmental responses, particularly in cities.

The aim here is to understand why reflections undertaken on environmental inequalities and injustices could, under certain conditions of action, generate a new perspective of the sustainable city in European regions, by repositioning the terms of the debate, linking social, spatial and environmental justice. Going out from findings and examples of scientific studies from several European countries (France, Germany, Netherlands, United Kingdom…), this chapter aims to highlight the scientific benefits of adopting a different approach to the environment, linking it to social situations and the construction of territories, in order to:

-   Not only provide different scientific findings on the state of environmental disparities, inequalities or even injustices, particularly relative to spatial injustices, singularly in cities;
-   But also invent other types of actions and means of intervention for sustainable development at urban or regional scales in Europe.

With this aim in mind, the chapter will be divided into two main parts.

The first will propose a few findings and an oriented synthesis of scientific research, based on 50 studies, both French and international (USA, the United Kingdom, Belgium, the Netherlands…). In 2008, the objective was establish, for the French Center for Scientific Research, an international and trans-disciplinary report on the state of the art on environmental justice (main topics, issues and purposes), in order to better identify scientifically relevant issues in France comparatively to other European countries, as assets or limitations, even as hot spots in public and private decision making supports and processes (Faburel, 2010a).

Two integrated hot spots and topics have been particularly explored. Traditionally, in France, environmental issues are viewed through an institutional lens which emphasizes technology and bureaucratic tools of assessment and action. Thus, the historical and legal spatial approach to justice (e.g.: land use and city planning, housing policies…) uses a technocratic and normative conception of the environment to face up to environmental challenges in innovative ways (such as environmental segregation in large cities). However, recent research projects carried out in several countries as well as in France stress the fact that environmental justice should take a more dynamic approach, for instance accounting for local and historic dimensions. So, considering the logic of decision makers and the cultures in the urban field, it has been proposed to explore new ways of thinking that would improve the inclusion of environmental inequalities from the perspective of sustainable development. One way would be to focus on lifestyles and people's experiences linked to the environment, and their attachment to a particular place. Another way would be to adopt a participatory rather than a structural approach to the investigation of exclusion and capacity forms of involvement (i.e. *capabilities*, in Sen, 1993 and 2009) instead of more conventional behavioural markers of urban inequality (such as moving house, for example).

The second part of the study proposes an empirical approach which applies these orientations towards environmental perceptions, representations and local experiences, such as:

-   Pertinent issues that provide an interesting scale for the observation or the highlighting of certain other factors that determine urban inequalities in cities;
-   Thus orienting both the evaluation (generally based on static and descriptive nomenclatures) and territorial decision making directed by sustainability.

It was conducted for the French Ministry of Ecology, Sustainable Development, Transport and Housing in close cooperation with the Ile-de-France (Paris) region (Faburel et Gueymard, 2008). On the one hand the study confronted so-called objective environmental data (geophysical indicators usually employed to characterize resources and harms: degrees of pollution, noise levels, density of green areas…) with classical socio-economic information (indicators on income, employment, housing…) in 1300 municipalities in the Paris region in order to pinpoint the major types of disparities in the environmental quality of the living environment. Thus, after identifying, on this basis, 6 municipalities close to Paris considered representative of different disparity situations, we conducted a survey in order to confront their responses with the data generated in the first part.

These various linkages notably made apparent a list of environmental objects and factors that make a place attractive or undesirable. Our study also highlighted certain difficulties relative to environmental evaluation and monitoring in urban, suburban and even rural territories. Information on the living and felt environment, through local experiences, satisfaction, place attachment relative to the environment, generated additional elements for a finer assessment of local disparities, inequalities or even injustices (neighbourhood, municipality, inter-municipality), in a sustainable development perspective. The conclusion addresses the issue of the role of the living environment and social involvement in decision making processes, balancing between institutional and bottom up approaches to sustainability for European regions.

2. On several major findings and conceptual stakes for sustainable development: Towards new links between justice and the environment in public policies?

2.1 On observing environmental inequalities at different scales

Abroad, this approach linking living conditions and environmental quality is not new, if one considers the *Environmental Justice* movement in North America which goes back to the 1970s (*supra*); even in France, where it was more modest and used different reference terms, it goes back to the 1980s. At the end of the 1980s, for example, in France a suburban social housing development was four times more likely to have an expressway running through it; in 1986, low income populations were proportionally four times more exposed to annoying noise levels (French National Institute for Research on Transport and Security, 1988). However, ecological crises and environmental ordeals have generated new stakes in this field, at different scales:

-   from the international scale, with for example between 50 to 163 million climate refugees, fleeing desertification, deforestation, soil erosion, and disasters (partly also caused by large scale development projects: mines, dams, periurbanisation, biofuels, etc.); in wider terms, owing to the poverty gap between regions (e.g. access to drinking water, food shortages),
-   to the more local scale of energy precarity and insalubrious housing of low income populations in certain urban neighbourhoods (plus, in our regions, emerging problems relative to environmental health),
-   but also including environmental segregation in cities, with such issues as pollution, nuisances and urban risks which increasingly discriminate between social groups,

regardless of sometimes laudable policies: green taxation and energy measures, steps to protect the landscapes of historical city center neighbourhoods, projects of so-called sustainable/eco neighbourhoods (*supra*)...

184 Cité Pierre-Semard à Saint-Denis, 1956-60. Vue aérienne (depuis le nord)

Photo 1. Public responsibility: large scale housing developments in France (Ile-de-France region in 1950s)

A great number of data were generated only recently. The attention of the international community focuses on climate change and natural hazards. The 2007 report of the Intergovernmental Panel on Climate Change (IPCC) shows for example that in 2004 the poorest countries represented 37% of the world's population, but only 7% of CO_2 emissions, whereas the richest countries showed an inverse proportion of 15% to 45%. Similarly, as shown in the the table below, natural disasters imply different levels of damage.

Catégorie de revenu	Nombres de désastres	Population (millions)	PIB par habitant	Nombres de morts	Coût total, en % du PIB
Haut revenu	1 476	828	23 021	75 425	0,007
Bas revenu	1 533	869	1 345	907 810	0,55

Source: Stromberg, David, 2007, « Natural Disasters, Economic Development, and Humanitarian Aid, » *Journal of Economic Perspectives*, Vol. 21 (Summer), pp. 199–22.

Column titles: Income category, Number of disasters, Population (millions), GDP per inhabitant, Number of dead, Total cost in % GDP. Line titles: High income, Low income)

Table 1 Rich countries and poor countries in the face of natural disasters

At the national scale, industrial risks (chemical and other), polluted sites and soils have been the object of several recent studies. It has for example been shown that metropolitan France has a very unequal distribution of high risk sites (safe industrial waste dumps, waste incineration facilities, Seveso sites). 8 % of municipalities harbour two sites, 2.5 % three or more. The southeastern and northern Paris regions (along the old industrial valley of the Seine), the poorer regions around Marseille as well as the large "industrial" agglomerations of the North - Pas de Calais harbour (Laurian, 2009).

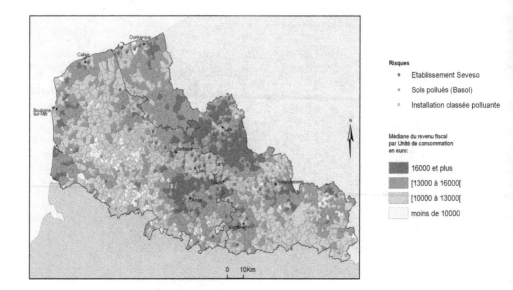

Source: IGN BD-CARTO, 2000
- INSEE, revenus_fiscaux_menages_par_commune_2002_nord_pas_de_calais, 2002
- DPPR, seveso, 2003
- DPPR, installations classées polluantes
- MEDD, basol, nov 2004
N.B. Document de travail

Legend: Seveso site (red), Polluted soil (blue), Site classified as polluted (green). Income category (brown)

Map 1. Inequalities relative to risks and polluted soils (Nord Pas-de-Calais region in 2000s)

Similarly, energy practices have begun to be analyzed from a social profile angle. In this register, the French Environment and Energy Management Agency (ADEME), for example, calculated that in France the part of energy expenditure of the 20% poorest households is 2.5 times higher than that of the 20% richest households.

Finally, on the urban scale, which up till now has certainly been the least studied, a differentiated offer of natural sites, unequal exposure to nuisances and the disparate quality of the living environment are attracting increased attention. Notably in the Ile-de-France region (Faburel, Gueymard, 2008), it has been shown that 2 750 000 persons were in a situation of environmental inequality, industrial decline and economic change, mainly concentrated in the northeastern departments of the "first ring" (e.g. Seine-Saint-Denis), with a historically low income population, or in more remote areas characterized by recent urbanization owing to poorer populations no longer being able to afford housing in the center of the agglomeration, accompanied by strong environmental impact (e.g. east of the Seine et Marne). We shall come back to this issue in the 3rd part.

And, at this more urban scale, environmental health is increasingly studied, throughout western European regions.

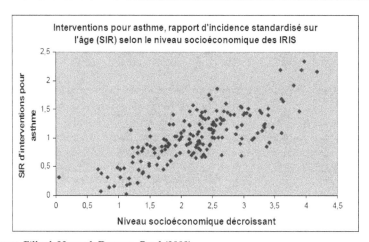

Source: Laurent, Filleul, Havard, Deguen, Bard (2008)
Titles: Interventions for asthma, standardized incidence report on age (SIR) according to socio-economic
level; SIR interventions for asthma; Decreasing socio-economic level

Graph. 1 Correlation between interventions for asthma and socio-economic status in
Strasbourg (France)

2.2 What justice do we mean when we speak of environmental inequalities?

2.2.1 Conceptions of the environment

These data, which we could easily extend to many geographic areas and countries, provide
us with several spatial findings on environmental disparity situations. However, they are
often still purely descriptive and static, and frequently address only pollution, nuisances
and risks. They express a conventional characterization of environmental inequalities:
proportionally higher physico-chemical exposure of low income populations to
environmental loads and sometimes to negative effects (on health, for example). They also
share one characteristic – they often only minimally address the socio-spatial dynamics and
segregation mechanisms that underpin the relative inequality in the environmental field,
particularly in cities where such mechanisms can be highly complex (Faburel, 2008). Thus,
they ignore possible connections between different types of social and environmental
inequalities. All this static information in fact ignores the dynamic nature of all inequality:
"differences that are the result of unequal access to the diverse resources offered by society".

How then, on the basis of only the exposure of populations, can we pinpoint the role of the
environment in mechanisms of segregation? How can one explain that although affluent city
centers are often subject to high noise and air pollution exposure levels, caused by heavy
automobile traffic, they are the object of urban requalification measures? Does this mean
that to ensure justice, everyone should get their equal "share" of exposure, regardless of the
socio-economic means available to avoid it? Could it be that environmental and/or
ecological inequalities are only social inequalities which, relatively to the physical, chemical
etc. attributes of the living environment, highlight other aspects of the historical production
of social divisions of and in places? On the other hand, are such inequalities not one of the

most difficult challenges we must face in view of their economic, cultural, social, psychological, environmental components? What then is their specific content? How should one view such inequalities in a more equitable urban perspective, in the name of sustainable development?

In fact, if these data generate other geographies and territorial characteristics at different scales, notably by means of mappings (*supra*), they above all question the concept of the environment that is involved. The statistics used are in fact generated by the historical assessment apparatus: nomenclatures, protocols and data. In France, this apparatus is inherited directly, as in many other countries, whether centralist or federalist (see '*materialist ethos of sustainability*' by Shirazi, 2011, from Germany), from a techno-centered approach to the environment (Theys, 2010), i.e. to a cognitive institutional rationale, "*conditioned by the possibility of aligning it (the environment) on a normative measure*" (Charvolin, 2003, p. 9). Expert and globalizing assessment criteria are often applied:

-    thresholds of physico-chemical exposure (for air quality, for example),
-    probabilities of the occurrence of official risks (for example to regulate housing construction in response to potential flood risks and hazards),
-    acoustic levels as predictors of annoyance (problems of noise nuisance),
-    distance for the accessibility of urban amenities (e.g. green spaces),
     ...

In our opinion, this very normative and thus objectifying approach to the environment is not appropriate for environmental policies, be they national or local, yet it influences all public policies. For example, the national observatory of so called Sensitive Urban Areas (Zones Urbaines Sensibles - ZUS)[1] recently showed that they suffer particularly from nuisances, pollution and environmental risks; it used approaches that were technical as well as surveys (Choffel, 2004): 38% of households living in ZUS areas declared that they were often bothered by noise, as against 20% of the inhabitants of low-rise residential areas; only 36% were satisfied with the abundance and quality of green areas in their neighbourhood, against 59% in non-ZUS areas. Other applications of these studies also indicate that children from families with a poor standard of living are overexposed to environmental nuisances (Rizk, 2003). However, although this qualitative opening is noteworthy, the psycho-sociological relations with the environment are viewed only within the strict perimeter of the neighbourhood. Also, the housing issue does not address all territorial aspects of the environment (access to nature, mobility, consumption attitudes).

The examples in the box below also illustrate the effects of such initiatives on the scientific understanding of environmental inequalities.

Why is this straitjacket imposed on official nomenclatures and institutional perimeters, including urban policies, although the latter are traditionally prone to opt for more qualitative and social approaches to the facts and mechanisms of inequality?

---

[1] Sensitive Urban Areas (ZUS) are infra-urban areas (e.g. neighbourhoods) which French public policy makers have defined as a priority target for urban policies, in view of the difficulties which their inhabitants encounter constantly (increasingly important fiscal and social provisions). There were 640 of them in metropolitan France in 2005.

Box 1. Some surprises and difficulties when approaching situations of environmental inequality: the case of large-scale transport equipment and infrastructure

Kruize analyzed environmental equity on the scale of the Netherlands and of two strongly urbanized regions, including the Amsterdam-Schiphol airport zone (2007). Environmental inequalities were analyzed according to the distribution of environmental "minuses" ("*bads*"), i.e. situations that did not comply with statutory norms, and of environmental pluses ("*goods*"), i.e. those that complied with the norms or fixed objectives, by income categories. As environmental indicators she used: noise levels (as defined by the statutory indicator), azote oxides rate (compared to thresholds of concentration in the air), official risks (planned zones) and distance to green areas.

The study shows that modest income populations usually live in slightly less environmentally friendly neighbourhoods, with stronger disparities relative to green spaces. The differences observed primarily concern areas in which noise and azote oxide emissions are low. But, surprisingly, the highest income populations are more exposed to noise (i.e. level of acoustical intensity) than populations with the lowest incomes. The author decided to couple this observation with a survey on perceptions and opinions.

Another study that goes back to 2004 (Faburel, Maleyre, 2007) involving eight municipalities in the vicinity of Orly airport (2nd airport in France) made use of the Hedonic Pricing Method to analyze the determinants of the property values of 688 accommodations, selected in the data base of the Paris chamber of notaries public (Chambre des Notaires de Paris). Property value depreciation is observed in the municipalities suffering the highest levels of noise generated by air traffic, with a *Noise Depreciation Index* of 0.96 % of the value, by decibel. This rate concords with what is stated in the literature on the subject, and the municipalities concerned are the poorest in the analyzed sample.

However, thanks to the segmentation of the value bases into several significant periods, one may observe that depreciation increased during the period from 1995 to 2003, going from 0.86% of the price of the property per decibel between the reference municipality and the three municipalities identified at 1.48 %... while noise level remained stable according to official indicators, due to a limitation (cap) on air traffic introduced in 1994.

Indicators based on physico-chemical exposure do not suffice – on the contrary – to explain the dynamic character of non-egalitarian phenomena. "*To draw conclusions with regard to the status of a person's health and well-being, the perception of exposure may be as important as or even more important than objectively measured exposure*"(Mielck, 2004, cité par Kohlhuber et al, 2006).

2.2.2 From concepts of the environment... to concepts of justice

Certainly the environment is still to a large extent viewed in total and universal terms, with prophecies based on technical mastery and the normed reduction of environmental "impacts" feeding into many areas. But above all – and we think this may be the most fundamental reason – any concept of the environment carries with it a concept of justice, since – as demonstrated notably by Peter Wenz (1988) – the environment is specifically linked to such reflection (Faburel, 2010b).

As an example, the *Environmental Justice* trends in the English speaking countries have developed consubstantially:

- a more individualized approach to the environment (at often primarily the local scale), and an essentially distributive justice (based on the measure of environmental values - *preferences-based approaches*), and its theoretical evolution (Rawls, 1971),
- with a few participatory (*Voice* in Hirschman model, 1970) though institutional, aspects (e.g. environmental self-determination, as *class action*), and in a vaster sense, on the capacities to defend, adapt and protect households, as in the Tiebout model (*"feet voting"*, 1956).

We also find this aspect in the definition of environmental inequalities that was officially formulated in 1995 by the US Environmental Protection Agency in a first handbook, *Environmental Justice Strategy*; this included a *toolkit* (indicators and quantitative tools), that was updated in 2004: *"Environmental Justice is the fair treatment and meaningful involvement of all people regardless of race, colour, national origin, or income with respect to the development, implementation, and enforcement of environmental laws, regulations, and policies"*. To this day, because it is grounded on regulation environmental studies, this framework remains highly relevant for waste storage depots and recycling sites, chemical plants, transport infrastructure (roads and airports), almost exclusively from the point of view of the potential or actual pollution they emit, as well as other risks and nuisances.

Most prevalent in European countries, especially the UK and Ireland, this working definition has moved away from racial discrimination to concentrate on social exclusion and environmental issues (Fairburn, 2008) with a specific focus on industrial polluters and clean air campaigns. But these slight differences between countries, for example in how social and ethnic divisions are measured, cannot hide a common factor: the way a concern for equity at the local level tends to strongly influence how we think about environmental issues. Again, in the UK, for instance, there is a tendency to privilege health and epidemiology. And the examples we have cited above at national scale are representative of the production of approaches of this type.

Similarly, with the approach via ecological inequalities of development, which positions itself at the global scale of development models (production conditions, technical systems, forms of social organization) to observe the ecological consequences of inequalities (internal) and disparities in poverty (external), another concept of the environment unfolds, more oriented towards ecological rights and obligations of societies (*rights-based approaches*, in Martinez-Alier, 2002). So by focusing on economic phenomena such as environmental dumping as a by-product of free trade policies (see Baumol and Oates, 1988) and more recent political defeats (such as in some cases a lack of regulation policies), the links between social inequalities, poverty and environmental disasters become clearer. The examples cited above at international scale (e.g. the 2007 Report of the Intergovernmental Panel on Climate Change - IPCC) perfectly illustrate this approach.

Moreover, it is more open to the diversity of lifestyles of populations. We can cite Pye et al's work (2008) which shows empirically how poor Europeans (single parent, low income or unemployed households) have a far lower carbon footprint than others. In this vein the work of Diamantopoulos, Schlegelmilch, Sinkoviks and Bolhen (2003) illustrates at the same scale the decreasing relevance of socio-demographic factors in green consumption habits.

This focus on how much waste our consumerist lifestyles generate also appears in the work of Dozzi, Lennert and Wallenborn (2008) carried out in Belgium at a microlevel: they looked at energy consumption and household spending on water and food including production and delivery costs. Other qualifications are thus given to environmental inequalities, such as those that Pye et al offer in their European Commission report (*op. cit.*): these include discrimination in terms of how different residents are able to access a green lifestyle (where social exclusion exists) and the uneven effects of environmental policies on these same residents.

Above all, this approach brings with it a conception of justice which is somewhat different (Dobson, 1998): much more social and openly procedural (focused on citizen involvement) than strictly (re)distributive at the economic level (via economic compensation for the weakest, for example). And, on this dual basis, the second approach pleads for the need of public action that is more re-founding than simply corrective or compensatory (as in the *Environmental Justice* approach), in order to more effectively face environmental inequalities.

Thus, over and above the common terms they use (inequalities, injustices, vulnerability), these two approaches differ greatly; the 1st focuses on epidemiological studies of risks, the 2nd more on social or ecological aspects. The second generates much more will for political change, although with undeniably different positions concerning the distribution of rights and duties. These differences both express and feed relatively different conceptions of the environment (and of justice): in the time scale they imply, notably for the no less diverse modalities of the regulations they propose; in the spatial frame of reference, much more micro-spatial for the 1st, revolving around individuals and their local collectivities, more macro-spatial for the 2nd, implying other forms of social organization and related conceptions of justice (more social and procedural).

Despite those approaches, in France, as in many other European countries, except for those cited above, the issue of environmental inequalities apparently suffers from a lack of political focus (Theys, 2007), and continues to be dealt with mainly in scientific publications. It is true that these, present also in Germany, seem to point towards a socio-urban opening: a meso-spatial reading (see for instance De Palma, Motamedi, Picard and Waddell, 2007). However, in the socio-urban and regional approach, frequent overlappings confirm that the content is far from stable, for example for such terms as risk, vulnerability, territorial disparities, environmental justice, spatial equity, ecological inequalities. To the point that we do not really seem to know the real specificity (does it exist?) of environmental as against social inequalities. Thus, things could be qualified in much broader terms, for example environmental inequalities could be described as follows: "*A difference in the situation between individuals or social groups that may be noted not only with reference to "ecological" considerations strictly speaking (pollution, public hygiene, natural environment), but also in terms of living space, accessible renewable resources, quality of human places, living conditions, landscapes, etc., this difference being seen as contrary to the rights and respect for the individual, and moreover likely the generate an imbalance that is harmful to the satisfactory functioning of the community*" (French committee for the Sustainable Development World Summit in Johannesburg, 2002, p. 164).

Yet it still fails to contribute to the nascent debate on sustainable development, although environmental inequalities are among the few issues that truly combine environmental, economic and social stakes (and "pillars"). It should certainly be viewed as a political aporia relative to sustainable development - which could, in theory at least, make public some of its aspects. Promoted by some as the nascent rationale for public action, notably in Europe (Beatley, 2000), in the area of urban planning and design (Riddell, 2004; Wheeler and Beatley, 2004; Ascher, 2004), even environment (Mazmanian and Kraft, 1999), it is frequently criticized in France for its empty eloquence which makes it possible to institutionally avoid essential reflections on the measures that must be taken in the face of economic, ecological and food crises (Lascoumes, 2001 ; Puech, 2011). While some authors view it in terms of a pragmatic construction of a meta-narrative (Rumpala, 2010), others report highly unequal territorial experiences, in which once again physico-chemical approaches to the environment (*supra*) or values defended, play an essential differentiating role (cf. notably, for the United States, Portney, 2003). And above all, many criticize its incapacity born of its generalization (every sector now boasts of its sustainability), its failure to prove its specificity and convincingly argue its fundamental and concrete contributions required to meet recognized challenges (e.g. climate change).

However, since this lack of ambition relative to environmental inequalities applies above all to France (cf. approaches abroad, *supra*), and since there can be no doubt that the links between conceptions of the environment and of justice, the republican tradition relative to the social pact and equality of treatment, the forms of injustice to which it also may have contributed (environmental?), have also marked it heavily. Thus, what conceptions of justice and of the environment should be debated in France? On what knowledge basis concerning environmental inequalities? For what view of urban sustainability?

2.3 The primary forms of environmental injustice: social inequity in the commitment to socio-ecological change

2.3.1 Towards a cosmopolitical approach

Certain economists consider that due to the vital questions relative to social justice in terms of environmental inequalities, we dispose of a first lever to socialize the environment via its (un)egalitarian aspects, as well as via a nascent perspective of a social ecology, given more egalitarian democracies (Laurent, 2010). This is certainly the case. But thanks to a reading that draws upon a cosmopolitical approach to the environment (see for instance debate between U. Beck and B. Latour in 2004), notably in its links with land use planning (Lolive and Soubeyran, 2007), the interest of this subject (but also certainly its failure to generate political reflection) is – we think - a different one. The issue is not to uniquely revise the founding myths (e.g. egalitarian), thus advocating compromises between economical progress and environmental conservation (*op. cit.*), but to fully establish them anew by means of:

- cornerstone questions which this subject would address consubstantially with concepts of justice and the environment,
- but also to be addressed to our 'governmentality' (e.g. the exercise of democracy in our liberal/free-market societies),
- in order to become fully aware of the means provided by the environment (human and non-human) to change our societies, their development models and modes of government, i.e. of a number of values and principles that have been advocated until now.

Everyone knows that over the last thirty years the environment has everywhere imposed itself as one of the most powerful filters for the understanding and interpretation of the

The Environment as a Factor of Spatial Injustice: A New Challenge for the Sustainable
Development of European Regions?

223

living environment, and thus as one of the primary operators of our reflections on modernity:

- the finite nature of resources and ecological irreversibility,
- the desynchronization of environmental time with time in terms of development,
- the growing distance between the spaces where problems occur and where decisions are made,
- with for example a growing lack of predictability relative to the effects of the "rationale" of modernist planning on places and its societies.

In France particularly, this change is observable in a certain number of recent programmatic aspirations or, essentially, watchwords which are often adequate in urban planning or urbanism fields: territorial energy transition, dense/slow city/short distance cities... and the post-Kyoto "paradigm" (cf. Greater Paris[2]). However, these aspirations do not compete with other aims relative to change, notably community or affinity-group-based solutions in the United Kingdom (e.g. *Cities in Transition*). In France, such bottom-up initiatives are still few and far between and rarely popularized (e.g. *Relocalisons !* movement).

Embodying values (esthetic, heritage-based, symbolic...), "*environmental situations*" and their "*qualitative variations*", terms which though dynamic are present everywhere in the literature on environmental inequalities[3], and more and more often mediate our relationship to (the) world(s). The growing importance of environmental considerations in the residential choices of households, in individuals' choices of transportation mode, in nutritional practices and individual energy choices... and even in our lifestyles and involvements in associations and local communities, shows this every day. Thus the environment contributes to a gradual re-founding of the joint government of humans and nature, reviewing certain values and action principles (Boltanski and Thévenot, 1991), particularly for policies with a strong territorial basis (land use planning, urban design, nature protection). According to Beck (1995), Latour (2004b)... this conception even announces, in different ways, a new age of politics, an age in which relations to identity, notably spatial identity, are being composed anew, to the extent that they shake up the historic chain of the construction of public action, above all in countries with a centralist tradition: a certain production of the rationalities (techno-scientific) for a certain exercise of democracy (delegative) (Stengers, 1997).

This larger purpose may even be found in certain recent French studies of environmental inequalities (below).

Box 2. When a more dynamic view of the environment raises the issue of time and space scales in the apprehension of environmental inequalities (French cases)

Having noted a lack of prospective and dynamic approaches to territories, Laigle (2005) proposes a territorial analysis of the urban dynamics that generate environmental inequalities, based on four cases: regions/territories characterized by a heavy industrial

---

[2] A choice which is apparently justified by environmental inequalities as evoked in the presidential discourse when the different architectural projects were presented in 2009.

[3] "*Environmental inequalities are inequalities of situation (...) resulting from qualitative variations of the urban environment*" (Inspection Générale de l'Environnement, French Ministry of Ecology and Sustainable Development 2005, p. 11).

past (Lille agglomeration – North of France), regions that are attractive economically and residentially (the Mediterranean agglomerations of Aix-en-Provence and Toulouse), territories or regions characterized by multipolar expansion (Strasbourg agglomeration, in the Rhein region).

Globally, the analysis generated two types of configurations, which according to the author encourage cumulative links: " *configurations in which past urbanization overlapping with industrialization resulted in: social deterioration, a degraded living environment making economic and urban reconversion difficult* "; " *configurations characterized by attractive economic and residential conditions, based on the quality of the living environment, which may strengthen selective factors of access to urbanity and – paradoxically – damage the quality of the environment.*" (p. 11).

Thus, local pathways, trajectories, heritage, as well as priority orientations and the dynamics of contemporary territorial action should be placed squarely in the center of the analysis of environmental inequalities. Further proof of this is supplied by the studies on the industrial heritage in the Seine-Saint-Denis department, to the northeast of Paris (from 1850 to 2000, cf. Guillerme, Jigaudon and Lefort, 2004), which gave rise to a historic phenomenon of discrimination and environmental and social segregation, notably due to choices made by public authorities, in spite of several recent large-scale requalification programs.

Picking up on the idea of cumulative disparities, Deboudt, Deldrève, Houillon and Paris (2008) examined a narrower, coastal territory: the Chemin Vert neighbourhood in Boulogne-sur-Mer (coastal industrial municipality in northern France). It was marked by its connection with the development of sea transport, tourism, port and residential economies, and thus by spaces with a high ecological value.

Their findings demonstrate firstly that social inequalities are cumulative (over-representation of unemployment, single parenthood, low income), and marginalization that is also geographic (remoteness to city center, topographic disparities, cuts in the urban tissue, few public spaces). Above all, there are few nuisance factors and the area is not vulnerable to natural hazards, with even a potential for amenities and enhancement. Consequently, urban policies wish to make use of this potential, notably by valorizing the "maritime" aspect.

However, according to a survey of the inhabitants, if the coastal environment is certainly seen as an element identifying and enhancing the living environment and a source of amenities, the inhabitants do not think that it should be preserved, since memories of the maritime past are not very strong, and the maritime professions are not in high regard. Thus, over and above the single issue of amenities and environmental practices, the study proposes to approach the subject from the point of view of the social value(s) ascribed to the environment. "*In a situation in which the inhabitants do not directly identify with the "maritime" concept, massive and qualitative public intervention leads to a paradoxical syndrome in certain individuals who ask themselves if they are "worthy of these new homes"* (p. 189).

This leads to a proposed analysis: should the analysis of inequalities, cumulative effects, and vulnerability aspects not be oriented more towards the spatial scale of ecosystems and human settings, as the historic crucible of the environmental offer and the social values attached to it?

Thus, more than just crossing static data, should not any investigation of environmental inequalities position itself with respect to privileged time scales (local itineraries and heritage, public and private arbitration in the past, current territorial strategies), and to the observed spatial scales (ecological or territorial ones, areas of practices, historic districts and divisions...).

2.3.2 The individual involvement capacities at the heart of environmental issues

Thus we think that the very first "disturbance" introduced by the subject of environmental inequalities, particularly in an urban analysis, is that in theory it makes possible a much more dynamic and active screening of a model of social equality, and its spatial correlations in land use planning, urban design, environmental protection policies. Here, beside the social aspect with the revitalization/reconfiguration of links (e.g. the importance of nature for local forms of solidarity in cities, in the North as in the South), or that of the economy of the new trends/sectors of locally-oriented production (ecological housing, local consumption of agricultural and cooperative products...), it first examines this model from the point of view of the *"myth of the passive citizen"* which makes this model operational (Rosanvallon, 2008). Individuals as subjects aspire more and more often to different ways of life and commitments, often invoking nature and the environment (see for instance Haanpää, 2007, for the role of lifestyles or Jagers, 2009, for the role of perceived ideologies in commitments; see also Dobré and Juan, 2009, for French cases). Also, the constitution of new, more informal collective entities, increasingly underpins no less social forms of mobilization (Lolive, 2010), also via different relations to the environment and to nature (e.g. sustainable/ecological/green communities in Roseland, 1997; and the return to Urban Design in Beatley, 2010).

From the point of view of the relationship between society and the environment we are encouraged to consider the contribution of environmental inequalities to the debate on sustainable development, in terms of both individual and collective capacities of involvement, and to examine their non-egalitarian social distribution and the very scope of such inequalities in the capacity for change. Let us also note the presence across-the-board, though with very different modalities (sometimes strictly regulated) of so-called citizen participation in the approaches targeting environmental inequalities that were discussed above (*Environmental Justice*, Ecological inequalities of development). This contradicts the official report of the Inspection Générale de l'Environnement (Diebolt and al., 2005, for French Ministry of Ecology and Sustainable Development, Transport and Housing) which denies this participatory dimension as an integral part of the issue of environmental inequalities.

In fact, it is here that we would today place the primary forms of environmental injustice. No longer simply disparities of exposure (although this interpretation remains useful for the detection of long term sanitary impacts, cf. Roussel, 2010), but gross social injustice relative to more individualized forms of access to formal or informal involvement (lifestyle commitments, unaffiliated collectives...) in socio-ecological transition. For, even though studies, mainly conducted in English-speaking countries, tend to show that the poor are increasingly involved in local causes (cf. case studied by Corburn, 2005), such capacities to influence environmental situations and the mechanisms behind environmental inequalities are no less unequally distributed than other capacities (Beck, 2001), as stipulated in Article 3.9 of the Aarhus Convention (1998), of which the countries of the European Union are signatories.

This would imply placing the means for change (still inequitable) at the heart of the reflection on sustainable development, perhaps in greater measure than social equality as a finality, which we know to be globally non-environmental (e.g. redistributive approaches of social and urban policies). This could also generate other axiological pluses, bearing witness to the scope of socio-political implications of a collective examination of environmental justice. This more dynamic and active option lies in fact at the crossroads of the various dividing lines:

- from the individual freedom to act, which is certainly a fundamental right inscribed upon the pediment of our liberal democracies, but which also – due to their backing of free market societies - suffers from all the spatial divisions which they are subject to as a result of social inequalities... to the responsibility, not via environmental education but via accompanying the poor in the definition of the stakes and the improvement of their own disparate environmental situations[4],
- from social mixity - and intergenerational mixity, which is at the forefront of sustainable development – via quotas and regulatory provisions often still implemented topdown on the strength of norms that are taken for granted (concerning the proportion of subsidized housing, for example) to more fundamental forms of solidarity which are spreading notably for and through nature (since we know that living together does not necessarily mean exchanging, and even less sharing or helping each other).

Without including – always in terms of values and principles – this conception of the environment in the moderation and sobriety displayed by certain lifestyles, or in the self-sufficiency which is increasingly invoked by local economy projects.

2.3.3 Inhabitants, lifestyles, and their places as subjects of environmental inequality?

As a result we have at least one proposal on the subject of environmental inequalities in the perspective of sustainability. It advocates the use of other conceptions of both the environment and of justice in public policies, which a possible horizon of sustainability should address (Faburel, 2010b). Thus it would seem that the concept of the environment presented here focuses on the environment as it has become, i.e. "on the qualitative differences between situations" (supra), recognizing the links and perception relations of local societies to the environment. "To perceive an atmosphere as sustainable, the physical dimension must meet the expectations of our existential living body; otherwise, an individual never perceives the environment as "sustainable" and never achieves a "sustainable status" (Shirazi, 2011, p. 8). This would call for an egalitarian project that would finally be open to socio-environmental singularities, to the ways in which they are lived and experienced through the inhabitants' sensibility, and how they are recognized by local knowledge (Fisher, 2000)... in short how they are embedded in ecological ways of living, lifestyles and involvements, in an cosmopolitical perspective of sustainability.

---

[4] Rather than for example to simply let households change their environment by residential mobility and its market stimulations, thus negatively positioning certain settings (environments).

The concept of justice would thus move away from an interpretation based on only
(re)distributive justice (with an egalitarian motivation but liberal rationale), characterized
notably in France by its real estate (rehabilitation/renovation, housing offers) and urban
aspects (the *Promethean* approach of land use planning, uniformization of public spaces,
social insertion via state-imposed policies). It would be more procedural (e.g participatory)
than structural and merely (re)distributive, based on the capacity of poor populations and
their place to face up to dynamic and inherited contexts via their own local experience. It
would thus admit that citizenship can be differentiated (Young, 1990)[5], and therefore open
to other factors of inequality than only individual income, and above all mindful of the
rights of affinity-based groups (and not just community-based ones). In brief, following
Schlosberg (2004) and Jamieson (2007), environmental justice needs to address not only the
distribution of environmental harms and benefits, but also people's participation in
decision-making processes, including recognition of people's particular identities and
visions of a desirable life.

On this reflexive and conceptual basis which develops a cosmopolitical approach to
environmental stakes, a more phenomenological conception of the subject-individual, and a
critical reading of the consubstantially dominant accepted meanings of the environment and
of justice, in 2008 we conducted an empirical study of environmental inequalities in the Ile-
de-France, i.e. the capital region of France (11.6 million inhabitants). The realization and
results of this project are discussed below.

3. Lived environmental experience, satisfaction and quality of life in the Ile-de France region. A different regional geography of environmental inequalities

3.1 A pluridisciplinary approach and a multi-scale procedure

As already stated in the previous section, several statistical observations tend to
demonstrate the existence of environmental inequalities in France and abroad, both now
and in the past. However, we have also seen that when conducted at scale-level, these
studies generated numerous conceptions in which the environment and justice overlapped;
they were also less and less adequate to the development of other approaches, better
adapted to the changes that our societies are subject to as a result of the environmental
situation and the challenges it brings: a gradual reformulation of the joint government of the
human element and nature, revising certain values and action principles of our so-called
reflexive modernity (Giddens, 1991). The system of environmental evaluation that still
dominates worldwide, i.e. principally technical, physico-chemical approaches, to normative
ends for environmental protection, and their regulatory and operational relays
(*Environmental Impact Assessment, Strategic Environmental Assessment...*) is increasingly ill
adapted to disclose the scope of a territorialized phenomenon, which has at least as much to
do with the socio-environmental as the bio-physical domain: environmental inequalities and
injustices. From its strictly evaluative aspect, this system still strongly depends on the
segmentation of knowledge and scientific disciplines, on their disparate recognition by the
powers-that-be, and – not to say above all – on a vision of the inhabitants as "statistical
individuals". This gives rise to a lack of instruments of territorialized assessment,
particularly in the cities, where socio-spatial and segregatory mechanisms are particularly

---

[5] For an application to urban policies, cf. Harvey (1992).

powerful, and old (Faburel, 2008). Such methodological obstacles or even limitations both contribute to and embody the deficits in the scientific recognition (techno-scientific production of rationalities) and the political action targeting such inequalities (delegative exercise of democracy).

The current scientific literature increasingly calls for pluridisciplinary, or even interdisciplinary approaches, in the attempt to integrate at least some elements of the inhabitants' living experience, complementing or contradicting existing observation and information systems. Since, where socio-environmental issues as well as others are concerned, the gap between what is given by so-called objective environmental data and what the population feels and experiences constantly widens. And, as already noted (cf. Box 2.), concerning the question of environmental inequalities, "*studies to clarify the relationships between objective and perceived exposure and the influence of social status on the perception of environmental exposures are still necessary*" (Kohlhuber et al, 2006, p. 254).

Conducted between 2006 and 2008 (Faburel and Gueymard, 2008) for the French Ministry of Ecology, Sustainable Development, Transport and Housing, for its *Territorial policies and sustainable development* research program (2005-2009), in close cooperation with the Ile-de-France (Paris) region, the resarch, a synthesis of which is presented here, had the primary objective of establishing a different geography of environmental inequalities.

On the one hand this geography confronts environmental disparities made apparent by the crossing of physico-chemical data[6] with no less institutional data relative to official socio-economic spatial characterization (income levels, proportion of subsidized housing, unemployment rates). More importantly, these observations of disparity were then compared with information on the living and felt environment, by means of local experiences, satisfaction, place attachment and political expectations relating to the environmental qualities which generated these observations. The aim was therefore to implement a perceptual and well-founded observation of "objectively" described socio-environmental situations, while opening oneself to the symbolic and identity factors that are at the basis of the attraction, attachment to or refusal of certain places by the populations. Within this framework, a further aim was to improve the understanding of operative mechanisms, notably residential ones, in the phenomena of spatial polarization for environmental reasons at a regional scale.

Several specific questions guided this work:

-   How do people perceive and judge environmental quality, and what experiences and expectations ground their points of view, notably during residential arbitration procedures?
-   How far do conventional indicators make it possible to register real satisfaction or dissatisfaction, when taken out of a given environment?
-   How then can one imagine a system of observation and measure that could best account for the influence of the quality/non-quality of the environment on individual decisions, and explain certain phenomena of inequality and segregation, and the resulting territorial dynamics?

---

[6] Thresholds of chemical exposure for air quality; probabilities of risks occurrence, flood risks and hazards for instance; acoustic levels for noise nuisance; distance for the accessibility of urban amenities, of green spaces...

In fact, we think that, due to its territoriality and resulting transversality, the register of the personal lived experiences and of environmental satisfaction constitutes a non-negligible source of information, which could prove essential to:

- (re)define the analytic frameworks of these situations which until now have been mainly perceived as "objectively" unequal, often presented as a "combination" of environmental degradation and socio-historical spatial disqualification (i.e. disparities),
- shed a light on potential levers for sustainable action, thus contributing to the entry into politics of a fully socio-environmental set of problems which are still rarely viewed from the perspective of public intervention and change (i.e. injustices),
- for example, by observing the aptitude of the current environmental evaluation system to describe a fully territorialized phenomenon, defined at least as much by felt, symbolic and axiological relations of local societies with their living space, as by largely accounted for physical or social characteristics (i.e. inequalities).

This was our first working hypothesis. The second resulted from it: the subject-individual, via his lived environmental experience and the cognitive and social transactions he operates, constitutes together with his immediate living environment, a pertinent scale of observation. Unlike the "statistical individual", this scale enables to both "territorialize" environmental quality, and to highlight certain determining dynamic factors of inequalities in this area, in order to perhaps differently ground no less territorial decision making.

Exploring the two dimensions of environmental inequalities, which are usually called "objective" and "subjective", first raised the question of the reference scale for observation. Working on the Ile-de-France[7] region, we opted for different, though complementary scales. This confrontation and overlapping of scales of analysis is also part of an approach underpinned by the territorialization of public action, particularly with reference to sustainable development: the progressive structuring of areas of competence (subsidiarity principle) and decision making levels (territorial governance) around the reality of phenomena and pertinent new scales of observation.

Two successive stages at two scales defined our empirical work. First, we made a conventional reading of environmental disparities, at regional scale, by spatializing so-called objective environmental data and crossing them with classical socio-economic and demographic data. The second step was to select six municipalities in the different environmental situations identified, with the aim to analyze inequalities of lived environmental experience. A survey was conducted with 600 inhabitants, face to face. However, in view of the size of the sample (600 questionnaires) and the various criteria which defined the choice of our sites as well as of our groups of individuals, we did not aim for representativity at a scale of a region with a population of 11.6 million. We thus adopted an essentially exploratory perspective, with a view to preparing the ground for a different system of observation, fully focused on environmental inequalities as linked to individuals' lived experience, in order to understand certain phenomena and mechanisms of dynamic socio-environmental spatial polarization. With this exploratory view, we developed and adopted a dual approach, referring to both spatial analysis (quantitative)

---

[7] Capital region of France, the Ile-de-France is the most densely populated with 11.6 million inhabitants, 90 % of whom live in the (Paris) agglomeration which covers 20 % of the regional territory.

and socio-cognitive investigation (qualitative), highlighting inhabitants and the socio-cognitive transactions with their living environment and the environment as such. Thus, our work closely combined geographical, economic, sociological and psychological knowledge.

3.2 A static reading of environmental inequalities in the Ile-de-France (Paris) region

The first stage of our work was to draw up a geography of environmental disparities at the regional scale, by setting up two typologies (environmental and socio-economic).

3.2.1 Construction of two multi-criteria typologies: choice of indicators and statistical method

To set up the environmental typology, we selected both classical criteria and indicators, but also such as are liable to interact with lived environmental experience and the environmental satisfaction of populations. We thought it important to address several thematic environmental registers by taking an interest in diverse environmental objects, referring certain of them to the sensitivity register (e.g.: noise) and above all such as could have contrasting effects (some perceived as agreeable, others as disagreeable). Twelve indicators, grouped into two families, which for clarity's sake we designated as resources and harms, were noted at the scale of the 1 300 municipalities in the Ile-de-France region.

In a next step, the environmental typology was established on the basis of discretization between 3 average classes (+/- standard deviation) for each of the variables. The different environmental parameters were then aggregated by calculating two weighted multi-criteria averages – average resources and average harms; based on certain findings concerning residential choices and on a conventional hierarchy of nuisances and risks in the Paris region (cf. 2.2.1). This calculation generated nine possible combinations, depending on different resource and harm levels, and 9 environmental groups, with at the two extremes: environments designated as very favourable or very degraded. For greater clarity, these different groups were then combined within three great environmental categories: good, average, bad.

This general map of environmental categories establishes a geography of disparities by clearly emphasizing areas of so-called "objective" good or bad environmental quality. These major disparities are generated by structural factors which have been known for a certain time, notably:

- the center, which corresponds to the heart of the Paris region, with mediocre environmental quality (density of infrastructures and of centers of economic activity, lack of vegetation...),
- municipalities that are environmentally and traditionally the most disadvantaged are mainly located in northeastern Paris, owing to an industrial past, but also to political choices to concentrate infrastructure and equipment, above all relative to traffic: in the Seine-St-Denis (93), in the northern Hauts-de-Seine (92), in southeastern Val-d'Oise (95) – along the "francilienne" (by-pass motorway for the agglomeration) and close to Roissy Charles de Gaulle airport (2nd airport in Europe),

Environmental variables	
Resources	Green*surface areas with possible landscape value (in % of municipal area)
	Population living close to green spaces open to the public (within a perimeter of 250 meters to 1.2 kilometers, depending on size of the space, in % of the municipal population)
	Surface of listed areas** (in % of the municipal surface area)
	Population living close to waterways and bodies of water (within a perimeter of 100 to 500 meters, in % of the municipal population)
Harms	Annual average nitrogen dioxide (NO2) level (2005)
	Population potentially concerned by local pollution *** (in % of municipal population)
	Population living in the flooding zone (in % of the municipal population)
	Population living close to a Seveso II**** class industrial site (within a radius of 500 meters, in % of the municipal population)
	Population exposed to aircraft noise caused by traffic at major airports***** (in % of municipal population)
	Population exposed to aircraft noise caused by traffic at small airports****** (in % of municipal population)
	Population living within railway traffic noise "hot spots" (in % of municipal population)
	Number of road segments with noise emissions higher than the hot spots daytime noise threshold (in % of the studied road area)

* Notably includes natural and agricultural lands, open urban gardens (allotment gardens, private family gardens), hippodromes, golfs and cemeteries.

** Designates listed sites and historic monuments, protected urban areas, protected urban architecture and landscape heritage areas (Zones de Protection du Patrimoine Architectural Urbain et Paysager, ZPPAUP).

*** Population living close to (100 meters) road segments with annual average NO2 levels higher than the annual quality objective, established by the air quality protection plan (Plan de Protection de l'Atmosphère, PPA) (2005-2010) and taken up by the air quality monitoring program for the Ile-de-France (Programme de Surveillance de la Qualité de l'Air en Ile-de-France, PSQA) for 2004.

**** The so-called Seveso directive or directive 96/82/EC is a European directive that imposes the obligation upon all EU member states to identify all industrial sites presenting major risks of accident. The directive, which was made official on 24 June 1982, was modified on 9 December 1996 (Seveso II) and amended in 2003 (2003/105/EC). Companies are listed according to the quantities and types of hazardous products they handle.

***** Populations included in the nuisance mitigation schemes (Plans de Gêne Sonore, PGS) for soundproofing grants, of Charles de Gaulle and Orly airports (1st and 2nd in France, 16th in Europe), or flown over at an altitude of less than 1000 meters.

****** Populations living in impact areas of other small airports, included in a land use compatibility noise program (Plan d'Exposition au Bruit, PEB) or, if no such program exists, within a radius of 1000 meters around the operator's infrastructural impact.

Table 2. Environmental variables selected to establish a descriptive geography of environmental disparities in the Ile-de-France region Source: Faburel et Gueymard (2008)

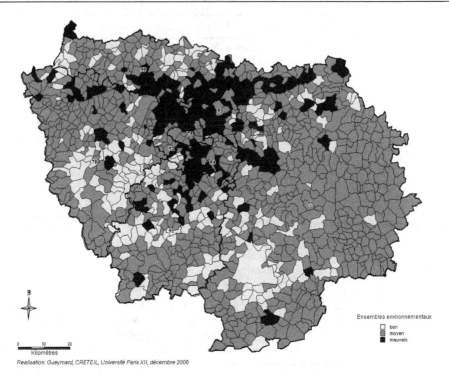

Realisation: Gueymard, CRETEIL, Université Paris XII, décembre 2006

Map 2. Distribution of Ile-de-France municipalities among 3 environmental categories

-   other smaller but also degraded sub-areas are located in the Val-de-Marne (94), close to Orly airport (2nd airport in France) and in the vicinity of the major motorways (A6 and A10), but sometimes also in the "second ring", at the peri-urban border of the northeastern agglomeration, notably in the Seine et Marne (77), owing to an influx of populations that can no longer cope with the cost of living in the center,
-   and, at the opposite, the most environmentally favoured munipalities, located more in the west and the south of the agglomeration, mainly in the departments of the "outer ring", with a major focus here on municipalities close to woodlands and the Regional National Parks (Parcs Naturels Régionaux PNR).

In other words, this first general illustration casts a light on certain structuring oppositions at regional scale (east/west, center/periphery) which are well known to geographers and urban planners. However, a conventional reading of environmental inequalities makes it necessary to cross given environmental characteristics and socio-urban data that are specific to the areas.

So, in parallel, and in the same spirit, we established a socio-economic typology of Ile-de-France municipalities, crossing information with a view to undertaking a first descriptive reading of environmental disparities at regional scale, before rigorously delimiting study areas (*supra*). Always going out from past findings, and in close cooperation with the Ile-de-France (Paris) region, we decided to opt for five variables, accounting for both the socio-economic characteristics of households, and housing stock.

Variables	Sources
Proportion of management level and higher level intermediary professionals	RGP*, 1999
Gross municipal income per inhabitant	DGI**, 2003
Unemployment rate	RGP, 1999
Proportion of tenants in social housing HLM	RGP, 1999
Proportion of social housing (2006)	DGI/DGCL, 2006

*General Census of the Population (National Institute National for Statistics and Economic Studies)
**General Tax Office (Ministry for the Economy and Finance)

Table 3. Variables to establish a socio-economic typology of municipalities in the Ile-de-France region, Source : Faburel et Gueymard (2008)

The discretization method adopted for these variables was the same as above. For each of the included variables, 3 classes (weak, average, strong) were defined on the basis of the average and the standard deviation. A municipal average was computed for all concerned ranks. Here too, we decided to distinguish between certain variables, by allotting a differentiated weight coefficient when calculating the average. This calculation generated 3 groups (low income, average, well-off).

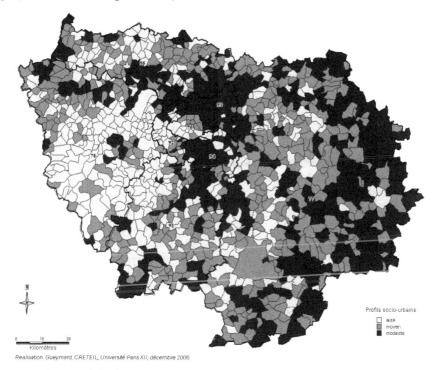

Profils socio-urbains
☐ aisé
▨ moyen
■ modeste

0   10   20
Kilomètres

Realisation: Gueymard, CRETEIL, Université Paris XII, décembre 2006

Map 3. Distribution of Ile-de-France municipalities among 3 socio-urban groups

This other general map comes as no surprise. The regional distribution of socio-urban groups again expresses the industrial past of certain central and peri-central areas of the agglomeration (e.g.: along the Seine Valley, in southeastern and northeastern Paris), with a social composition that is very different from the munipalities marked by the development of the tertiary sector in the western and southwestern sectors. Here we find again the usual separation between the Seine-Saint-Denis (93), poorest department in the region, and its opposite, the Yvelines (78). It largely coincides with the geography of income (Saint-Julien, François, Mathian, Ribardière, 2002), as well as with other socio-economic large studies (Berger, 2004; Préteceille, 2003). Above all, this map illustrates the growing recent impoverishment of more remote areas, first of all of the Seine et Marne (77), which since the mid 90s has absorbed households that can no longer afford to live in the agglomeration, and a decline of certain agricultural activities.

3.2.2 A first reading of crossed regional environmental disparities, confirming our intuitions

Crossing environmental and social typologies generated a first reading of this phenomenon at regional level. The table below presents the crossed selection generated for the environmental types grouped into 3 categories (good. average, low) and the 3 socio-economic groups.

Environmental category	Socio-economic profile			
	Affluent	Average	Low income	Total
Good	45.53	31.49	22.98	100
Average	33.57	36.15	30.28	100
Low	17,84	32.39	49.77	100

Table 4. Socio-economic profile of the three major environmental categories, Source: Faburel et Gueymard (2008)

Unsurprisingly at this stage, we observe an increasing relation between environmental and social characteristics of municipalities in the region:

- 45.5 % of the municipalities in the good quality environmental category are municipalities with the highest socio-economic profile in the Ile-de-France,
- Symmetrically, almost 50 % of municipalities in the low quality environmental category are municipalities that are home to deprived populations.

Seemingly, these first, general results, confirm the existence of environmental disparities at regional scale, in conventional terms: proportionally more of the poorest households live in environments of low or mediocre quality, according to the standard indicators used to characterize these situations.

Another approach to this phenomenon was to cross it with the presence of Sensitive Urban Areas (Zones Urbaines Sensibles – ZUS). As a reminder - these areas (ZUS) are infra-urban areas (e.g. neighbourhoods) which French public policy makers have defined as a priority target for urban policies, in view of the difficulties which their inhabitants encounter constantly (increasingly important fiscal and social provisions). There were 640 of them in metropolitan France in 2005; of these 138 in the Ile-de-France, with a population of 1.1 million. Here, too, we see a strong link, confirming the several findings that already exist in this field (cf. 2.2.1). The proportion of municipalities in which there are no ZUS is almost of 100 % in the "good" environmental category. On the other hand, ZUS are over-represented

in municipalities with a high level of environmental harms; 20 to 30 %, depending on the
type of environment, while the proportion of municipalities with ZUS in the Ile-de-France
lies under 10 %.

Environmental category	Presence of ZUS on municipal territory		
	No	Yes	Total
Good	95.74	4.26	100
Average	92.96	7.04	100
Low	73.71	26.29	100

Table 5. Proportion of municipalities with ZUS in the three environmental categories,
Source: Faburel et Gueymard (2008)

Nevertheless, if the existence of a global correspondence between socio-economic and
environmental characteristics is here clearly apparent at the aggregate scale, what about
each environmental factor? Do the environmental factors investigated in our typology
confirm this link when taken one by one? Could there be environmental factors which, on
the basis of this static "objective" reading, are more likely to feed the disparities we have
already noted at this stage?

3.2.3 Socio-spatial distribution of environmental objects and the environmental profile of social groups: The structuring role of factors of degradation

The table below gives an example of a crossing between environmental factors and different
socio-economic groups. This was in particular established for class 3 of environmental
objects, where they are present in greater proportion than in the regional average, and –
since they illustrate a "caricature-like" situation - better highlight the specificities of each
group's spatial distribution.

Environmental objects (class 3)	Socio-economic profile (in %)			
	Well-off	Average	Low income	Total
Green spaces	42.42	25.76	31.82	100
Green components	29.88	42.07	28.05	100
Listed spaces (e.g.: ZPPAUP)	45.3	38.46	16.24	100
Waterways and bodies of water	34.2	29.97	35.83	100
Overall pollution (average N02)	3007	21.57	48.37	100
Local pollution	3571	21.43	42.86	100
Flooding zone	2597	19.48	54.55	100
Seveso industrial risk site	1176	47.06	41.18	100
Aircraft noise (major airports)	14.06	40.63	45.31	100
Aircraft noise (small airports)	21.43	57.14	21.43	100
Railway traffic noise	20	31.43	48.57	100
Road traffic noise	31.37	21.57	47.06	100

Table 6. Spatial distribution of social groups according to the above environmental factors,
Source: Faburel et Gueymard (2008)

The first global reading of this table confirms the first conclusions. There is indeed a rising linear correlation between environmental factors and socio-economic situations. We observe (gray boxes) that for almost the totality of positive environmental factors, so-called well-off municipalities are much better represented. Symetrically the same observation can be made for poor municipalities.

To validate and further investigate these findings, we then extended these crossings to all the different classes of environmental objects, corresponding to the three classes of distinction (*supra*). In order to easily spot the constitutive environmental factors of the various socio-economic categories, we decided to think in terms of the representation interval (under or over-representation), with reference to the weight of each of the groups in the sample. When assembled, these representation intervals enable the establishment of a hierarchy of the most structuring objects for each group, thus highlighting, via comparison, the environmental factors that appear as the strongest vectors of socio-spatial differentiation.

The table below presents this hierarchy of objects, generated by the digressive classification of representation intervals of groups, from the strongest over-representation to the strongest under-representation.

Socio-economic groups		
Well-off	**Average**	**Low income**
Listed sites (e.g.: ZPPAUP) (+) 3	Green components (+)	Seveso (+) 1
Green spaces (+) 4	Air traffic noise (small airports) (+)	Railway traffic noise (+) 2
Overall pollution (+)	Waterways and bodies of water (-)	Local pollution (+) 3
Waterways and bodies of water (-)	Green spaces (-)	Road traffic noise (+) 4
Green components (-)	Listed sites (-)	Air traffic noise (major airports) (+) 5
Road traffic noise (-)	Overall pollution (-)	Flooding zones (+)
Local pollution (-)	Air traffic noise (major airports) (-)	Overall pollution (+)
Flooding zones (-)	Seveso (-)	Green spaces (+)
Air traffic noise (small airports) (-) 6	Flooding zones (-)	Green spaces (+)
Air traffic noise (major airports) (-) 5	Railway traffic noise (-)	Air traffic noise (small airports) (-)
Railway traffic noise (-) 2	Road traffic noise (-)	Listed sites (e.g.: ZPPAUP) (-) 7
Seveso (-) 1	Local pollution (-)	Green components (-) 6

Table 7. Environmental profiles of socio-economic groups, Source : Faburel et Gueymard, 2008

In view of this classification and also going out from the strongest absolute difference in terms of representation (gray boxes), we observe differentiated environmental profiles, profiles with several characteristics. The first characteristic confirms noted disparities, by making them explicit:

- The group of well-off municipalities is defined primarily by a strong under-representation of Seveso class industrial risks, and of railway noise. Only then do we find a strong over-representation of listed sites (listed sites and historic monuments, protected areas, protected urban architecture and landscape heritage areas - Zones de Protection du Patrimoine Architectural Urbain et Paysager), and green spaces; this is followed by a strong under-representation of aircraft noise (from both major airports - Roissy CDG and Orly - and small ones, commercial for instance), all variables taken together.
- On the other hand, the group of municipalities designated as low income is above all affected by an over-representation of harms: Seveso class industrial risks, railway traffic noise, local pollution (nitrogen dioxide levels close to roads) and by road traffic noise. In a smaller measure, this group is also characterized by the presence of aircraft noise generated by the major airports. Only then is it characterized by an under-representation of green components (natural and agricultural spaces, allotment and private family gardens, hippodromes, golf courses, etc) and of listed heritage sites.
- The group of municipalities designated as average is mainly characterized by a much smaller number of discriminating factors, be they positive or negative.

The second characteristic, and perhaps the newest one, is that in fact, at the scale of the three groups, four environmental objects powerfully structure the expected difference between the environmental offer of the most well-off municipalities and the poorest: listed heritage sites, Seveso class industrial risks, railway noise, noise generated by the major Parisian airports.

Finally, a third major characteristic: environmental degradations (expressed in technical and normative terms) appear to be the primary structuring factors of the general assessment, whether positive (due to absence) or negative (due to presence). These factors confirm the results of past studies on environmental issues, notably designating transportation noise as the first source of environmental disqualification; we think they shed another novel light, and perhaps an essential one: at regional scale, and above all with reference to the "extreme" social groups, the presence or absence of degradations seems to play a more important role in structuring and social differentiation than the presence or absence of amenity factors. Thus, it would seem that the repulsion caused by environmental nuisances and degradations may give us a better understanding of environmental disparities mechanisms at the scale of the Ile-de-France region than the attraction operated by certain settings, notably those we designate as natural (green spaces, waterways).

This said, how does the environment intervene concretely in household choices and strategies? Is avoidance of nuisances and pollution actually more important than the search for amenities? How do environmental experiences and satisfaction, and more generally the living environment, intervene? Do they confirm or invalidate the geography we have devised? And on the basis of what other factors, indicators and methods?

3.3 Towards environmental inequalities in terms of lived experience and satisfaction: the structuring role of felt environmental experience and the capacity to act at the local scale

3.3.1 A population survey: presentation of the method, the thematic fields and the locations selected

Six municipalities were selected on the grounds of the regional environmental typology, ensuring that the number of questionnaires were kept at a minimum (100). We privileged the choice of municipalities of clearly differentiated environmental types (good, average, bad). While globally environmental criteria predominated, we took care to retain for each environmental category binomes of municipalities equally close socially, guaranteeing a certain internal comparability for each of the groups. But we also sought to vary the history and the dynamics proper to each of these territories, by selecting municipalities from different departments in each category.

In this case we opted for the three "first ring" (première couronne) departments of the Paris agglomeration. In 2008, these departments were home to 37 % of the regional population (11.6 million people). Together, they represent the diverse social and environmental situations encountered throughout the region: residential areas that are sometimes identical to those of the peri-urban areas of the "outer ring"; an environmental offer (e.g.: woodlands) that is to a certain extent comparable to municipalities more remote from Paris; or – as a last example – certain links with or proximity to agricultural areas. Moreover, these three "first ring" departments differ clearly as to their trajectories:

- economic (type of activities and development, for example a very different industrial past),
- social (socio-professional aspects, for example municipalities that may be situated at the extremes in terms of tax base),
- urbanistic (morphotypes, with for example very variable proportions of collective and social housing),
- and thus environmental (amenities/disamenities, with for example strong differences in terms of protected areas, or transport-related nuisances, industrial risks, etc.).

On the basis of these diverse crossed criteria we retained:

- for municipalities of good environmental quality: Sceaux (Hauts-de-Seine - 92[8] department) and Vincennes (Val-de-Marne – 94 department),
- for municipalities of low environmental quality: Asnières-sur-Seine (Hauts de Seine - 92) and Noisy-le-Sec (Seine-St-Denis - 93),
- for municipalities of mixed environmental quality: Choisy-Le-Roi (du Val-de-Marne - 94) and Epinay-sur-Seine (de Seine-St-Denis - 93).

The questionnaire addressed the inhabitants (average length 45 minutes), and consisted of 75 questions, 23 of which were open (verbal qualification). It was structured around our queries on the satisfaction and lived experience of the environment and established an analytic register for a different geography of environmental inequalities, notably pointing towards two major explanatory dimensions:

---

[8] Cf. Maps 1 and 2 to localize precisely the different departments involved in the survey.

Municipalities	Asnières-sur-Seine	Noisy-le-Sec	Choisy-le-Roi	Epinay-sur-Seine	Sceaux	Vincennes
Population 2005	82 800	38 600	36 300	50 800	19 400	47 200
Gross income per inhabitant 2003	1403,16	899,58	1038,98	837,94	2136,68	1806,92
Unemployment rate 1999 (in %)	11.6	15.56	13.76	18.84	6.88	9.51
Property owners 1999 (in %)	37.86	33.26	37.90	34.91	46.75	43.84
Social housing 2005 (in %)	21.89	42.05	34.44	38.06	22.55	6.07
Presence of a Zone Urbaine Sensible	yes	yes	yes	yes	yes	no

Table 8. Socio-economic characteristics of the municipalities surveyed, Sources: RGP (1999 et 2005), DGI (2003), DGI/DGC (2006)

- the felt environment, including affective relation to place, territorial anchoring... notably via residential trajectories or sensible perceptions of the near environment;
- people's involvement and their willingness/capacity to take action at local level for example, or via their attitude towards public action, the public authorities or provisions for participation.

We must also specify that the questionnaire was established after a preliminary phase of 50 exploratory interviews conducted in the 6 selected municipalities; these enabled us to fine-tune the potential role of certain local factors, and to test the wording of some questions.

Finally, eight major thematic headings structured the questionnaire (Appendix 1 presents the overall structure of the information collected and the variables tested):

- Residential trajectory of the person and the household, length of residence (seniority) and assessment of the neighbourhood
- Motivations and criteria for the decision to live in given municipality
- Representations of the quality of the environment and of the living environment (at different scales)
- Environmental experience, perception and satisfaction levels (at local and urban scale)
- Projects of residential moblity, motivations and conditions
- Spatial practices (work, services/equipment, tourism)
- Opinions on territorial action at different scales and points of view about relations with the public authorities

These seven headings were completed by a further one establishing the major socio-economic characteristics of individuals and their households (professions and socio-professional categories, educational level, age, sex, type of housing, occupational status, length of residence – seniority - in the municipality).

As stated in the introduction to this third part, these eight headings and the whole survey as such call upon and cross contributions from several scientific disciplines:

- from cognitive psychology, notably for the parameters of satisfaction and mastery of the private character of the environment,
- to political sociology, in order to grasp the social relations to territorial action, the modalities of its construction, and the criteria that legitimize it,
- via psychology and social geography, for the analysis of the weight of representations, but also of the identity factors of attachment to the living space,
- and via spatial economy to assess the structuring role of socio-economic factors in the distinctive construction of urban areas.

We opted for a quota-based sampling method, with three criteria: distribution by professions and socio-professional categories (in French: PCS); distribution by age, and by gender. For these three criteria, the 600 persons surveyed are representative of the municipal populations. Moreover, various filtering criteria were applied (sampling objectives):

- the age of surveyed individuals: only persons aged 18 and over were selected, in order to guarantee a certain stability relative to choice and particularly to residential choice;
- a minimal length of residence (seniority): individuals with less than one year of residence in their current home were excluded from the sample, in order to ensure that they had a certain experience of the environment, the neighbourhood and the larger living environment;
- homogeneous infra-municipal distribution, in order to ensure complete coverage; to this aim, quotas were established by sector (function of the number of sectors to be surveyed and the environmental charactiristics of the neighbourhood)

Appendix 2 presents an example of municipal breakdown by surveyed neighbourhood and socio-environmental characteristics.

3.3.2 Between environmental repulsion and attraction: the weight of sensory interpretative environmental filters for residential choices

Firstly, 58.2% of the persons surveyed declared that they were attentive to the quality of the environment when choosing their home. This means that the environment is the 4th major criterion, behind the internal characteristics of housing and the price variable, but more important than the offer of services, shops and facilities, than the neighbors and/or nearness to family/friends, or parameters relative to the general atmosphere in the neighbourhood. This result fully complies with what numerous surveys conducted in several European regions over the past 15 years have demonstrated (see for instance Bonaiuto, Fornara, Bonnes, 2003), and confirms the argument voiced in the previous part of this chapter: the environment is increasingly important for lifestyle choices of European populations.

In a next step, an analysis of environmental factors for these criteria enabled a more precise view of what makes up the environment and its quality.

Setting aside the financial and material constraints liable to influence their choice, we see that persons are above all likely to avoid disagreeable factors (e.g.: repulsive effects of pollution, nuisances and risks). Firstly, this validates the observation made previously on the basis of standard indicators and the disparities thus observed. Here too we must note, with reference to factors of attraction, that parameters relating to the sensory atmosphere (ex: tranquility) constitute another privileged register: the presence of nature in the neighbourhood, the view, cleanliness, architectural quality and low building density. Thus, perceptual operations, the dimensions of experience and sensibility do indeed operate as primary interpretative filters of environmental quality, at least when planning to move house. This is certainly one of the first contributions of the survey to our overall issue. We shall return to this question later.

	Number	Proportion
Housing quality	392	65.33
Price of housing or rent	381	63.50
Nearness to collective transport	377	62.3
Environmental quality	349	58.17
Presence of shops and services	307	51.17
Nearness to place of work or study	290	48.33
Image and mood of the neighbourhood	266	44.33
Building density and quality of architecture	243	40.50
Safety	204	34.00
Proximity to good schools	175	29.17
Presence of friends or family	143	23.83
Neighbours	129	21.50
Presence of sports and cultural facilities	111	18.50
Total / surveyed	600	

Table 9. Criteria privileged by households in the choice of their current home, Source : Faburel et Gueymard (2008)

	Number	Proportion
No traffic noise	297	52.2
No factories in the vicinity	280	49.2
Green spaces	270	47.5
Cleanliness	268	47.1
No air traffic noise	264	46.4
Presence of trees, vegetation in the neighbourhood	258	45.3
View	197	34.6
No rail traffic noise	192	33.7
No flooding risks	183	32.2
Quality of local architecture	164	28.8
Low building density	164	28.8
Air quality	135	23.7
Presence of waterways and bodies of water	78	13.7
Total / respondents	569	

Table 10. Environmental criteria linked to the choice of future housing, Source: Faburel et Gueymard (2008)

From the point of view of the environmental inequalities problem we address, we prolonged the analysis by crossing future criteria of residential choice with socio-professional categories of households (PCS). The strongest attraction and repulsion between on the one hand the different social categories, and environmental objects on the other hand, were recorded using the Correspondence Factor Analysis, with the Maximum Percentage Deviation as indicator (PEM). Two response modalities (yes/no) were systematically associated with criteria that proved significant: yes designating a significant relation to choice, no to non-choice.

Profession and socio-professional category	Question	Modality	Effectifs	Khi2	PEM	Test Khi2
Craftsmen/tradesmen, shopkeepers, heads of businesses	No railway traffic noise	Yes	14	3,325	29	••
Managers and higher level intellectual jobs	No railway noise	Yes	43	4,909	19	•••
	No road traffic noise	Yes	56	1,513	17	•
	Low building density	Yes	37	4,412	15	•••
	Quality of local architecture	Yes	36	3,63	14	••
Intermediary professions	No flooding risk	No	85	2,808	45	•••
	No railway noise	No	78	1,076	26	•
	Low building density	Yes	41	6,174	18	•••
	No air traffic noise	No	65	1,087	18	
	Quality of local architecture	Yes	37	2,983	12	••
Employees	Presence of green spaces	Yes	53	2,223	19	••
	No factories in the vicinity	Yes	53	1,501	16	•
Workers	Quality of local architecture	No	68	1,188	38	••
	Cleanliness	Yes	52	6,453	34	•••
	No road traffic noise	No	55	4,46	33	•••
	No factories in the vicinity	No	52	1,563	22	•
	Presence of green spaces	Yes	43	1,008	14	
	No flooding risk	Yes	32	1,954	12	•
Retired persons	Low building density	No	128	1,697	32	•••
	Cleanliness	No	103	2,994	23	•••
	No flooding risk	Yes	65	6,117	16	•••
	No factories in the vicinity	Yes	82	1,041	10	

Table 11. Environmental objects designated as important in the choice of new housing, according to socio-professional category, Source: Faburel et Gueymard (2008)

These various relations highlight the oppositions between social categories, notably between
the categories that are most emblematic of problems of inequality. Among the objects that
embody these oppositions, we note that:

-   Road traffic noise seems to structure a difference between the richest and the poorest.
    Avoidance factor for managers and upper level intellectual professions, it has no
    repulsive effect on workers.
-   Railway traffic noise, and less importantly air traffic noise, distinguishes between the
    middle classes (intermediary professions) and the richest (managers).
-   The quality of the local architecture (and in a lesser measure building density)
    distinguishes, via its attractivity or lack of it, management and upper level intellectual
    professions, as well as the intermediary professions, from workers.
-   Finally, the presence of a factory here differentiates employees from workers, both low
    income groups (PCS). In fact, since the factory is part of the worker's social universe,
    this result appears highly plausible.

Thus, though the "objects of opposition" generated here are not all identical to those
identified previously as structuring the social composition of space at regional scale (a
reminder: listed heritage sites, Seveso type industrial risk, railway traffic noise, airway
traffic noise from major Paris airports), we observe:

-   an a priori global correspondence between objectivized environmental quality and
    motivational objects,
-   and above all, a strong structuring of social distribution due to repulsion, i.e. the
    avoidance of certain potentially disagreeable environmental factors.

It would therefore seem that negative environmental factors do indeed introduce greater
spatial social distinctions than positive objects, actively contributing to selection
mechanisms and consequently to the construction of the geography of environmental
inequalities. However, and this is the third point, and doubtless the most important one, we
must again admit that the parameters of sensory atmosphere (sight and sound) constitute an
important register, an interpretative filter for the assessment of the environment – both
positive and negative. Above all it distinguishes – via its statistically validated presence or
absence - between the rich and the poor. For us this was a strong incentive to pursue our
analysis by examining environmental satisfaction, and understanding the social factors of its
construction.

3.3.3 On the inequalities of felt and lived experience: The primary role of local attachment, of sensible operations and of political involvement

Several questions attempted to evaluate the environmental satisfaction of households:
variables of numerical assessment on a scale from 0 to 10, addressing a list of environmental
objects (positive and negative), but also open questions on what was perceived as agreeable
or disagreeable (cf. Appendix 1).

Following a factorial analysis (AFC) based on the responses to these questions, we used
classification methods enabling us to establish sub-populations, depending on how close
their responses were to each other. This generated three sub-populations, homogeneous in
size: the dissatisfied (A : 24.3%, n=146), the more or less satisfied (C : 39.8%, n=239), the very

satisfied (B : 35.8%, n=215). We then operated various crossings with the explanatory dimensions established by the body of information (corpus) generated by the survey. Numerous differences appear behind these 3 great levels of satisfaction: different residential trajectory/seniority (with as a modulating factor the degree of attachment to the municipality), different modalities and factors of residential choice (for example the choice or rejection of a home), different representations of the environment (positive/negative, local/global, bio-centred/anthropo-centred), different spatial and leisure practices (e.g.: use of green spaces), different relations to public involvement (confidence in elected representatives, memberhip in or cooperation with an association) and different socio-economic characteristics (for details cf. Faburel and Gueymard, 2008).

Above all, the three categories of satisfaction (and their associated socio-spatial profiles) with professions and socio-professional categories (PCS), enabled us to note the existence of environmental inequalities of the lived environmental experience. First of all, there are indeed notable social differences of the felt experience depending on socio-professional category. A priori, the most affluent social categories are proportionally much more satisfied with their environment than the poorest categories, and this within the same municipality. It appears that the socially most vulnerable are proportionally the most dissatisfied.

	Dissatisfied	Moore or less satisfied	Very satisfied
Craftsmen/tradesmen, shopkeepers, heads of businesses	25.9	33.3	40.7
Managers and upper-level intellectual jobs	14.6	44.8	40.,6
Intermediary professions	20.6	47.1	32.4
Employees	33.3	40.6	26.0
Workers	48.8	32.9	18.3
Retired persons	15.9	36.3	47.8
Other persons with no professional activity	17.5	40.0	42.5

Table 12. Level of environmental satisfaction by socio-professional category, Source: Faburel et Gueymard (2008)

However, beyond generally confirming the overall link between environmental satisfaction and the usual social indicators, certain results generated by the measure of satisfaction directly question the conventional measure of environmental inequalities (i.e. mainly technical physico-chemical approaches aiming for normative action for protection). If a priori the most vulnerable socially are proportionally the most likely to be dissatisfied with their environment, some questions still have to answered. Notably, the distribution among the different satisfaction levels of professions and socio-professional categories highlights a large diversity of situations, with strongly contrasting felt experiences which did not allow us to establish a clear (unequivocal) relation to the environment by social category. In other words, satisfaction may vary strongly at an identical social level and at a comparable educational level. At municipal and infra-municipal scale, and again going out from the 3 different categories of environmental quality:

-   45.6 % of very satisfied individuals do not live in a municipality designated as having very good environmental quality,

- 41.2% of persons living in municipalities of good environmental quality appear to be not fully satisfied with their environment, and 6% are totally dissatisfied.

These fine distinctions put the explanatory scope of "objective" environmental satisfaction characteristics into a more relative perspective, and queried the instruments used to measure environmental inequalities.

Continuing the analysis of the different explanatory dimensions of environmental satisfaction, we finally attempted to establish a hierarchy of the variables which, by crossing all the explanatory dimensions arising from the questionnaire's thematic headings (particularly the relation to the living environment and the ways of life), appear to structure and discriminate between the different groups most strongly. Again using the Maximum Percentage Deviation (PEM), we then established a decreasing classification of the variables most strongly associated with the environmental satisfaction of the persons surveyed.

Variable	Deviation	Khi2	Test Khi2	PEM
Expectations as to improvement of environmental quality	19	14	•••	42
Feeling of being "at home"	80	63	•••	40
Regret at having to move	78	36	•••	35
Confidence in the municipal elected representatives	106	73	•••	35
Regret at having to leave the neighbourhood	87	82	•••	32
Criteria of residential choice: environmental quality	80	32	•••	32
Municipality of residence	101	72	•••	30
Living in a ZUS classified neighbourhood	20	18	•••	30
Reference to ideal living environment: here	42	18	•••	30
Municipal environmental characteristics (3 categories characterized "objectively", see above)	96	46	•••	29
Attachment	75	44	•••	29
Membership in association	45	12	•••	27
Confidence in local public authorities	61	41	ooo	27
Frequency of use of green spaces	62	38	•••	26
Criteria relative to residential choice: the image and atmosphere of the neighbourhood	39	10	•••	25

Table 13. Classification of explanatory variables of environmental satisfaction (global PEM), Source : Faburel et Gueymard (2008)

Once again, environmental satisfaction seems to imply objective environmental endowment, here expressed by the variables "municipality of residence" and "municipal environmental characteristics", and we note a certain correspondence between such endowment and its lived experience; once again, declared satisfaction appears to be socially anchored; nonetheless this classification casts a few interesting lights, above all putting the weight of strictly socio-economic criteria into a more relative perspective, and generating other, potentially more promising types of determinants within detailed local situations.

1. As a matter of fact, it appears that environmental satisfaction is above all strongly linked to the emotional identity-related aspects that accompany a living environment – including so-called low-quality environments: to regret moving from the house and the neighbourhood, the chosen environment, the current place of residence considered as the ideal living environment, the strength of the attachment to it. All these elements are expressed above all in terms of being "at home".
2. In this register of lifestyles, sensory parameters that qualify the surrounding atmosphere and the perceptual operations and dimensions of experience are the primary filters to interpret environmental quality and the resulting satisfaction or dissatisfaction (frequency of using green spaces, criteria of atmosphere in the residential choice).
3. Finally and above all, this satisfaction seems to depend on the confidence of the persons in collective means of action and particularly in the elected local representatives, and their capacity to respond to expectations relative to environmental matters and the living environment. While association-based involvement, and in consequence the evaluation of provisions for public participation, seem to express a commitment that seemingly unfolds in a more political dimension relative to environmental satisfaction and social inequalities as they are experienced, it also appears to ground it.

Thus, it appears that environmental satisfaction depends less directly on socio-economic variables, or on "objective" environmental characteristics, than on the differentiated capacities and aptitudes of persons (who are, let us remember, socially unequal, cf. 2nd part) to control their local environment and act upon it, thus confirming a number of relevant findings from cognitive psychology and political sociology.

Once crossed, these three types of results make apparent the strength of affective mechanisms and political involvement to influence the relation to the environment. They also point towards the growing weight, running across all social categories, embodied at the local scale by the "universe of what is near", in the assessment of the environment and the desired changes (cf. 2nd part).

4. Some practical indications for evaluation and implementation, with a view to the sustainable development of European regions: From socio-spatial disparities to territorialized environmental injustices

The research summarized here confronted statistical data on so-called environmental inequalities on the scale of the Ile-de-France region with the environmental experience of the region's inhabitants. The research aimed to build a different geography of environmental inequalities, taking into account the lived and felt environment, through local experiences, satisfaction, and place attachment relative to the environment. A further aim was to improve the understanding of the operative mechanisms, notably residential ones, in the phenomena of spatial polarization for environmental reasons at regional scale. Our two working hypotheses were:

- the register of personal lived experiences and of environmental satisfaction constitutes a non-negligible source of information which, due to its territoriality and resulting transversality, distinguishes between environmental qualities, thus pinpointing disparities, inequalities and even injustices in this area;
- the subject-individual, via his lived environmental experience and the cognitive and social transactions he operates, constitutes together with his immediate living environment, a pertinent scale of observation to highlight certain relevant determining dynamic factors of inequalities, in order to perhaps differently ground territorial decision making.

The first stage was based on crossing two typologies, one environmental, the other social, going out from previously existing statistical data. This led – classically – to the observation of a growing correlation between the environmental and social characteristics in the Ile-de-France. This distribution confirms the situation of certain areas in the nearer suburbs, which used to be industrial, but also that of peri-urban areas absorbing the dispersion of low-income populations to areas which may have been subject to recent deterioration (e.g. certain parts of the eastern Seine-et-Marne). Above all, as of this stage, it became clear that it would be easier to understand residential choices and the resulting geography, more via the repulsive effects of environmental damage and deterioration than via the attractiveness of certain elements, notably those called natural here (green spaces, waterways). It also generated a list of environmental objects and factors that make a place attractive or undesirable.

The second step was to select six municipalities close to Paris considered as representative of the different environmental disparity situations identified. A survey based on a semi-open questionnaire was conducted with 600 inhabitants, face to face (average length 45 minutes), to gather information concerning the environmental experiences and satisfactions of the households concerned, their real life situations and perceptions of environmental quality and of their living environment, their residential itineraries and attachments, places, practices, and relations to public action. The survey confirmed our argument that people are more likely to make their residential choices to avoid nuisance factors; with traffic noise or the bad quality of local architecture (and to a lesser extent, the presence of an industrial sites) as the major arguments. It also showed that environmental satisfaction is probably strongly linked to territorialized experiences and expectations relative to the lived and felt environment: the capacity of the near environment to provide a feeling of *"being at home"*, and confidence that elected representatives (above all municipal) will do something about these expectations.

These results elucidate the strengths and weaknesses of the conventional approach to environmental inequalities, founded (if we remember) on a static reading of quantified physico-chemical (e.g.: exposures) and socio-economic facts (e.g.: income level). The situations described are situated at least as much in the domain of sensible, symbolic and axiological relations and transactions of local societies to their living environment, as in the more conventional domain of the physical or social components of local places, which are often largely accounted for: thresholds of chemical exposure for air quality (with, for instance, short interest in health effects) ; data probabilities of risks occurrence, flood risks and hazards for instance (whit, for example, no more interest in local history, social habits or economical activities linked to rivers); acoustic levels for noise nuisance (with well-known gaps between doses and annoyance); distance for the accessibility of urban amenities, of green spaces…

To this aim, considering the logic of decision makers and the cultures in the urban field, we now wish to propose a few approaches that would improve the inclusion of environmental inequalities from the perspective of sustainable development (Faburel, 2011). One way would be to focus on lifestyles and people's experiences linked to the environment, and their attachment to a particular place. Another way should adopt a participatory rather than a structural approach to the investigation of exclusion and capacity forms of involvement (i.e. Sen's *capabilities*) instead of more conventional behavioural markers of urban inequality (such as moving house, for example). How can this be done? First empirically, then politically.

Certainly one must be careful when generalizing these results. There is no way in which a local survey of 600 persons could be representative of a population of over 4 million (3 "first ring" departments), and even less of the 11.6 million inhabitants of the Ile-de-France region. As the objective of our study was exploratory, it became imperative to structure spatial scales to account for the role of ecological dynamics and social transitions in shaping environmental challenges and their differential impact within society (for this see for instance Marcotullio and McGranahan, 2007, or World Bank, 2007). Moreover, we must admit that certain standard indicators have undeniable predictive power, for example when evaluating the increasingly significant weight of so-called environmental disqualifications on the repulsive effect of certain environmental situations, e.g. in the residential choices of households.

But, as for the less static and descriptive, more dynamic and "pro-active" interpretation of our approach, it addresses both the production of scientific knowledge and its usual divisions/habits by scientific discipline, as well as the still dominant worldwide system of environmental evaluation, i.e. mainly technical approaches aiming for normative and strictly protective action, usually at project, national or continental levels (*Environmental Impact Assessment, Strategic Environmental Assessment…*). Following in the steps of Krieg et Faber on the subject of environmental inequalities, who proposed some interesting views on the cumulative indicators of social vulnerability inspired by the *capabilities* of Sen and on environmental hazards (2004), and in the wake of Bonaiuto et al. (2003) on the importance of place attachment in households' residential choices, let us cite two examples taken from our work.

Like others, we have stated that the registers of perceptions, of the sensible, and of involvement were a powerful force structuring the lived environmental experience, to the extent that in adapting to great environmental disparities (and to an environment of so-called bad quality), the resulting appropriation ("to make it one's own") may play an essential role. Here appropriation implies mechanisms which in certain cases could easily be defined by already existing classical indicators, or as readily grasped via certain adjustments. For the first, the length of residence (seniority), which is often included in surveys on social issues (for instance housing and environment surveys) may reveal the attachment to the place of residence and a grounding in it; given the confidence granted to territorial players (confirmed by numerous official barometers) this generates means of action seen as addressing environmental problems and allowing for an assessment of the level of personal involvement. For the second, the variable "presence of a garden or terrace", for example, constitutes a true environmental relay for certain people, whereas for others it

acts as a compensatory factor within the domestic sphere. This may be more important than
the distinction by type of housing (house/apartment) or by status (owner/tenant/rent-free)
which surveys habitually use to distinguish socially between populations and/or to typify
relations to the environment.

Similarly and perhaps even more central to the issue of environmental inequalities, or at
least to the various aspects addressed in the 2nd part, a gap revealed in the survey enabled
us to query the pertinence of the official statistical classifications generally proposed or
used. The analysis shows that the rich are not systematically the most satisfied with their
environment. Our results allowed us to cast a light on a social category which is often
ignored in socio-economic approaches to the environment: non-working persons (retired
persons of all social origins and others not gainfully employed). In fact, the differentiation
relative to environmental satisfaction may have less to do with differences between
professions and socio-professional categories (PCS), or between managers and workers, for
example, than with the opposition between non-working/working persons, with the retired
dominant in the first group, and workers in the active population. We will have to
understand how time set free by retirement, or links between age and local attachment, may
generate possibilities of involvement in environmental issues and challenges.

As we can see, information on the living environment, through local experiences,
satisfaction, place attachment relative to the environment enabled us – under the condition
of using complementary indicators - to obtain additional elements for a finer assessment of
local disparities (neighbourhood, municipality, inter-municipality). The type and nature of
environmental objects in these contexts, the importance of certain morphological and socio-
economic factors, as well as the environmental perceptions and beliefs that underlie
relations with local policies also provide a basis for action addressing environmental
inequalities in a sustainable development perspective.

The information arising from the population's *on site* felt experience raises queries that are
pertinent to an empirical measure of environmental inequalities (e.g.: what observation
indicators) and for analytic frameworks (e.g.: what conceptions of the environment, of
justice, of the individual, should be privileged). To a large extent, these questions still have
to be resolved. In a wider sense, it addresses the systems of knowledge of the environment
and their capacity to perceive what makes people "inhabit" a given place, their sensible
lived experience, attachments, involvement, and thus the types of inequalities in this field.

It is true that, as stated by Charles: *"Although the environment is recognized as an object of
universal concern, concrete measures relative to it, its consideration at a finer scale and the subjective
dimensions that constitute it are largely underestimated and ill perceived."* (2001, p. 21). In fact,
although it has shown itself to be effective to a certain extent (see predictive power above),
the environment is still viewed in terms of overlapping technical and legal norms, which do
not disclose the ways in which it is "lived', nor its interfaces with other territorial
characteristics. The instruments to measure these aspects are still inadequate.

Last but not least, via this cosmopolitical approach to the environment and the socio-spatial
changes that accompany it, we move away from the dominant approach to both inequalities,
the environment and justice, i.e. from a strictly egalitarian reading of social disparities in
terms of the environmental endowment of places, and towards a more dynamic analysis of
inequalities, which are (as already stated) *"the result of unequal access to the diverse resources*

offered by a society", and thus to apprehend by means of a survey and qualify the territorialized phenomenon they constitute via lived experience.

In so doing, we move from only "combining" environmental degradation and socio-historical spatial disqualification (disparities) to what we see as a first evaluation of injustices through the different ways in which the inequalities thus evaluated make their entry into politics. Doubtless, as we have shown, because of the vital queries addressing the inaptitude of the current and official environmental assessment system to describe a fully territorialized phenomenon, i.e. the shortage or inadequacy of evaluation tools. But also because the factors and mechanisms used in in our work refer directly to action and its recent changes and trends via the symbolic and axiological dimensions thus highlighted.

For instance, in this new geography the structuring role of the repulsive nature of certain damages is particularly linked to the installation of so-called high-impact equipment (industrial sites, transportation infrastructures…). Let us assume that attractivity founded on exceptional elements (sites and monuments, green components) perhaps less explains the inequalities that have been pointed out than the repulsive effect of certain degradations. Are not then public and private policies responsible for the installation of this equipment and above all for monitoring compliance with the relevant environmental standards? Are they not directly implicated, owing to their history, notably that of the State with its enterprises and services? This makes clear the impact of past arbitration, and the responsibility of public and private authorities, with their underlying conceptions of social and spatial justice, for these decisions.

Along the same lines, the influence on environmental satisfaction arising from differentiated judgments, expectations and capacities for commitment to local action, could involve inhabitants in novel ways, both via their own experience of unequal environmental situations, via forms of involvement which such situations increasingly generate, or as vital resources for participatory projects in local forms of action. In short, environmental satisfaction also addresses new forms of institutional and territorial governance of projects, and their regulation.

In fact, municipal collective bodies are confronted and must often manage growing spatial inequalities which mix socio-urban and economic stakes with increasingly environmental ones (concentration of economic activities, the social specialization of space, social disqualification and environmental degradation). Thus, today they must take a greater interest in instruments of evaluation and intervention to counter the mechanisms of socio-spatial segregation and reverse the lot assigned to certain portions of the territory which cumulate economic and social problems and environmental handicaps.

All this reinforces the idea that environmental injustice might well represent, over and above merely social disparities relative to exposure, *"the social and territorial inequalities of capacities and means given to populations and local public authorities to act in view of improving their lived and felt environment"* (Faburel, 2008). We are close here to the readings proposed by Schlosberg (2004) or Jamieson (2007): environmental justice needs to address not only the distribution of environmental harms and benefits, but also people's participation in decision-making processes, including recognition of people's particular identities and visions of a desirable life. This is also an extension of the interpretation of inequalities given by Mitchell and Walker (2003), and born of the *Environmental Justice* movement in the

English speaking countries: *"Unequal capacities to act upon the environment and address public
authorities in order to change the living environment"*.

Such an extension would in fact mark the appearance of an updated conception of the
environment (and of justice), accounting for the importance it assumes for our social cognition,
practices and projections. As shown in the first part of our chapter this qualification is better
adapted to the changes that the environment now imposes upon our societies, and might offer
a more perceptive view of action, particularly in the urban regions where socio-spatial
dynamics and segregation mechanisms are particularly strong and go back a long time.

Without this growing awareness one may well wager that the question « Evaluate, but for
what purpose? » remains unanswered. For example, without fine-tuning the noted
disparities highlighted by the first stage of our study (according to which 2 750 000 persons
in the Île-de-France are victims of such situations), the costs of public and private
intervention will act as an obstacle to action for a long time. As we see, the increasingly
frequent current efforts to define and observe environmental inequalities are not able to
counteract the objectives of targeted action in multi-player systems, nor their underlying
conceptions of the environment (and of justice).

Here is where this more pro-active qualification sheds a light on potential levers for
sustainable development for European regions, balancing between institutional and bottom up
approaches to sustainability. However, this is perhaps less a question of sustainable
development consisting of themes and pillars side by side, than of increasingly inclusive and
plural, mixed dimensions, i.e. of a *"conceptual framework within which the territorial, temporal, and
personal aspects of development can be openly discussed"*. These would include 'Place',
'Permanence', and 'Persons' (Seghezzo, 2009). In an intersecting perspective combining
different sciences and policies (Stengers, 1997), cross-disciplinary research should contribute to
defining integrated, locally based issues relative to social and spatial aspects.

These could be used by decision makers in the field of environmental justice: experimental
knowledge (for instance landscaping experiments), real participatory approaches (ex: diversity
of collaborative initiatives with inhabitants and their empowerment, see for instance
Cruikshank, 1999), and new subjects (well-being, sustainable/eco neighbourhoods, ecological
housing). Let us also note as a last example in this perspective, that health progressively
imposes itself as a paramount subject in the analysis of environmental injustice. Far from its
purely biomedical and quantitative (epidemiological) aspects, this approach is evolving
rapidly to view health primarily as well-being in the larger sense (e.g. emotional dimensions).
Crossing it with ecological findings (Corvalan et al, 2005), it thus emphasizes its fundamental
and qualitative links with poverty, participation, or the sustainability of territories (Sen, 2002).

In fact, if within the framework of the territorialization of urban action via sustainable
development, as well as within that of a democracy willing itself to be more participatory
(notably owing to environmental stakes, see for instance Dietz and al., 2008), poorer
populations are not given the capacity of involvement, notably to make a political issue of
environmental inequalities (cf. sanitary whistle blowers), certain well-known socio-
economic mechanisms and the non-environmental character of dominant conceptions of
social justice will continue to segregate populations and spaces, notably due to residential
mobility, competitive policies, property or finance based reasons behind the installation of
equipment generating negative external effects.

In consequence, the sustainable city should take a real interest in the long-term dynamics, past and future, of the environmental marginalization of certain of its places and populations, and protect them: against spatial fragmentation, social segregation, environmental gentrification. In any case this could prove more useful than uniquely institutional answers which have in the end doubled environmentally based social vulnerability; imposing limits on action in favor of mixity in housing policies, and enhancing the weight of the market in the attractiveness competition between various places or regions…

5. Appendix 1

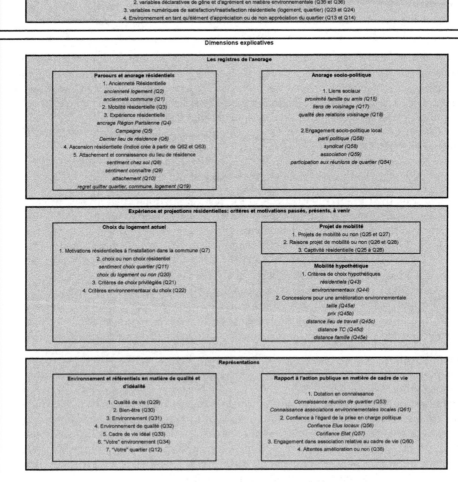

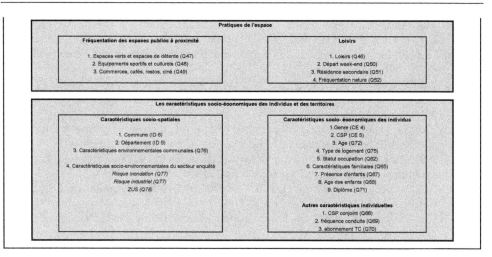

6. Appendix 2

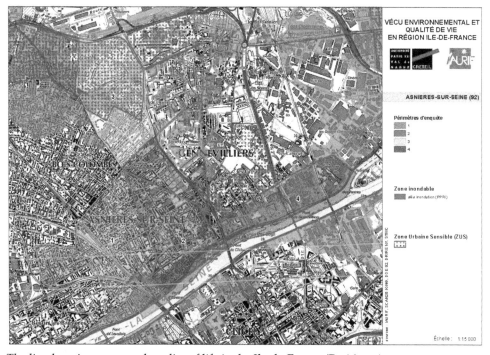

The lived environment and quality of life in the Ile-de-France (Paris) region

7. References

Ascher F., 2004, Les nouveaux principes de l'urbanisme, L'Aube, 110 pages

Baumol W.J. and Oates W.E., 1988, The Theory of Environmental Policy, Cambridge University Press, Cambridge, UK, 299 pages.

Beatley T., 2010, Biophilic Cities: *Integrating Nature Into Urban Design and Planning*, Island Press, 200 pages

Beatley T., 2000, Green urbanism: *learning from European cities*, Island Press, 2000 - 491 pages

Beck U., 2004, "The Truth of Others: A Cosmopolitan Approach", Common Knowledge, 10(3), pp. 430-449

Beck U., 2001, La société du risque. Sur la voie d'une autre modernité. Editions Aubier, Coll. Alto, 521 p. Traduction de World Risk Society, Cambridge, Polity Press (1999).

Beck U., 1995, Ecological politics in an age of risk, Wiley-Blackwell, 216 pages (see chapter 7, Technocratic challenge to Democracy, pp 158-184).

Berger M., 2004, Les périurbains de Paris. De la ville dense à la métropole éclatée ?, CNRS Editions, 176 p.

Boltanski L., Thevenot L., 1991, De la justification. Les économies de la grandeur, Paris, Gallimard, Coll NRF Essais, 483 p.

Bonaiuto M., Fornara F., Bonnes M., 2003, "Indexes of perceived residential environment quality and neighbourhood attachment in urban environments: a confirmation study in the city of Rome", in Landscape and Urban Planning, n°65, pp. 41-52.

Bullard B. (ed.), 1994, Unequal protection: Environmental Justice and Communities of Color, San Francisco, Sierra Club Books.

Bullard R., 1990, Dumping in Dixie: Race, Class, and Environmental Quality, Boulder, CO: Westview Press.

Bullard R.D., 1983, "Solid Waste Sites and the Black Houston Community", Sociology Inquiry, 53, pp. 273-288.

Charles L., 2001, « Du milieu à l'environnement », in Boyer M., Herzlich G., Maresca B. (coord.), L'environnement, question sociale. Dix ans de recherche pour le ministère de l'environnement, Paris, ed. Odile Jacob, 229 p.

Charvolin F., 2003, L'invention de l'environnement en France, La découverte, 127 p.

Choffel P. (coord.), 2004, Observatoire national des zones urbaines sensibles, Paris, Editions de la Délégation interministérielle à la ville, 252 p.

Corburn J., 2005, Street Science. Community Knowledge and Environmental Health Justice, MIT Press, Cambridge, London.

Corvalan C., Hales S., McMichael A., 2005, Ecosystems and human well-being: health synthesis. A report of the Millennium Ecosystem Assessment, World Health Organization, Geneva.

Cruikshank B., 1999, The Will to Empower: Democratic Citizens and Other Subjects. Cornell University Press, Ithaca.

Deboudt P., Deldrève V., Houillon V. et Paris D., 2008, « Inégalités écologiques, inégalités sociales et territoires littoraux : l'exemple du quartier du Chemin Vert à Boulogne-sur-Mer (Pas-de-Calais, France) », Espace, Populations, Sociétés, n°1, pp. 173-190.

De Palma A., Motamedi K., Picard N., Waddell P. (2007), "Accessibility and environemntal quality: inequality in the Paris housing market", European Transport, 36, pp. 47-74

Diamantapoulos A., Schlegelmilch B., Sinkoviks R. et Bolhen G. (2003), "Can socio-demographics still play a role in profiling green consumers ?", Journal of Business Research, 56, pp. 465-480.

The Environment as a Factor of Spatial Injustice: A New Challenge for the Sustainable
Development of European Regions?

255

Diebolt W., Helias A., Bidou D. et Crepey G., 2005, *Les inégalités écologiques en milieu urbain*, Rapport de l'Inspection Générale de l'Environnement et du Conseil Général des Ponts et Chaussées, 68 p.

Dietz Th., Stern P. C., National Research Council (U.S.), Panel on Public Participation in Environmental Assessment and Decision Making, 2008, Public participation in environmental assessment and decision making, National Academies Press, 305 pages

Dobré M., Juan S. (dir.), 2009, Consommer autrement – La réforme écologique des modes de vie, Paris, l'Harmattan, 312 p.

Dobson A., 1998, Justice and the environment, Oxford University Press, Oxford.

Dozzi J., Lennert M. et Wallenborn G. (2008), « Inégalités écologiques : analyse spatiale des impacts générés et subis par les ménages belges », Espace, Populations, Sociétés, n°1, pp. 127-143.

Environmental Protection Agency – EPA, 1995, Environmental Justice Strategy, Note, EPA, Washington, 41 p.

Faburel G., Gueymard S., 2008, Vécu environnemental et qualité de vie en région Ile-de-France. Une approche des inégalités environnementales, Rapport final du CRETEIL pour le Programme Politiques Territoriales et Développement Durable, Ministère de l'Ecologie, du Développement Durable, des Transports et du Logement, mai, 133 p.

Faburel G., Maleyre I., 2007, « Dépréciations immobilières, polarisation sociale et inégalités environnementales pour cause de bruit des avions », Revue Développement Durable et Territoires, mai, 17 p. (http://developpementdurable.revues.org/document2775.html)

Faburel G., 2011, "How to green our urban spaces (people participation, experiences and attachment to place)? Environmental conceptions in public policies and local perspectives for spatial justice in France", 10th Conference on Environmental Justice and Global Citizenship, Mansfield College, Oxford, july, 16 p.

Faburel G., 2010a, « Inégalités et justice environnementales », in O. Coutard et J-P. Lévy (coord.), Ecologies urbaines, Economica, coll. Anthropos, Chap. 13, pp. 214-236.

Faburel G., 2010b, « Débats sur les inégalités environnementales. Une autre approche de l'environnement urbain », Justice spatiale/spatial justice, n°2, Justice spatiale et environnement, octobre, pp. 102-132.

Faburel G., 2008, « Les inégalités environnementales comme inégalités de moyens des habitants et des acteurs territoriaux. Pour que l'environnement soit un réel facteur de cohésion urbaine ? », Revue Espace - Populations – Sociétés, n° 2008-1, pp. 111-126.

Fairburn J., 2008, 'Adressing environmental justice : a UK perspective', Colloque Inégalités environnementales et risques sanitaires, Agence Française de Sécurité Sanitaire, Environnementale et du Travil, Paris, 10 avril, 9 p.

Fisher F., 2000, Citizens, experts and the environment. The political of local knowledge, Duke University Press, 328 p.

Giddens A., 1991, The Consequences of Modernity, Stanford University Press, 188 pages.

GIEC - Intergovernmental Panel on Climate Change, 2007, Climate Change 2007: Synthesis Report, Intergovernmental Panel on Climate Change, 32 p.

Gueymard S., 2009, Inégalités environnementales en région Ile-de-France : répartition socio-spatiale des ressources, des handicaps et satisfaction environnementale des habitants, Doctorat d'aménagement, d'urbanisme et de politiques urbaines, Institut d'Urbanisme de Paris, Université Paris XII, 387 p.

Guillerme A., Jigaudon G. et Lefort A-C., 2004, Dangereux, insalubres et incommodes. Paysages industriels en banlieue parisienne XIXe-XXe siècles, Champ Vallon, Collection Milieux, 343 p.

Haanpää L., 2007, 'Consumers' green commitment: indication of a postmodern lifestyle', International Journal of Consumer Studies, Volume 31, Issue 5, pp. 478–486.

Harvey D., 1992, "Social justice, postmodernism and the city", International Journal of Urban and Regional Research, 16, pp. 588-601

Hirschman Albert O., 1970, Exit, Voice and Loyalty. Responses to Decline in Firms, Organizations and States, Cambridge, MA, Harvard University Press, 284 p.

Jagers S. C., 2009, 'In search of the ecological citizen', Environmental Politics, Volume 18, Issue 1, pp. 18-36.

Jamieson D., 2007, 'Justice: the heart of environmentalism', In Sandler, R. and Pezzullo (eds) Environmental justice and environmentalism: the social justice challenge to the environmental movement. MIT Press, 352 p.

Kohlhuber M., Mielck A., Weiland S. K. and Bolte G., 2006, "Social inequality in perceived environmental exposures in relation to housing conditions in Germany", Environmental Research, Vol. 101, Issue 2, pp. 246-255

Krieg E.J., Faber D.R., 2004, "Not so Black and White: environmental justice and cumulative impact assessments", Environmental Impact Assessment Review, Vol.24, pp. 667-694.

Kruize H., 2007, On environmental equity – exploring the distribution of environmental quality among socio-economic categories in the Netherlands, KNAG / Copernic institute, Utrecht, Pays Bas, 219 p.

Laigle L., 2005, Les inégalités écologiques de la ville : caractérisation des situations et de l'action publique, Rapport Ministère de l'Ecologie, programme Politiques territoriales et développement durable, 122 p.

Lascoumes P., 2001, « Les ambiguïtés politiques du développement durable », Université de tous les savoirs, La nature et les risques, Vol. 6, Odile Jacob, pp. 250-263.

Latour B., 2004a, "Whose Cosmos, Which Cosmopolitics? Comments on the Peace Terms of Ulrich Beck", Common Knowledge, 10(3), pp. 450-462.

Latour B., 2004b, The Politics of Nature, Cambridge MA: Harvard University Press, 326 p.

Laurent E., 2010, Social – Ecologie, Flammarion, 230 p.

Laurent E., 2009, « Ecologie et inégalités », Revue de l'OFCE, Presses de Sciences Po, 2, n°109, pp. 33-57.

Laurian L., 2008, " Environmental injustice in France", Journal of Environmental Planning and Management, vol.51, n°1, janvier, pp. 55-79

Lolive J. et Soubeyran O. coord., 2007, L'émergence des cosmopolitiques, La Découverte, 383 p.

Lolive J., 2010, « Mobilisations environnementales », in Ecologies urbaines, Coutard O. et Lévy J-P. (coord.), Economica, coll. Anthropos, pp.

Marcotullio, P. and McGranahan, G. eds., 2007, Scaling urban environmental challenges: from local to global and back, Earthscan, London, 212 pages.

Martinez-Alier J., 2002, The Environmentalism of the Poor, Edward Elgar, Northampton.

Mazmanian D. A., Kraft M. E., 1999, Toward sustainable communities: *transition and transformations in environmental policy*, MIT Press, 323 pages.

Mitchell G., Walker G., 2003, Environmental Quality and Social Deprivation, technical report E2-O67/1/TR, The Environment Agency, Bristol, 61 p.

Préteceille E., 2003, La division sociale de l'espace francilien. Typologie socioprofessionnelle 1999 et transformations de l'espace résidentiel 1990-1999, Rapport de recherche, Observatoire Sociologique du Changement, Paris, 146 p.

Portney Kent E., 2003, Taking sustainable cities seriously: *economic development, the environment, and quality of life in American cities*, MIT Press, 284 pages.

Puech M., 2010, Développement durable : un avenir à faire soi-même, Le pommier, coll. Mélétè, 226 pages.

Pye S., I. Skinner, N. Meyer-Ohlendorf, A. Leipprand, K. Lucas et R. Salmons, 2008, Addressing the social dimensions of environmental policy – A study on the linkages between environmental and social sustainability in Europe, European Commission Directorate-General "Employment, Social Affairs and Equal Opportunities, 148 p.

Riddell R., 2007, Sustainable urban planning: *tipping the balance*, Wiley-Blackwell, 335 pages

Rizk, 2003, « Le cadre de vie des ménages les plus pauvres », Insee Première n° 926, octobre, 4 p.

Rawls J., 1971, A Theory of Justice, Cambridge, Massachusetts: Belknap Press of Harvard University Press.

Rosanvallon P., 2008, La contre-démocratie : la politique à l'âge de la défiance, Seuil, Paris, 344 p.

Roseland M., 1997, Eco-city dimensions: *healthy communities, healthy planet*, New Society Publishers, 1997 - 211 pages.

Roussel I., 2010, « Les inégalités environnementales », Air Pur 76 - Inégalités environnementales, inégalités de santé, pp.5-12.

Rumpala Y., 2010, Développement durable ou le gouvernement du changement total, Editions Le Bord de l'eau, collection « Diagnostics », 450 pages.

Saint-Julien T., François J-C., Mathian H., Ribardière A., 2002, Les disparités de revenus des ménages franciliens en 1999. Approches intercommunales et infracommunales, et évolution des différenciations intercommunales (1990-1999), Paris, Direction Régionale de l'Equipement d'Ile-de-France, 108 p.

Schroeder, R., St Martin, K., Wilson, B., Sen, D., 2008, "Third World environmental justice", Society and Natural Resources, 21, pp. 547-555.

Sen A.K., 2002, "Why health equity?", Health Economics, Vol. 11, pp. 659-666.

Sen A.K., 2009. The Idea of Justice, Harvard University Press, 468 p.

Sen A.K., 1993, "Capability and Well-being", in Martha C. Nussbaum and Amartya K. Sen (eds), The Quality of Life, Oxford: Clarendon Press, pp. 30 - 53.

Schlosberg D., 2004, 'Reconceiving environmental justice: global movements and political theories', Environmental Politics 13, pp. 517-540.

Shirazi R., 2011, 'Sustainability and the Hegemony of Technique. Towards a new approach to cultural sustainability', Inter-disciplinary.net *10th Global Conference:* Environmental Justice and Global Citizenship, Mansfield College, Oxford, United Kingdom, july, 10 p.

Stengers I., 1997, Sciences et pouvoirs. La démocratie face à la technoscience, Paris, La Découverte, Coll. Sciences Sociétés, 116 p.

Theys J., 2010, « Les conceptions de l'environnement », in Ecologies urbaines, Coutard O. et Lévy J-P. (coord.), Economica, coll. Anthropos, pp. 16-34

Theys J., 2007, "Pourquoi les préoccupations sociales et environnementales s'ignorent-elles mutuellement. Essai d'interprétation à partir du thème des inégalités écologiques », in P. Cornut, T. Bauler et E. Zaccaï (coord.), Environnement et inégalités sociales, Editions de l'université de Bruxelles, pp. 24-35

Tiebout C., 1956, "A pure theory of local expenditures", Journal of Political Economy, vol. 64, pp. 416-424.

UK Environment Agency (2007), Adressing environmental inequalities: cumulative environmental impacts, Science report, 81 p.

Villalba B. et Zaccaï E., 2007, « Inégalités écologiques, inégalités sociales : interfaces, interactions, discontinuités », Revue Développement durable et territoires, Dossier 9 Inégalités écologiques, inégalités sociales.
http://developpementdurable.revues.org/document3502.html.

Wheeler S. M., Beatley T., 2004, The sustainable urban development reader, Routledge, 2004, 348 p.

Wenz, P.S., 1988, Environmental Justice, State University of New York Press, Albany, 217 p.

World Bank, 2007, Poverty and the environment: understanding the linkage at the household level, World Bank, Washington, DC.

Young I.M., 1990, Justice and the Politics of Difference, Princeton University Press, 294 p.

12

Unraveling Stakeholders' Discourses Regarding Sustainable Development and Biodiversity Conservation in Greece

Evangelia Apostolopoulou[1,*], Evangelia G. Drakou[1,2]
and John D. Pantis[1]
*[1]Department of Ecology, School of Biology,
Aristotle University of Thessaloniki, Thessaloniki,
[2]Current Address: Global Environment Monitoring Unit, Institute for Environment and
Sustainability, Joint Research Center, European Commission, Ispra,
[1]Greece
[2]Italy*

1. Introduction

The designation and implementation of adaptive conservation strategies able to respond to changing socio-ecological conditions, requires understanding protected areas as complex, interconnected social-ecological systems able to reconcile human needs with biodiversity conservation (Davidson-Hunt & Berkes, 2003). This consideration leads to perceiving ecosystems involved in biodiversity conservation and the social, political and economic processes and structures behind their management, as interrelated. Sustainable development has been considered, at least the last two decades, as an integrative concept aiming at combining ecological, economic and social issues. However, the concept of sustainable development has received much criticism, whereas the outcomes of successfully combing economic development, social welfare and ecological sustainability can be characterized as quite mixed.

Focusing on the relationships between conservation and development we should refer to Blaikie & Jeanrenaud (1997) who describe three distinct intellectual paradigms, which also entail fundamentally different approaches to human welfare and assume different set of relations between civil society, the market and the state: the classic/authoritarian, the neo-populist (people-oriented conservation programs such as integrated conservation and development projects - ICDPs, joint or co-management schemes) and the neo-liberal. Similarly, Salafsky & Wollenberg (2000) analyze three types of linkages between livelihoods and conservation: no linkage, indirect linkage and direct linkage, whereas Nygren (1998) analyzes four ideological perspectives dominant in the current discourse of sustainable environmentalism in the "Third World": (i) environmentalism for nature, (ii)

---

* Corresponding/Invited Author

environmentalism for profit, (iii) alternative environmentalism, and (iv) environmentalism for the people. Finally, Adams et al. (2004) offer a conceptual typology of the relationships between poverty reduction and conservation, which is quite relevant to the discussion regarding sustainable development and biodiversity conservation: (i) poverty and conservation as separate policy realms, (ii) poverty as a critical constraint on conservation, (iii) conservation as a process which should not compromise poverty reduction, and (iv) poverty reduction as depending on living resource conservation.

This chapter presents the different discourses between sustainable development and biodiversity conservation strategies, using as an example that of protected areas while presenting at the same time the actions these discourses are promoting. Given also that the research design was based on a multilevel governance approach, a significant aim of this research is to investigate the perceptions of stakeholders acting at different governance levels regarding development and conservation. Towards this goal and in order to disentangle the multiple myths surrounding conservation and development discourses we try to contextualize the latter in their specific institutional fields as well as to reconceptualize (Nygren, 1998). We furthermore investigate how the perverse understanding of the sustainable development concept could cause scale mismatches between ecological and social systems and thus between natural and social processes (see also Cumming et al., 2006).

The analysis followed in this chapter uses primary data obtained from the authors through qualitative research methods. Following the grounded theory approach (Strauss & Corbin, 1998), 87 semi-structured interviews were conducted in Greece in order to analyze current policy and governance discourses as well as management strategies on sustainable development and biodiversity conservation.

2. Literature review: About sustainable development and conservation

A brief description of the historical development of conservation policy is essential to unravel the links between sustainable development and nature conservation and understand the broader context in which these concepts have emerged. The establishment of protected areas has been the leading conservation strategy since the late 19th century (Adams et al., 2004). The different purposes of the wilderness movement and the wise use movement raised the focus of conservation and utilization – oriented dialogues (Kalamandeen & Gillson, 2007). These two opposing perceptions of nature have always reflected a wavering between idealistic and mechanistic representations of nature (see also Foster, 2002). The latter has been evident in the history of the establishment of protected areas which has been based on aesthetic, moral (Thiele, 1999), political and economic criteria as well as on the displacement of indigenous people (Abakerli, 2001).

The gradual increase in the number and extent of protected areas during the 20th century led to the establishment of networks of protected areas during the '90s. The Habitats Directive and the establishment of the Natura 2000 Network were influenced by the political and economic context of the period, especially the "meteoric rise of sustainable development'" (Peterson et al., 2005). This has been associated with a shift towards consensus-driven policies based on the belief that longstanding conflicts between development and conservation could be resolved through collaborative governance (Apostolopoulou & Pantis, 2010). Similarly, the ecological modernization theory emerged during the '80s and remains until today dominant in environmental sociology, fact that can become evident by the prevalence of the opinion that environmental protection, including biodiversity conservation, can be a potential source of economic or developmental growth (Clark & York, 2005), something in line with the core principles of sustainable development. On the other hand, the role of the market as a tool against biodiversity loss has been linked to private property rights over natural resources (Mukhopadhyay, 2005), an expanded role for non state economic and development actors, and results-based regulatory approaches, typical characteristics of neoliberal governance (McCarthy, 2006).

In current discourses the terms of sustainable development, ecological modernization and collaborative governance constitute the core ideas around which the dominant conservation policy is being framed. The new approach to sustainable development and management of protected areas which became institutionalized in the European policy of protecting the natural environment with the Habitats Directive (92/43/EEC) reflects this broader understanding of the relationship between society and nature. To understand what is involved in this perception it is crucial to consider the context in which it was formed (see also Apostolopoulou, 2010).

The sustainable development concept first appeared in the early '70s. At that time, the capital accumulation crisis combined with a major ecological crisis, lead to the emergence of the ecological issue as an independent and important face of social and political struggle. Amid the increasing intensification of environmental problems and the increasing inability of the environmental policy to solve them (Weale, 1992) and sufficiently respond to radical ecological movements (Hajer, 1995), the issue of environmental protection has officially -i.e. institutionally- emerged. The emergence of measures for the protection of the natural environment by state policies as a separate issue came after a relatively long period during which EU's environmental policy had not set specific rules for protecting the natural environment. On the other hand, EU's policies included from the beginning a variety of social, economic and technological objectives the achievement of which would further enhance growth. Specifically, the treaty on European unification in 1957 did not contain elements of environmental policy and the First Action Programme for the Environment was launched in 1972 (Naxakis, 1997). Furthermore, the sustainable development concept also served as a reaction to the growing literature which addressed the need to set limits on development and has been initially set with the publication of the Club of Rome in 1972.

The overall ideology of that era, and particularly the fact that the end of the three postwar decades of dynamic economic growth has been followed by the entry into a period of crisis and recession, played a major role in the autonomous emergence of ecological issues. The official entry into the global scenery of the concept of sustainable development has been followed by the emergence of the concept of participation of "civil society" in environmental governance initiated by the United Nations and especially through the Brundtland report (World Commission on Environment and Development [WCED], 1987). The Brundtland report was practically the formal adoption of a system of ideas, which was substantially based on the concept of "ecological modernization" (see also Hajer, 1995; Weale, 1992).

Without ignoring the fact that sustainable development and ecological modernization do not have the exact same meaning and interpretation, nevertheless it is considered that they largely overlap and coincide (see Blowers, 1998; Dryzek, 1997; Jänicke, 1997). The theory of ecological modernization has its roots in the work of the German sociologists Joseph Huber and Martin Jänicke (Spaargaren, 1997) and began to emerge more clearly in Western countries and international organizations in the early '80s. In particular, by the mid-80s it has been widely recognized as a promising alternative policy while following the general acceptance of Agenda 21 in 1992 it began to be the dominant approach to environmental policy (Hajer, 1996).

Sustainable development is often defined as the development that meets present needs without jeopardizing the ability of future generations to meet their own needs (according to the Brundtland report, see WCED, 1987). However, most neoclassical economists understand sustainable development as the development in which consumption remains undiminished by time (Vlachou, 2005). Specifically, they suggest that the concept of sustainability should focus on maintaining the productive opportunities of future generations, without specifying if it is determined by physical capital or natural boundaries (Vlachou, 2005). Similarly, the theory of ecological modernization seeks to broadly analyze the way in which today's societies organize their economic, political and cultural institutions to cope with environmental crises. Based on this logic, ecologically modernized societies are those that incorporate environmental principles into the design of institutions to regulate human interactions with nature. The concept of democracy and constitutionally guaranteed rights and freedoms are the necessary institutions for ecological modernization in the sense that they operate as self-regulating mechanisms that have the potential to alleviate the human impacts on the planet. Alongside the market, further industrialization and technology are considered to be the main forces of modernization that will lead to ecological sustainability (Hajer, 1995; Mol, 1995, 2002; Mol & Sonnenfield, 2000; Mol & Spaargaren, 2000; Spaargaren, 1997, 2000).

If the concept of sustainable development has lost some of its original "glory" the same does not apply to the ecological modernization theory, which is a prominent neo-liberal theory and one of the leading theories in environmental sociology (York & Rosa, 2003). Its widespread prevalence in shaping the current environmental policy is evident in the dominant approach that environmental protection should not be treated as an obstacle to economic development, but rather as a potential source for future growth (Weale, 1992, p. 75), a view highly related to the concept of sustainable development. This perception is accompanied by the assumption that the "ecological rationality" will emerge from already existing institutions rather than from radical environmental movements. As Clark & York (2005) argue the theory of ecological modernization is basically a functionalist theory, in the sense that it does not see the emergence of ecological rationality as derived primarily from social conflicts but rather from ecological enlightenment within the key institutions of modern society (Mol, 1995). As a response to the increasing criticism regarding the intensification of environmental problems in the existing socio-economic system and delays in the emergence of this "rationality", supporters of the theory consider it to become a prevalent trend in the near future.

In particular, they argue that in the early years of modernization states degraded the environment, while in its last stages environmental concerns will diffuse through society, leading to the restructuring of major political, economic and social institutions towards ecological sustainability and social welfare (Mol, 1995). It is important to note that discussions of these concepts are mainly oriented to performance issues (efficiency), a focus that is not at all accidental. The criterion of efficiency is fully accredited by various public and private economic actors and establishes a common ground within the theory of ecological modernization that manages to combine the concerns of environmentalists, businesses and states. It is typical that many environmental organizations have embraced this consensual framework for discussion because it promises to make environmental protection "attractive" to governments and businesses whose cooperation is needed for environmental organizations to achieve specific reformist objectives (Hajer, 1995). The collaborative governance concept is also based in the same core ideas by supporting the participation of a variety of stakeholders in a context of apparent equity and fairness.

In general, strategies promoting sustainable development and ecological modernization goals in the context of collaborative governance do not address environmental degradation as an inherent characteristic of the current socio-economic system. Moreover, these perceptions often get to support that the forces of modernization will lead to the de-materialization of society and will succeed to decouple the economy from energy and material consumption, thus allowing the human society to overcome the environmental crisis within the present socio-economic system (Mol, 1995; Spaargaren, 1997). Nevertheless, so far, and despite the numerous signed conventions and directives, the various policy objectives and legislative measures towards the conservation and protection of the natural environment and biodiversity, the protection of human health, the rational use of natural resources etc., economic development remains the primary objective at both international and EU levels, thus weakening the effectiveness of existent environmental measures (Naxakis, 1997). After all, as Brand and Görg (2003) argue, politics on the conservation of biodiversity focus more on the creation of a stable political institutional framework for its commercialization, rather than on its actual conservation.

The above insights are essential in order to disentangle the variety of approaches regarding sustainable development and biodiversity conservation. Even though debates around the issues described above may not be transferred as such in the case of biodiversity conservation and protected areas they are relevant in order to contextualize the latter. Moreover, it is now obvious that debates regarding the relationship between biodiversity conservation and sustainable development are closely related to different approaches regarding the relationship between society and nature. These debates are of fundamental importance for the designation and implementation of management strategies and more generally for the direction and content of future conservation policy and governance.

3. Methodological approach

In our research we adopted a grounded theory approach. Grounded theory has been widely used in environmental research (e.g. Berghoefer et al., 2010; Kittinger et al., 2011) and produces theories that are likely to offer insight, enhance understanding and provide a meaningful guide towards action (Strauss & Corbin, 1998).

During the process of grounded theory building, four analytic and not strictly sequential phases were identified: research design, data collection, data analysis and literature comparison. In the initial phase (sampling) of the research, when the major purpose was to generate as many categories as possible, we gathered data using the method of snowball sampling, asking our initial list of individuals who they thought would be good informants based on their experience and then tried to gather data in a wide range of pertinent areas. We also followed the method of purposive sampling that allowed us the use of personal judgment for the selection of the respondents who had the knowledge and the experience to cover the topics of the research. Once the initial categories were formed, then the sampling became more specific. The sample was then selected according to the "theoretical sampling" method, based on analytic questions and comparisons, pinpointing places, people or events to maximize the chances of discovering variations among concepts (Corbin & Strauss, 1990). This meant that the sample definition and requirements evolved during the research process itself. The criterion for stopping the research of a certain category was based on the category's "theoretical saturation", a major component of our research methodology since without this the theory would be unevenly developed and lacking density and precision (Strauss & Corbin, 1998).

Overall, we conducted 87 face-to-face semi-structured interviews with actors acting at several governance levels between January of 2007 and September of 2008 (see Table 1). Research questions were open, exploring issues regarding the relationships between sustainable development and biodiversity conservation with specific reference to the role of local community, the state and the market. The interviews were then transcribed and the transcribed passages were labeled with codes. In grounded theory data analysis is a well-defined process that begins with basic description and moves to conceptual ordering and then on to theorizing (Patton, 2002). Thus, the main analytical techniques include three types of coding: open, axial and selective coding (for further details see also Apostolopoulou & Pantis 2009, 2010). The main purpose of coding is the same as in other types of qualitative research (Padgett, 1998; Patton, 2002) but its level of specificity is what distinguishes grounded theory from other qualitative methods.

Interviewees were also asked to give specific examples in line with our approach to disentangle confusing and contradicting discourses and investigate what actions they actually support in practice. We selected as case studies Zakynthos Marine National Park and Schinias National Park as in both cases previous research has already been conducted by the authors, thus a relatively good knowledge on the actual situation served as a basis for the interpretation of research outcomes. Moreover, both cases are exceptional examples, since the history of their establishment as national parks has been characterized by significant conflicts between development and conservation.

We should clarify that the formulation of the research questions was based on an extensive literature review and on the described theoretical section (see section 2), in order to ensure that they will serve to investigate the relationship between sustainable development and biodiversity conservation. Similarly, given that the objective of this research is to use empirical research as a start for a broader discussion, the findings of the analysis are linked to relevant existing scientific literature and in particular to research on human-environment relationships. This is in line with the focus of coding in grounded theory which is not based only on the opinion of the individual interviewees but also on the core emerging concepts which can guide researchers from "description to conceptualization and from the more specific to the general or abstract" (Strauss & Corbin, 1998).

Stakeholders participating in Greek biodiversity governance	Number of interviews
Central administration	
Ministry of the Environment	14
Ministry of Rural Development and Food	5
Ministry of Development	3
Ministry of Economics	1
Ministry of Tourism	1
National Center for the Environment and Sustainable Development	1
Council of the State	1
Total	26
NGOs	
World Wide Fund for Nature Greece	5
The Sea Turtle Protection Society of Greece	3
Hellenic Ornithological Society	2
Hellenic Society for the Study and Protection of the Mediterranean monk seal	2
Mediterranean association to save the sea turtles	1
Hellenic Society for the Protection of Nature	1
Pan – Hellenic Network of Ecological Organizations	1
Hellenic Society for the Protection of the Environment and the Cultural Heritage	1
Total	16
Management agencies and local administration	
Management agency of National Park of Schinias – Marathon	9
Management agency of National Marine Park of Zakynthos	5
Municipalities and Regions	5
Central Union of Municipalities and Communities of Greece	1
Total	20
Other key stakeholders	
Companies providing consulting and assessment services in the field of nature conservation	3
Greek Biotope/Wetland Center	2
Greek General Confederation of Labor	2
The centre of Athens labor unions – Department of environment and international relations	1
Hellenic Federation of Enterprises	1
Pan – Hellenic Federation of Tourism Enterprises	1
Technical Chamber of Greece	1
The Mediterranean Initiative of the Ramsar Convention on Wetlands	1
Total	12
Scientific community	
Aristotle University of Thessaloniki	4
National & Kapodistrian University of Athens	4
Scientific institutions	2
National Centre for Social Research	3
Total	13
TOTAL	87

Table 1. The sample of interviewees.

During data analysis we chose to categorize *discourses* (see Phillips & Jorgensen, 2006) and not categories of stakeholders as such, a choice based both on theoretical arguments as well as on empirical findings. Regarding the theoretical arguments, following Blaikie & Jeanrenaud (1997, p. 47) we argue that different actors tend to use different parts of specific paradigms and approaches in eclectic and contradictory ways in order to support their projects, policies and often their special interests. Therefore, a focus on ideas and paradigms is valuable in order to disentangle dominant policies and practices. As far as empirical findings are concerned, the analysis of our qualitative empirical data confirmed that the majority of stakeholders tend to combine different arguments and conceptualizations of the linkages between conservation and development in order to support different claims. Even if specific stakeholder groups, such as state officials, adopted a more specific standpoint reflecting a specific ideology or even if in some cases one approach was more dominant, e.g. in the case of private economic actors, in many occasions stakeholders were using different arguments in order to support these approaches.

Therefore, the process of data analysis resulted in the construction of a conceptual model according to which the variety of stakeholders' discourses relating sustainable development to biodiversity conservation is categorized in the three main approaches that are described in the results and discussion section, as well as in several subcategories regarding the role of local community (including the role of NGOs), the role of protected areas, the role of the market and the state as well as the explanation of (natural resource) conflicts between development and conservation with reference to the two case studies.

4. Results & discussion

4.1 Biodiversity conservation and sustainable development as incompatible discourses

This approach could be divided in two different but interlinked discourses, in the sense that they result to the same conceptualization of the relationship between nature and society: in the first one the priority lies explicitly on biodiversity protection and in the second one on development. Each discourse, even if it often has completely different foundations and arguments, by putting either development or protection as a priority, leads to the reproduction of a dichotomy between nature and society.

4.1.1 Development as a barrier to conservation

One of the two dimensions of this approach, the conceptualization of development as a barrier to conservation, was dominant mainly between stakeholders from NGOs and Universities. Underneath this approach lies a strong moral imperative regarding nature's intrinsic value as well as a strong belief that development should be understood as the main cause of current biodiversity loss. The latter was grounded in the chronic failure of Greek state conservation policy to ensure the conservation of biodiversity which was mainly attributed to the explicit prioritization of development and public works. Sustainable development was interpreted as mainly based on rhetorical arguments in order to support actions and initiatives that are not actually "sustainable", more or less as a term used in official reports or policy documents without being based on concrete actions that take seriously into account biodiversity conservation concerns.

Interviewees adopting the above arguments often failed to distinguish between corporate developmental interests and local community livelihoods and thus to explain the roots behind the current direction of Greek economic development. Conflating sustainable and non-sustainable human activities has in its core the idea that humans are an *a priori* threat to biodiversity conservation leading to calls for strict nature protection excluding any type of human activities including those of the local community.

The above approach is quite similar to what has been characterized as the classic or authoritarian paradigm to biodiversity conservation (Blaikie & Jeanrenaud, 1997) and one significant element of this discourse was the explicit support for the establishment of more protected areas as the dominant proposed strategy for biodiversity conservation. The effect of the presence of local communities and in particular of indigenous people around protected areas (Chatty & Colchester, 2002), along with the economic and social impacts of these areas have been widely acknowledged and investigated (Adams & Hulme, 2001; Igoe, 2006; McNeely, 1993), mainly highlighting the tendency of conservation policy to act against the economically weaker groups of local communities. The establishment of protected areas has often been accompanied by the financial exploitation of these areas and the degradation of local communities. One of the most significant consequences is related to the displacement of local populations with direct impacts on their survival, as well as on livelihood provision (Brown, 1998; Cernea, 1997; Cernea and Schmidt-Soltau, 2006; Chatty & Colchester, 2002; Geoghegan & Renard, 2002; Gjertsen, 2005; Harper, 2002; Knudsen, 1999; Nygren, 2005). Even if the majority of the research on the unequal costs and benefits of conservation policies and on their increasing economic and social consequences have focused on "developing" countries (e.g. Sodhi et al., 2010; Swetnam et al., 2011) the last decades research on "developed" countries is also increasing (Apostolopoulou & Pantis, 2010; Foster, 2002; Haggerty, 2007).

However, interviewees who considered development as a barrier to conservation tended to underestimate the social dimensions of conservation policies. The latter was linked to the general tendency towards blaming humanity or mankind as a whole for environmental degradation. Moreover, usually interviewees mainly blamed the economic "weaker" social groups following the dominant ideology that poverty and environmental degradation are interlinked, considering poverty as a cause of environmental degradation and not both as a result of the existing socio-economic structures. Interviewees often considered that local communities by prioritizing their individual welfare could easily adopt short-term developmental goals and initiatives without taking into account environmental concerns. The conceptualization of local community as a group of people caring only for short-term profit from rapid development has been often associated with a more general understanding of society as a unified entity without significant differentiations (e.g. economic, social, political, gender), but as divided between two poles: the people that are environmental conscious and those that are not. It is worthy to note that many interviewees considered the latter category as a typical example of Greek local communities, attributing major responsibilities for biodiversity loss to local people, thus arguing that non-conservation is not only a state choice but also a social demand. Interestingly, this claim was in accordance with arguments of private economic actors and state officials who considered conservation as a barrier to development (see §4.1.2. and also Apostolopoulou & Pantis, 2010).

The way that the above argumentation has been used in order to explain the conflicts between development and conservation that have emerged in the two case studies, the National Park of Schinias and the National Marine Park of Zakynthos, is remarkable. The following quote from a non-governmental organization (NGO) representative regarding the conflicts between development and conservation in National Marine Park of Zakynthos is indicative:

"The best way to end these unresolved conflicts between biodiversity and development is to buy the beaches that are the matter of debate. This has already happened –from an NGO- for one beach with very positive results. Sustainability is not an issue any more in these areas; they have abandoned it forever when they uncritically adopted mass tourism. Local people create problems each summer and they will continue to do the same because they understand the protected area as a barrier to their welfare so the only realistic way to proceed and ensure biodiversity conservation is unfortunately without them…"

The conceptualization of local community through the adoption of the term "civil society" proved in many cases as a crucial factor leading to the homogeneity of local people's interests. The determinative role of "civil society" in environmental policy has firstly emerged during the international meeting in Rio (1992) and today the term is more or less established. Civil society includes all the organizations and institutions which, at least theoretically, are located outside direct state control like associations, community groups, corporations, NGOs as well as business interests (Scholte, 2004). Prominent role in environmental and conservation policy is given to NGOs at international, national and local levels. It is characteristic that many interviewees consider that NGOs' participation is equal to public participation stemming from the belief that NGOs are the main representatives of common opinion. It is crucial to notice that the establishment of NGOs has been increased approximately 400% the last twenty years. These organizations are quite heterogeneous not only regarding the scale of their activity (local, national, international) but also in the forms of their organization, their goals, as well as their general standpoint towards political processes (De Angelis, 2003). However, in many cases they share common economic goals with business cycles while in other cases they purposefully promote the values and policies of neoliberal state and market (De Angelis, 2003).

A crucial point of this research is that representatives of NGOs tended to consider themselves during interviews as representatives of society without clarifying if they represent the interests of a specific part of society, whilst in many cases the only role that they were acknowledging for local people was the need to include them in environmental education activities. This seems to be a general trend especially if we take into account the fact that none of the international organizations which are promoting the establishment of protected areas has adopted and published explicit policies and official principles which would forbid the displacement of local people from these areas (Cernea & Schmidt-Soltau, 2006). However, big international organizations which are active in biodiversity conservation and which are lobbying for more protected areas, mainly at international and national levels, are receiving significant economic support from states, public and business whereas small-scale and mainly local NGOs working along with local communities in order to actively support the combination of sustainable development and biodiversity conservation are mainly based on the voluntary work of activists (Chapin, 2004).

Another issue raised here is that as in the past the movement of strict nature protection found common ground with the utilitarian movement in the creation of national parks, nowadays this seems to happen again. In particular, the fact that the idea that environmental and conservation initiatives and goals come first has often been used as an excuse for excluding local people and promoting private economic interests revealing that often in this classic approach the problem is not people in general, but local people and especially indigenous people. It is indicative that many interviewees were highly critical towards local practices but at the same time they were supportive of private market-based initiatives for conservation. Not only powerful economic interests are not distinguished as potentially harmful for biodiversity, but to add to that, local people are blamed for unsustainable behavior with the common arguments that they are either not educated or due to their small-scale and short-term economic interests they do not take seriously into consideration environmental impacts. Thus we could argue that this approach is partly an inverted image of the "human-in-nature" approach that reproduces the division between human society and environment leading to serious mismatches between social and ecological systems.

4.1.2 Conservation as a barrier to development

This discourse was dominant in a small group of state officials and private economic actors and has been largely based on the criticism regarding unsatisfactory current developmental trends in Greece. Interviewees adopting this view argued that current environmental laws are very strict hindering opportunities for real development in Greek rural, marine and coastal areas. Particular emphasis was put on Natura 2000 network and the large percentage of Greek land (27,13% of national terrestrial area) that it covers. Many interviewees were adopting a mix of different arguments in order to support this approach whereas local community has mainly been portrayed as poor people-victims of strict biodiversity policies. However, the real focus of this approach was related to concerns regarding the fact that corporate interests do not select Greece for their investments because of the dominance of strict conservation measures. The following quotes from a private economic actor are quite indicative:

"Natura 2000 network has been designed without taking into account the developmental opportunities that existed in these areas which are now "trapped" inside the boundaries of the network. The result is that more than the 50% of areas ideal for tourist, residential and energy development are now Natura 2000 sites. Inside these areas, but also outside them because of the strict legislation, an investor should wait four years or more in order to receive a "yes" or a "no" from the Greek state regarding the authorization of his project. Thus, in this era of competition you should wait at least four years for investment in Greece! You can understand that this is detrimental for any kind of development and completely contrary to sustainable development: sustainability is supposed to be taking into account economy-society-environment; If it considered only the last two things then it wouldn't be called development".

And:

"Sustainable means "viable" and this clarifies current extremities in the interpretation of the term in Greek discourses considering that sustainable is equal to "inheriting" resources to the next generations. But who can guaranty to us that the future societies of our children or grandchildren will have the same needs with us? Technology continuously invents and develops new materials and substitutes for the rare ones".

It is rather remarkable that interviewees adopting this discourse had a completely different opinion regarding biodiversity policy in Greece. In contrast with all other interviewees (see also Apostolopoulou & Pantis, 2009), they argued that there is a national biodiversity strategy and a clear priority from Greek state to promote and support the establishment of protected areas and especially Natura 2000 network.

It is important to keep in mind at this point the huge struggle regarding the issue of arbitrary building, which has emerged during the voting of the new national (Greek) biodiversity law in 2011. Even though this happened after the period that this research was carried out, it has been quite indicative of the predominance of this minority discourse during interviews in real biodiversity politics in Greece.

Simultaneously, the local community has been portrayed in quite contradicting ways. The statement of K. Brown (2002) that "development perspectives have often argued that conservation is a threat to human welfare and highlight the exclusion of local people from protected areas as a denial or rights to resources and as undermining livelihoods" was dominant in this discourse, which on the other hand was combined with the conceptualization of local people as environmentally uneducated. The "value" of local people seemed to be considered as increased because of their role as voters at least for the ministries' and local administration's political leadership. As a ministry representative stated:

"Sustainable development is good in theory. But can it be really implemented in practice? Big interests will always prioritize development. [...] EU is trying to reach consensus between environment and society and is willing to take into account citizens' opinions. They were imagining sensitive citizens though, but in reality citizens use environmental protection as a "flag" while claiming other things. Therefore, state leadership will conflict only for something of huge importance. And since people are those voting for the existing governments we should prioritize people's preferences –which are clearly development and profit-, especially in a small underdeveloped country like Greece".

This has been more clearly illustrated in stakeholders' opinions about the conflicts in the two case studies. The following quote is indicative:

"The participation of local community in combining sustainable development and biodiversity conservation is risky. People who have interests will try to promote their interests and just name this "sustainable development" and in this way personal interest for profit will dominate, exactly as it happened in the cases of Schinias and Zakynthos. But on the other hand these people are those who really know these areas, own land there, thus you cannot exclude them from decision-making or decide the transformation of these areas to protected parks, which stands for actually taking over their land based on some international agreements or NGOs strategies".

However, the fact that the above argument if expressed in a different way can be used for exactly the different purpose from populist discourses or from the "human-in-nature" approach on biodiversity conservation should sensitize researchers to focus both on the social construction of nature and on the politics of conservation and development including the actual practices that each discourse supports.

4.2 Biodiversity conservation and sustainable development as compatible discourses: Green economy, ecosystem services and "win-win" scenarios

A different stakeholders' discourse, the dominant one in comparison with all others analyzed in this chapter, consisted of the main idea that biodiversity conservation and sustainable development could and should be combined with the main goal to support development through biodiversity conservation. Interviewees adopting this approach, tended to emphasize that environment is a common good, from the conservation of which every individual member of society could significantly benefit. In this context, ideas of collaborative and multilevel governance were dominant in the discussions. Even though the term civil society was the main term used for the local community also in this discourse, in contrast with the previous cases, this time local community was portrayed as consisting of more or less equal groups of actors between which "win-win" solutions could be reached. Behind this approach lies the ideology that social groups with different interests could co-decide and reach consensus through negotiations. A central argument supporting this potential has been based in the importance to render biodiversity protection politically viable through the development of new partnerships between various actors in the context of the common expectation for economic development, something in line with ecological modernization theory (Fisher & Freudenburg, 2001).

The idea of green economy is placed at the core of this discourse as well as the attempt to explicitly link economy to biodiversity conservation. The core issue here proved to be not the conservation of biodiversity as such but as a potential strategy that would benefit economic development. This was a dominant discourse among interviewees from different organizations and sectors, even though there was a continuum of approaches, from the more explicitly economic to the more socially conscious. In particular, interviewees located in the first end of this continuum, explicitly referred to the necessity to include nature into market through its valuation. This general proposal was further supported by more specific ones in the context of the popular motto, "the polluter pays" or even "the user pays". In the first case, it was argued that environmental degradation, including biodiversity loss, has been primarily caused by polluting activities, infrastructure or overexploitation of natural resources and thus should be arranged through negotiations among relevant actors and compensations should be made mainly from the responsible enterprises. In the second case, it was argued that each person visiting a national park or swimming in a protected beach should pay for using this natural "service". This is in line with what Naxakis (1997) explains as the dominant response to environmental problems: "to give prices to nature, to consider that natural resources have value, they are commodities, since the prices are those regulating the changes of available quantities of goods. The economic valuation of natural resources –from the air that we breath, the oceans, the forests, the fauna and flora- and their exchange in the market will determine according to neoliberals their demand and consumption rates, will thus regulate their exhaustion (destruction) rate".

In particular, interviewees from central and local administration, NGOs as well as private economic actors argued that the transition to a green economy, including a green tourist industry, green investments and banking, as well as green products, by regulating supply and demand will lead to the increase of the price of products and services which are rare, including ecosystem services, endangered species, habitats and landscapes and will accelerate their protection or in some cases even their replacement through technological innovation. This approach by adopting the core arguments of the ecological modernization theory is in line with the main arguments of the classical theories of ecological economics. In particular, the latter theories support that increased demand for environmental quality, expressed mainly through the preference for green commodities and services, would force governments and business to invest in ecologically friendly technologies and practices. Financial support for these technologies would in turn be possible because of the increased profit that these would bring. However, interviewees tended to underevaluate the fact that if value and scarcity are inversely related then species recovery and relative abundance could paradoxically result in reduced support for conservation (Vira & Adams, 2009).

Simultaneously, interviewees tended to support that the causes of biodiversity extinction are strongly related to the communal property of natural resources. As an interviewee noted:

"If a rare species belongs to everybody then it belongs to no one and therefore nobody has the responsibility for its conservation and management. To put it simply: nobody cares if it disappears because it won't cost anything to him".

This has been strongly attributed to the incapacity of Greek state to effectively conserve valuable ecosystems and combine in practice sustainable development with biodiversity conservation. The absence of explicitly defined property rights has been considered as a main factor, which helps and legitimates state intervention while at the same time does not allow the market to take an active role in managing conservation problems. These arguments have been further supported through the open support for free businesses and in general for a free market unobstructed from both governments and employees (see also Jamal et al., 2003). The market was portrayed here as rational, fair and representative of social interests whereas according to more extreme opinions it was considered as able to ensure the democratic distribution of ecosystem services and natural resources.

Following the above line of argumentation, interviewees explained that they consider as necessary the further privatization and commercialization of natural resources, while arguing that private property rights in individual parts of natural environment could partly deal with many current threats to biodiversity. As Fraser (1996) explains, a neoliberal approach aligns government and capital more directly, thus leading to a variety of services and goods which are neither public nor commodities, but more or less hybrids combining characteristics of both forms. This was expressed during interviews while discussing the role of protected areas on conservation and development where it was argued that initiatives towards the commercialization of landscapes and environmental "experience" could be good solutions in order to make protected areas economically viable and a core element of tourist and residential development. This is actually a very popular trend, which has led to many partnerships between parks administration, especially national parks, and mainly tourist industry in the face of constantly decreasing state funding for nature conservation (Searle, 2000). The following quotes from a representative of the tourist industry and a member of local administration regarding the two national parks are indicative:

"Eco-development and thus ecotourism set as the main economic touristic value, the preservation of innocence, wilderness and of the broad variety of nature that is essential for the modern societies that are trapped in the large cities, the fast working conditions, even in the large luxurious apartments of the big cities. Volunteering, extreme sport activities and generally the "mother nature" package that is considered as a shelter from the wild city life, is rapidly evolving as the new way of commercializing sustainability, thus enhancing the huge benefits one can gain from the non-monetary economic development. This model if adopted in Schinias and especially in Zakynthos would transform these areas into green and expensive paradises".

And:

"Many experts agree that nature which is pristine can remain like this only if it brings money. Local people in Zakynthos, and generally in Greek rural areas, are still trying to gain money from the exploitation of natural resources but they could achieve the same, if not more, by guiding tourists into their beautiful and valuable areas. Without tourists and their money neither local economies nor endangered species have the prospect to survive".

The dominance of these approaches has also been evident in the fact that the last 30 years the valuation of ecosystem services has been proved to be one of the faster developing research areas in environmental economics (Jenkins et al., 2010; Jim & Chen, 2009; Lange & Jiddawi, 2009). This has been a core argument of many interviewees, especially those more familiarized with environmental research, and it was widely used in order to support that these ideas have somehow "naturally" evolved in the era of modernity and have now become a highly respected scientific endeavor. Reference to ecosystem services, mostly seen as economic benefits provided by natural ecosystems, was not only a dominant theme during interviews but it can also be considered as the dominant trend in conservation science (MA, 2003; McCauley, 2006). However, research focusing on these issues tends to support approaches which are based in consumers' preferences and which are compatible with the usual monetary system of valuating competitive products and services (e.g. Jenkins et al., 2010).

Supporters of these approaches defending themselves towards criticism for attempts to costing and selling natural environment, argue that the main goal it is not to select a specific price («$ price tag») for natural environment or its components, but to express in economic terms the result of a change in the benefits of ecosystem services in relation to other services that people are willing to pay for (e.g. Jenkins et al., 2010), thus to calculate the potential, as well as the amount, that "consumers" are willing to pay for conserving natural environment in comparison to other "products" (Hanley & Shogren, 2002; Randall, 2002). As Vira & Adams (2009) argue "the ecosystem services approach may provide a useful additional argument for conservation, but practitioners should be cautious about the potential pitfalls of utilizing economic metaphors that are not always perfectly related to the biological systems that are the subject of conservation interest". They furthermore explain that "while natural capital is a useful economic concept, it does not capture the full complexity of relations between genes, species, and ecosystems that is associated with the term biodiversity (cited Wilson, 1992)". This is of particular importance for conservationists, who seem to adopt ecosystem services as the new "win-win" strategy, which in contrast to sustainable development puts the emphasis of this dual relationship on conservation and not on development. On the other hand, the concept of ecosystem services is essentially based on human valuation systems, which are based on changing consumer preferences, willingness to pay, and technological advances (Vira & Adams, 2009). As Tallis et al. (2008) explain "if policy and financial incentives for conservation of ecosystem services are to be successful and equitable, we will also need a solid scientific understanding of how services flow from one region to another, what human groups benefit from ecosystem services, and what groups or populations would need to be compensated for protecting those services". Moreover, there is "a strategic risk in justifying biodiversity conservation primarily in terms of ecosystem services", as McCauley (2006) points out. One should thus be aware of the potential risk that economic benefits from services that are valued by human society will overwrite and outweigh noneconomic justifications for conservation (Redford & Adams, 2009).

In the above context, proposals for economic benefits from the establishment of protected areas dominated interviews. The most dominant one concerned proposals for investigating visitors' willingness to pay for visiting a protected area or for establishing a small market based on souvenirs sold in the protected areas. Simultaneously, proposals for partnerships with the tourist industry proved to be quite popular. The conceptualization of public access to recreation as a commodity has been a constantly emerging topic in scientific literature and since the early '90s the optimal way to "charge" people for recreational purposes has been considered to be a market under constant development (see Bishop & Phillips, 1993). As Kiss (2004) notices "ecotourism represents one facet of the sustainable use approach, in which biodiversity is regarded as a product to be sold to consumers (using the terms broadly)". As a representative of an NGO put it:

"There have been many thoughts about promoting ecotourism and partnerships with the private sector, but in Zakynthos –as well as in Schinias and in Greece in general- we are too far from these approaches – the problem is not the protected area but the investment projects. We do not have the necessary development law, which would help each local landowner or business owner to think in an environmentally friendly way and start acting towards this direction [...] It is a matter of time that these people realize that environment and biodiversity conservation can give major job opportunities and bring money to their areas. Similarly, it is a matter of time for administration to realize that valuation of ecosystem services is crucial for the survival of Greek protected areas".

The latter proposals were often related, in economic actors' discourses, to proposals for promoting the importance of protected areas for residential development. This is in line with peer-reviewed publications (Pejchar et al., 2007), where authors argue that the most direct benefit from such initiatives would be the decrease in the total amount (and therefore cost) of necessary infrastructure in order to support development, assuming that almost the same number of houses is built in a smaller area. Pejchar et al. (2007) notice that the National Association of Home Builders calculated that a medium developmental complex in a protected area costs 34% less in order to be developed compared to a conventional area (citing Thomas, 1991). They also add that there are plenty of proofs indicating that vicinity to open space, like protected areas, increase the value of a property whereas the bigger increases in values concern houses located in a area of approximately 455 m from permanent protected natural areas. Therefore, development based on the conservation of natural environment could potentially give competitive advantage to those who would choose it given that it would offer them the opportunity to differentiate their houses in relation to those which are part of the "classic developmental paradigm" which tends to offer limited variations in a rather common subject (Pejchar et al., 2007). However, as Pejchar et al. (2007) argue, it is estimated that development based on conservation entails a degree of risk and under specific circumstances could be considered as less advantageous because for instance, "the identification and protection of important ecological assets could eliminate the best potential home sites on a property [sic]".

A major characteristic of this discourse is the way that different and conflicting arguments co-existed in the same sentence reflecting a tendency to resolve real conflicts around development and conservation in theory. Although each discourse was not expressed in a concise way, as explained earlier in this chapter, it was especially in this approach that policies and goals not sufficiently combined in current practices were presented as totally compatible.

4.3 Biodiversity conservation and sustainable development as complementary discourses: "Human-in-nature" approaches

This discourse proved to be dominant among scientists, NGOs representatives and state employees. Nevertheless we must emphasize that this idea has been more directly expressed through individual arguments and has not, at least for most interviewees, been presented as a clear and concrete standpoint reflected in all individual questions. An important feature of this view is that is was expressed through calls for redefining or reconceptualizing both sustainable development and biodiversity conservation. Again, we could place these discourses in a continuum from approaches that argued for redefining both concepts through criticizing current practices, to more holistic approaches that were based on specific proposals towards adopting a new more integrative approach.

At the core of this discourse lies the understanding of society and environment as co-evolving social-ecological systems. Interviewees argued that sustainable development should be interpreted as the type of development that encompasses both biodiversity conservation and human welfare, primarily defined as explicit support for environmental friendly activities, resilient local livelihoods and increased quality of life for the majority of local people. Therefore, current developmental trends were considered as non-compatible with conservation. Interviewees emphasized that those who support the need for conservation of biodiversity and sustainable development should become more critical against the wider-scale policies that threaten it and the specific actors who promote and actually profit from these (see also McAfee, 1999). They should also consider more carefully the connections between individual acts and the wider structures and processes that drive social and environmental practices and changes (see also Adams & Hutton, 2007). It was argued that for the latter it is crucial to acknowledge the differentiation within local community groups as well as the variety of activities through which they interact with natural environment. Clearly taking into account the role of power and productive relations and the way that these influence human metabolism with nature was considered as crucial towards the above acknowledgement as well as towards resolving scale mismatches between natural and social systems. As a state employee explained:

"In natural resource conflicts all social groups are not sharing equally either the costs or the benefits of conservation policy. This has been obvious in both cases, I mean in both Schinias and Zakynthos, and it was further aggravated by the fact that many people in these areas were obliged to make sacrifices whereas at the same time others remained unaffected from regulations despite the environmental disastrous character of their activities. [...] And moreover there is no official policy or strategy trying to explain and deal with the reasons behind non-sustainable activities of local people. Maybe because in this case state would have to blame itself and then who could ever be convinced for biodiversity conservation from a state which is mainly responsible for its loss?"

In this view sustainable development is closely related to social equity and local community livelihoods and therefore local perceptions regarding issues of social justice and improvement of life conditions are considered as crucial factors for the success of projects aiming at combining development and conservation. However, it was pinpointed that this approach in the Greek case is not an established trend and therefore besides reference to some specific cases where initiatives towards this direction have been adopted, this was mainly conceptualized as a proposal for future Greek policies and practices. As a scientist explained:

"The main goal of current practices of biodiversity conservation in Greece, along with the absence of references to a "societal economy" which could potentially distribute the benefits of the management agencies established in protected areas to local communities, are indicatory of the dominant direction

of present initiatives towards the commercialization of the natural environment and the exclusion of local people from their areas. They are reproducing a dichotomy between society and nature arguing at the same time that this is the only realistic approach in the era of modernity. Future integrative policies will be successful only if they escape from such dipoles. After all, our experiences with areas such as Zakynthos or Schinias confirm this argument".

Interviewees adopting this view emphasized the role of multiple types of knowledge in the process of designing, implementing and evaluating conservation projects. The role of scientific knowledge and monitoring proved to be of fundamental importance for assessing whether sustainable development is actually combined with conservation, by analyzing the impacts of development projects for ecosystems. Similarly, traditional and lay knowledge were considered as necessary tools for assessing the incorporation of socio-economic and cultural objectives in conservation-development projects as well as for resolving mismatches occurring from inaccurate analysis of the interactions between ecosystem processes and human activities.

It is remarkable that some interviewees while analyzing their proposals for improving the situation in Schinias, they focused on the need to realize that in order to minimize the degradation of ecosystems in the region due to a set of interrelated factors (fragmentation and habitat loss, hydrologic regime, residential and tourist development, etc.) an interdisciplinary designation of conservation policy that would be based on the interaction and interrelationship of social, political, ecological, economic and cultural conditions should be considered as a prerequisite. While assessing the situation in Schinias National Park interviewees explained that it would be critical to promote sustainable development by integrating the social dimension in biodiversity conservation to improve the living standards of local community with investments in areas such as infrastructure for local people (sewers, flood control, biological control of mosquitoes, etc.), works to strengthen the family income, agrienvironmental schemes, economic incentives for environmental protection and compensation for loss of income. These actions could significantly restore the chronic unequal distribution of costs and benefits that state policies have promoted. In this view, sustainable development was perceived as a strategy which could potentially lead to resilient social-ecological systems. Similar insights were documented for the National Marine Park of Zakynthos.

Similarly, the principle of "participation" has been dominant in discussions whereas the role of local communities has been described as crucial during the designation and implementation of environmental policy with the main goal to promote sustainable development. Elements of the neo-populist paradigm can be traced here in stakeholders' discourses, mainly in the quotes of NGOs representatives, especially regarding the role of protected areas (see also Brown, 2002). However, it is important to notice that the reference to ICDPs projects (for further details about ICDPs see Garnett et al., 2007), which has been a dominant strategy for combining development and conservation worldwide, was not considered as relevant for the Greek case, something also evident from the fact that this kind of projects are actually non-existent. The latter is also related to the fact that discourses about local communities leaving in harmony with nature tend to be popular in "developing" countries and not in "developed" ones.

Concurrently, some interviewees, mainly researchers, claimed that in order to make stakeholders networks to effectively work, research should be focused on the investigation of the socio-economic and political power relations within and among social groups (see also Paulson et al., 2003) across scales. The designation of a socially inclusive conservation strategy, including specific incentives, compensation measures and support for traditional human activities, at national level was considered as necessary in order to achieve real changes at both regional and local levels. Such initiatives can be based on the benefits from the establishment of linkages between humans and protected areas (Hoole & Berkes, 2010) and on schemes that would deal with the unequal distribution of cost and benefits that conservation policies produce (Apostolopoulou & Pantis, 2010).

Finally, interviewees argued that issues such as the definition of "local community" and the description of how societal participation in implementing conservation programs will occur in practice should no longer be located on the margins of the dominant approaches. It was argued that conflicts over natural resources cannot be treated as "technical" issues that need to be resolved by the appropriate "communication strategies".

The adoption of adaptive co-management strategies (see Armitage et al., 2009) was a dominant proposal in this discourse. The latter was perceived as a strategy which could lead to the improvement of the current situation through the transition to a comprehensive and long-term adaptive management plan where a variety of management measures will be implemented and tested in practice to achieve the integration of sustainable development and biodiversity conservation in a context of environmental justice. Similarly, a participatory and transparent decision-making process was considered by interviewees as necessary in order to implement integrative conservation and development strategies according to the social needs of the majority of local people.

5. Conclusion

It is evident that different discourses regarding sustainable development and biodiversity conservation have core differences on the way they interpret and frame the relationship between nature and human society. The conceptualizations of the role of local community, NGOs, the state and the market have fundamental consequences for the way that biodiversity conservation and sustainable development as well as potentials for their integration are being understood and defined. Undoubtedly, there is a huge confusion around all these terms evident in the apparent difficulty of the interviewees to explicitly explain their ideas and offer integrative approaches whereas in many cases same observations on several points were used to support very different arguments (see also Brechin et al., 2002). The fact that different agencies interpret the linkages between development and conservation in different ways, along with the different policy instruments implemented in protected areas, results in a range of prescriptions and management strategies (Brown, 2002). This confusion is directly related to the variety of cultural and ideological perspectives as well as to the influence of powerful economic interests and especially to the fact that "by no means all of these different interests and normative notions about biodiversity concern human welfare although they may be invoked in its name (Blaikie & Jeanrenaud, 1997)". The existence of many contradictory tendencies and rivalries in development strategies requires a thorough analysis of the social construction of nature especially given that the discourses regarding conservation and sustainability are directly linked to the broader systems of development and power (Nygren, 1998).

Towards this direction the role of the state, the market and local communities should be carefully analyzed. Apart from the overall role of NGOs, the "civil society" term is still unclear and problematic, since it encompasses the definition of society as a homogeneous entity. In the current reconstruction of the term, the "civil society" concept diminishes the structural conflicts that occur among different social groups (Meiksins Wood, 1998). The failure to recognize differences within local communities has been highly criticized by a broad variety of researchers (Agrawal & Gibson, 1999; Brosius et al., 2005; Brosius et al., 1998). Similarly, focusing on "actors" rather than state and market structures and processes tends to remove agents from structures forgetting that the central questions related to environmental degradation and rural deprivation are to be found in land tenure relations, market dependencies, organization of economies, and violence against local knowledge (Bebbington, 1993; Nygren, 1998). The latter are primarily responsible for serious mismatches between social and ecological sustainability whereas they are strongly related to the fact that individuals in fundamental societal roles have the power to influence ecological patterns and processes at scales beyond expected. This is highly apparent in the case of biodiversity conservation and sustainable development with the example of Protected Areas, to be of the most typical ones. This, in turn, causes several misconceptions, mismatches and conflicts of interest among the various administrative levels as well as between and within institutional and governance structures and processes.

Today, it is considered that due to scientific advances and new political coalitions new approaches are emerging that align development with conservation linking human and environmental well being (Daily & Matson, 2008). However, we could argue that combining nature protection with social justice has not yet been implemented as a general strategy whereas the role of sustainability remains complicated and quite ambiguous in current policies and practices (Apostolopoulou & Pantis, 2010; Apostolopoulou, 2010; Brechin et al., 2002). Therefore, it is of fundamental importance to unravel current stakeholders' discourses in order to analyze current deficits in both theory and practice. The latter is, in turn, critical for the designation and implementation of future integrating policies which would consider biodiversity conservation and sustainable development as complementary goals in the context of a new positive relationship between nature and human society.

6. Acknowledgements

An important part of this study was funded by the EU European Social Fund (75%) and the Greek Ministry of Development - GSRT (25%). Funding for a part of this study was also provided by the Large-Scale Integrating Project within FP7 SCALES (grant 226 852). We thank the many people and organizations for providing us with information necessary for our research.

7. References

Abakerli, S. (2001). A critique of development and conservation policies in environmentally sensitive regions in Brazil. *Geoforum*, Vol.32, pp. 551-565

Adams, W.M. & Hulme, D. (2001). Conservation and communities: Changing narratives, policies and practices in African conservation, in: *African Wildlife and Livelihoods: The Promise and Performance of Community Conservation*, D. Hulme & M. Murphree, (Eds), 9-23, James Currey, London

Adams, W.M. & Hutton, J. (2007). People, Parks and Poverty: Political Ecology and Biodiversity Conservation. *Conservation and Society*, Vol.5, pp. 147-183

Adams, W.M., Aveling, R., Brockington, D., Dickson, B., Elliott, J., Hutton, J., Roe, D., Vira, B. & Wolmer, W. (2004). Biodiversity conservation and the eradication of poverty. *Science*, Vol.306, pp. 1146-1149

Agrawal, A. & Gibson, C.C. (1999). Enchantment and disenchantment: The role of community in natural resource conservation. *World Development*, Vol.27, pp. 629-649

Apostolopoulou, E. & Pantis, J.D. (2010). Development plans versus conservation: explanation of emergent conflicts and state political handling. *Environment and Planning A*, Vol. 42, pp. 982-1000

Apostolopoulou, E. & Pantis, J.D. (2009). Conceptual gaps in the national strategy for the implementation of the European Natura 2000 conservation policy in Greece. *Biological Conservation*, Vol.142, pp. 221-237

Apostolopoulou, E. (2010). Critique of the dominant developmental ideology regarding the nature-society relationship: The case of conservation policies. *Outopia*, Vol.91, pp. 87-106 *(in Greek)*

Armitage, D.R., Plummer, R., Berkes, F., Arthur, R.I., Charles, A.T., Davidson-Hunt, I.J., Diduck, A.P., Doubleday, N.C., Johnson, D.S., Marschke, M., McConney, P., Pinkerton, E.W. & Wollenberg, E.K. (2009). Adaptive co-management for social-ecological complexity. *Frontiers in Ecology and the Environment*, Vol.7, pp. 95-102

Bebbington, A. (1993). Modernization from below: an alternative indigenous development? *Economic Geography*, Vol.69, pp. 274-292.

Berghoefer, U., Rozzi, R. & Jax, K. (2010). Many eyes on nature: diverse perspectives in the Cape Horn Biosphere Reserve and their relevance for conservation. *Ecology and Society*, 15(1): 18, Available from
http://www.ecologyandsociety.org/vol15/iss1/art18/

Bishop, K.D. & Phillips, A.A.C. (1993). Seven Steps to Market – the Development of the Market-led Approach to Countryside Conservation and Recreation. *Journal of Rural Studies*, Vol.9, pp. 315-338

Blaikie, P. & Jeanrenaud, S. (1997). Biodiversity and human welfare, In: *Social change and conservation: environmental politics and impacts of national parks and protected areas*, K. Ghimire & M.P. Pimbert, (Eds), 46-70, Earthscan, London

Blowers, A. (1998). Power, participation and partnership, In: *Cooperative Environmental Governance: Public–Private Agreements as a Policy Strategy*, P. Glasbergen, (Ed.), 229-249, Kluwer Academic Publishers, Dordrecht

Brand, U. & Görg, C. (2003). The state and the regulation of biodiversity. International biopolitics and the case of Mexico. *Geoforum*, Vol.34, pp. 221-233

Brechin, S.R., Wilshusen, P.R., Fortwangler, C.L. & West, P.C. (2002). Beyond the square wheel: toward a more comprehensive understanding of biodiversity conservation as social and political process. *Society and Natural Resources*, Vol.15, pp. 41-64

Brosius, J.P., Tsing, A.L. & Zerner, C. (2005). *Communities and Conservation: Histories and Politics of Community-Based Natural Resource Management*, Altamira Press, Walnut Creek, USA

Brosius, J.P., Tsing, A.L. & Zerner, C. (1998). Representing communities: histories and politics of community-based natural resource management. *Society and Natural Resources*, Vol.11, pp. 157-168.

Brown, K. (2002). Innovations for conservation and development. *The Geographical Journal*, Vol.168, pp. 6-17

Brown, K. (1998). The political ecology of biodiversity, conservation and development in Nepal's Terai: confused meanings, means and ends. *Ecological Economics*, Vol.24, pp. 73-88

Cernea, M.M. & Schmidt-Soltau, K. (2006). Poverty risks and national parks: Policy issues in conservation and resettlement. *World Development*, Vol.34, pp. 1808-1830

Cernea, M.M. (1997). The risks and reconstruction model for resettling displaced populations. *World Development*, Vol.25, pp. 1569-1589

Chapin, M. (2004). A challenge to conservationists. *World Watch Magazine*, Vol.11-12, pp. 17-31

Chatty, D. & Colchester, M. (2002). *Conservation and Mobile Indigenous Peoples: Displacement, Forced Resettlement and Sustainable Development*, Berghahn Press, New York

Clark, B. & York, R. (2005). Dialectical Materialism and Nature: An Alternative to Economism and Deep Ecology. *Organization and Environment*, Vol.18, pp. 318-337

Corbin, J. & Strauss, A. (1990). Grounded theory research: procedures, canons, and evaluative criteria. *Qualitative Sociology*, Vol.13, pp. 3-21

Cumming, G.S., Cumming, D.H.M. & Redman, C.L. (2006). Scale Mismatches in Social-Ecological Systems: Causes, Consequences, and Solutions. *Ecology and Society* 11(1): 14, Available from
http://www.ecologyandsociety.org/vol11/iss1/art14/

Daily, G.C. & Matson, P.A. (2008). Ecosystem services: from theory to implementation. *PNAS*, Vol.105, pp. 9455-9456

Davidson-Hunt, I.J. & Berkes, F. (2003). Nature and society through the lens of resilience: toward a human-in-ecosystem perspective, In: *Navigating Social-Ecological Systems*, F. Berkes, J. Colding & C. Folke, (Eds.), 53-82, Cambridge University Press

De Angelis, M. (2003). Neoliberal Governance, Reproduction and Accumulation. *The Commoner*, Spring/Summer 2003, Available from
http://www.thecommoner.org

Dryzek, J.S. (1997). *The Politics of the Earth: Environmental Discourses*, Oxford University Press, Oxford

Fisher, D.R. & Freudenburg, W.R. (2001). Ecological modernization and its critics: assessing the past and looking toward the future. *Society and Natural Resources*, Vol.14, pp. 701-709

Foster, J.B. (2002). *Ecology against capitalism*, Monthly Review Press, New York

Fraser, N. (1996). Clintonism,Welfare, and the Antisocial Wage: The Emergence of a Neo-liberal Political Imaginary, In: *Marxism in the Postmodern Age: Confronting the New World Order*, A. Callari, S. Cullenberg, C., Biewener, (Eds), 493-505, The Guildford Press, New York and London

Garnett, S.T., J. Sayer & Du Toit, J. (2007). Improving the effectiveness of interventions to balance conservation and development: a conceptual framework. *Ecology and Society* 12(1): 2, Available from
http://www.ecologyandsociety.org/vol12/iss1/art2/

Geoghegan, T. & Renard, Y. (2002). Beyond community involvement: lessons from the insular Caribbean. *Parks*, Vol.12, pp. 16-25

Gjertsen, H. (2005). Can habitat protection lead to improvements in human well-being? Evidence from marine protected areas in the Philippines. *World Development*, Vol.33, pp. 199-217

Haggerty, J.H. (2007). I'm not a greenie but …: environmentality, eco-populism and governance in New Zealand: experiences from the Southland whitebait fishery. *Journal of Rural Studies*, Vol.23, pp. 222-237

Hajer, M.A. (1996). Ecological modernization as cultural politics, In: *Risk, Environment and modernity: towards a new ecology*, S. Lash, B. Szerszynski, B. Wynne, B., (Eds), 246-268, Sage publications, London

Hajer, M.A. (1995). *The politics of environmental discourse: Ecological modernization and the policy process*, Oxford University Press, Oxford

Hanley, N. & Shogren, J.F. (2002). Awkward choices: economics and nature conservation, In: *Economics, Ethics and Environmental Policy: Contested Choices*, D.W. Bromley & J. Paavola, (Eds), Blackwell Publishing, Oxford

Harper, J. (2002). *Endangered species*, Carolina Academic Press, Durham, North Carolina

Hoole, A. & Berkes, F. (2010). Breaking down fences: Recoupling social-ecological systems for biodiversity conservation in Namibia. *Geoforum*, Vol.41, pp. 304-317

Igoe, J. (2006). Measuring the costs and benefits of conservation to local communities. *Journal of Ecological Anthropology*, Vol.10, pp. 72-77

Jamal, T., Everett, J. & Dann, G.M.S. (2003). Ecological rationalization and performative resistance in natural area destinations. *Tourist studies*, Vol.3, pp. 143-169

Jänicke, M. (1997). The political system's capacity for environmental policy, In: *National Environmental Policies: A Comparative Study of Capacity-Building*, M. Jänicke & H. Weidner, (Eds), 1-24, Springer Verlag, Berlin

Jenkins, W.A., Murray, B.C., Kramer, R.A. & Faulkner, S.P. (2010). Valuing ecosystem services from wetlands restoration in the Mississippi Alluvial Valley. *Ecological Economics*, Vol.69, pp. 1051-1061

Jim, C.Y. & Chen, W.Y. (2009). Ecosystem services and valuation of urban forests in China. *Cities*, Vol.26, pp. 187-194

Kalamandeen, M. & Gillson, L. (2007). Demything "wilderness": implications for protected area designation and management. *Biodiversity and Conservation*, Vol.16, pp. 165-182

Kiss, A. (2004). Is community-based ecotourism a good use of biodiversity conservation funds? Trends in Ecology and Evolution, Vol.19, pp. 232-237

Kittinger, J.N., Dowling, A., Purves, A.R., Milne, N.A. & Olsoon, P. (2011). Marine Protected Areas, Multiple-Agency Management, and Monumental Surprise in the Northwestern Hawaiian Islands. *Journal of Marine Biology* doi:10.1155/2011/241374

Knudsen, A. (1999). Conservation and controversy in Karakoram: Khunjerab National Park, Pakistan. *Journal of Political Ecology*, Vol.6, pp. 1-30

Lange, G.-M. & Jiddawi, M. (2009). Economic value of marine ecosystem services in Zanzibar: Implications for marine conservation and sustainable development. *Ocean & Coastal Management*, Vol.52, pp. 521-532

MA (Millenium Ecosystem Assessment) (2003). *Ecosystems and human well-being*, Island Press, Washington

McAfee, K. (1999). Selling nature to save it? Biodiversity and green developmentalism. *Environment and Planning D: Society and Space*, Vol.17, pp. 133-154

McCarthy, J. (2006). Neoliberalism and the Politics of Alternatives: Community Forestry in British Columbia and the United States. *Annals of the Association of American Geographers*, Vol.96, pp. 84-104

McCauley, D.J. (2006) Selling out on nature. *Nature*, Vol.443, pp. 27-28

McNeeley, J.A. (1993). Economic incentives for conserving biodiversity: Lessons for Africa. *Ambio*, Vol.22, pp. 144-150.

Meiksins Wood, A., 1998. The communist manifesto 150 years later. Monthly Review Press, New York.

Mol, A.P.J. & Sonnenfield, D.A. (2000). *Ecological modernization around the world: Perspectives and critical debates*, Frank Cass, London

Mol, A.P.J. & Spaargaren, G. (2000). Ecological Modernization Theory in Debate: A Review. *Environmental Politics*, Vol.9, pp. 17-49

Mol, A.P.J. (2002). Ecological modernization and the global economy. *Global Environmental Politics*, Vol.2, pp. 92-115

Mol, A.P.J. (1995). *The Refinement of Production: Ecological Modernization Theory and the Chemical Industry*, Van Arkel, Utrecht, The Netherlands

Mukhopadhyay, A. (2005). *Merchandising nature. Political ecology of biodiversity*, Working paper series, Indian Institute of Management, Calcutta

Naxakis, C. (1997). The environmental impacts of the Maastricht Treaty. *Outopia*, Vol. 26, pp. 107-118 *(in Greek)*

Nygren, A. (2005). Community-based forest management within the context of institutional decentralization in Honduras. *World Development*, Vol.33, pp. 639-655

Nygren, A. (1998). Environment as Discourse: Searching for Sustainable Development in Costa Rica. *Environmental Values*, Vol.7, pp. 201-222

Padgett, D.K. (1998). *Qualitative methods in social work research: Challenges and rewards*, Sage, Thousand Oaks, CA

Patton, M.Q. (2002). *Qualitative research and evaluation methods*, Sage, Thousand Oaks, CA. 3rd edition

Paulson, S., Gezon, L.L. & Watts, M. (2003). Locating the political in political ecology: an introduction. *Human Organization*, Vol. 62, pp. 205-217

Pejchar, L., Morgan, P.M., Caldwell, M.R., Palmer, C. & Daily, G.C. (2007). Evaluating the Potential for Conservation Development: Biophysical, Economic, and Institutional Perspectives. *Conservation Biology*, Vol.21, pp. 69-78

Peterson, M.N., Peterson, M.J. & Peterson T.R. (2005). Conservation and the myth of consensus. *Conservation Biology*, Vol.19, pp. 762-767

Phillips, L. & Jørgensen, W. (2006). *Discourse Analysis as Theory and Method*, Sage Publications, London

Randall, A. (2002). Benefit cost considerations should be decisive when there is nothing more important at stake, In: *Economics, Ethics and Environmental Policy: Contested Choices*, D.W. Bromley & J. Paavola, (Eds.), Blackwell Publishing, Oxford

Redford, K.H. & Adams, W.M. (2009). Payment for ecosystem services and the challenge of saving nature. *Conservation Biology*, Vol.23, pp. 785-787

Salafsky, N. & Wollenberg, E. (2000). Linking livelihoods and conservation: a conceptual framework and scale for assessing the integration of human needs and biodiversity. *World Development*, Vol.28, pp. 1421-1438

Scholte, J.A. (2004). Civil society and democratically accountable global governance. *Government and Opposition*, Vol.39, pp. 211-233

Searle, M. (2000). *Phantom Parks: The Struggle to Save Canada's National Parks*, Key Porter Books, Toronto

Sodhi, N.S., Lee, T.M., Sekercioglu, C.H., Webb, E.L., Prawiradilaga, D.M., Lohman, D.J., Pierce, N.E., Diesmos, A.C., Rao, M. & Ehrlich, P.R. (2010). Local people value environmental services provided by forested parks. *Biodiversity and Conservation*, Vol.19, pp. 1175-1188

Spaargaren, G. (2000). Ecological modernization theory and domestic consumption. *Journal of Environmental Policy & Planning*, Vol.1, pp. 323-335

Spaargaren, G. (1997). *The ecological modernization of production and consumption: essays in environmental sociology*, Wageningen Agricultural University, PhD thesis.

Strauss, A. & Corbin, J. (1998). *Basics of Qualitative Research, Techniques and Procedures for Developing Grounded Theory*, Sage Publications, London

Swetnam, R.D., Fisher, B., Mbilinyi, B.P., Munishi, P.K.T., Willcock, S., Ricketts, T., Mwakalila, S., Balmford, A., Burgess, N.D., Marshall, A.R. & Lewis, S.L. (2011). Mapping socio-economic scenarios of land cover change: A GIS method to enable ecosystem service modeling. *Journal of Environmental Management*, Vol.92, pp. 563-574

Tallis, H., Kareiva, P., Marvier, M. & Chang, A. (2008). An ecosystem services framework to support both practical conservation and economic development. *PNAS*, Vol.105, pp. 9457-9464

Thiele, L.P. (1999). *Environmentalism for a new millennium: the challenge of coevolution*, Oxford University Press, Oxford

Vira, B. & Adams, W.M. (2009). Ecosystem services and conservation strategy: beware the silver bullet. *Conservation Letters*, Vol.2, pp. 158-162

Vlachou, A. (2005). Debating sustainable development. *Rethinking Marxism*, Vol.17, pp. 627-638

WCED (World Commission on Environment and Development) (1987). *Our Common Future,* Oxford University Press, Oxford

Weale, A. (1992). *The new politics of pollution,* Manchester University Press, Manchester

York, R. & Rosa, E.A. (2003). Key Challenges to Ecological Modernization Theory: Institutional Efficacy, Case Study Evidence, Units of Analysis, and the Pace of Eco-Efficiency. *Organization & Environment,* Vol.16, pp. 273-288

Permissions

The contributors of this book come from diverse backgrounds, making this book a truly international effort. This book will bring forth new frontiers with its revolutionizing research information and detailed analysis of the nascent developments around the world.

We would like to thank Dr. Chaouki Ghenai, for lending his expertise to make the book truly unique. He has played a crucial role in the development of this book. Without his invaluable contribution this book wouldn't have been possible. He has made vital efforts to compile up to date information on the varied aspects of this subject to make this book a valuable addition to the collection of many professionals and students.

This book was conceptualized with the vision of imparting up-to-date information and advanced data in this field. To ensure the same, a matchless editorial board was set up. Every individual on the board went through rigorous rounds of assessment to prove their worth. After which they invested a large part of their time researching and compiling the most relevant data for our readers. Conferences and sessions were held from time to time between the editorial board and the contributing authors to present the data in the most comprehensible form. The editorial team has worked tirelessly to provide valuable and valid information to help people across the globe.

Every chapter published in this book has been scrutinized by our experts. Their significance has been extensively debated. The topics covered herein carry significant findings which will fuel the growth of the discipline. They may even be implemented as practical applications or may be referred to as a beginning point for another development. Chapters in this book were first published by InTech; hereby published with permission under the Creative Commons Attribution License or equivalent.

The editorial board has been involved in producing this book since its inception. They have spent rigorous hours researching and exploring the diverse topics which have resulted in the successful publishing of this book. They have passed on their knowledge of decades through this book. To expedite this challenging task, the publisher supported the team at every step. A small team of assistant editors was also appointed to further simplify the editing procedure and attain best results for the readers.

Our editorial team has been hand-picked from every corner of the world. Their multi-ethnicity adds dynamic inputs to the discussions which result in innovative outcomes. These outcomes are then further discussed with the researchers and contributors who give their valuable feedback and opinion regarding the same. The feedback is then collaborated with the researches and they are edited in a comprehensive manner to aid the understanding of the subject.

Apart from the editorial board, the designing team has also invested a significant amount of their time in understanding the subject and creating the most relevant covers. They scrutinized every image to scout for the most suitable representation of the subject and create an appropriate cover for the book.

The publishing team has been involved in this book since its early stages. They were actively engaged in every process, be it collecting the data, connecting with the contributors or procuring relevant information. The team has been an ardent support to the editorial, designing and production team. Their endless efforts to recruit the best for this project, has resulted in the accomplishment of this book. They are a veteran in the field of academics and their pool of knowledge is as vast as their experience in printing. Their expertise and guidance has proved useful at every step. Their uncompromising quality standards have made this book an exceptional effort. Their encouragement from time to time has been an inspiration for everyone.

The publisher and the editorial board hope that this book will prove to be a valuable piece of knowledge for researchers, students, practitioners and scholars across the globe.

List of Contributors

Oscar Fernández
University of León, Spain

Janie Liew-Tsonis and Sharon Cheuk
School of Business and Economics Universiti, Malaysia Sabah, Malaysia

Michael Zgurovsky
National Technical University of Ukraine "Kyiv Polytechnic Institute", Kyiv, Ukraine

Sasho Kjosev
University "Ss. Cyril and Methodius", Faculty of Economics, Republic of Macedonia

Mago William Maila
University of South Africa, Pretoria, South Africa

Wiebke Schone, Cornelia Kellermann and Ulrike Busolt
Hochschule Furtwangen University, Germany

Tereza Kadlecová and Lilia Dvořáková
Institute for Sustainability, United Kingdom
University of West Bohemia in Pilsen, Czech Repulbic

Miguel Rocha
ITESM Campus Querétaro, Mexico

Cory Searcy
Ryerson University, Canada

Francesco Fusco Girard
Seconda Università Degli Studi di Napoli, Italy

Kayano Fukuda and Chihiro Watanabe
National University of Singapore, Singapore

Guillaume Faburel
Urban Planning Department (Institut d'Urbanisme de Paris), University of Paris Est – Coordinator of Research Unit Aménités, France

Evangelia G. Drakou
Department of Ecology, School of Biology, Aristotle University of Thessaloniki, Thessaloniki, Greece
Current Address: Global Environment Monitoring Unit, Institute for Environment and Sustainability, Joint Research Center, European Commission, Ispra, Italy

Evangelia Apostolopoulou and John D. Pantis
Department of Ecology, School of Biology, Aristotle University of Thessaloniki, Thessaloniki, Greece

Printed in the USA
CPSIA information can be obtained
at www.ICGtesting.com
JSHW011500221024
72173JS00005B/1145